# Transmission Techniques for 4G Systems

# OTHER TELECOMMUNICATIONS BOOKS FROM AUERBACH

**Ad Hoc Mobile Wireless Networks: Principles, Protocols, and Applications**
Subir Kumar Sarkar, T.G. Basavaraju, and C. Puttamadappa
ISBN 978-1-4200-6221-2

**Communication and Networking in Smart Grids**
Yang Xiao (Editor)
ISBN 978-1-4398-7873-6

**Delay Tolerant Networks: Protocols and Applications**
Athanasios V. Vasilakos, Yan Zhang, and Thrasyvoulos Spyropoulos
ISBN 978-1-4398-1108-5

**Emerging Wireless Networks: Concepts, Techniques and Applications**
Christian Makaya and Samuel Pierre (Editors)
ISBN 978-1-4398-2135-0

**Game Theory in Communication Networks: Cooperative Resolution of Interactive Networking Scenarios**
Josephina Antoniou and Andreas Pitsillides
ISBN 978-1-4398-4808-1

**Green Communications: Theoretical Fundamentals, Algorithms and Applications**
Jinsong Wu, Sundeep Rangan, and Honggang Zhang
ISBN 978-1-4665-0107-2

**Green Communications and Networking**
F. Richard Yu, Xi Zhang, and Victor C.M. Leung (Editors)
ISBN 978-1-4398-9913-7

**Green Mobile Devices and Networks: Energy Optimization and Scavenging Techniques**
Hrishikesh Venkataraman and Gabriel-Miro Muntean (Editors)
ISBN 978-1-4398-5989-6

**Handbook on Mobile Ad Hoc and Pervasive Communications**
Laurence T. Yang, Xingang Liu, and Mieso K. Denko (Editors)
ISBN 978-1-4398-4616-2

**IP Telephony Interconnection Reference: Challenges, Models, and Engineering**
Mohamed Boucadair, Isabel Borges, Pedro Miguel Neves, and Olafur Pall Einarsson
ISBN 978-1-4398-5178-4

**LTE-Advanced Air Interface Technology**
Xincheng Zhang and Xiaojin Zhou
ISBN 978-1-4665-0152-2

**Media Networks: Architectures, Applications, and Standards**
Hassnaa Moustafa and Sherali Zeadally (Editors)
ISBN 978-1-4398-7728-9

**Multihomed Communication with SCTP (Stream Control Transmission Protocol)**
Victor C.M. Leung, Eduardo Parente Ribeiro, Alan Wagner, and Janardhan Iyengar
ISBN 978-1-4665-6698-9

**Multimedia Communications and Networking**
Mario Marques da Silva
ISBN 978-1-4398-7484-4

**Near Field Communications Handbook**
Syed A. Ahson and Mohammad Ilyas (Editors)
ISBN 978-1-4200-8814-4

**Next-Generation Batteries and Fuel Cells for Commercial, Military, and Space Applications**
A. R. Jha, ISBN 978-1-4398-5066-4

**Physical Principles of Wireless Communications, Second Edition**
Victor L. Granatstein, ISBN 978-1-4398-7897-2

**Security of Mobile Communications**
Noureddine Boudriga, ISBN 978-0-8493-7941-3

**Smart Grid Security: An End-to-End View of Security in the New Electrical Grid**
Gilbert N. Sorebo and Michael C. Echols
ISBN 978-1-4398-5587-4

**Transmission Techniques for 4G Systems**
Mário Marques da Silva
ISBN 978-1-4665-1233-7

**Transmission Techniques for Emergent Multicast and Broadcast Systems**
Mário Marques da Silva, Americo Correia, Rui Dinis, Nuno Souto, and Joao Carlos Silva
ISBN 978-1-4398-1593-9

**TV Content Analysis: Techniques and Applications**
Yiannis Kompatsiaris, Bernard Merialdo, and Shiguo Lian (Editors)
ISBN 978-1-4398-5560-7

**TV White Space Spectrum Technologies: Regulations, Standards, and Applications**
Rashid Abdelhaleem Saeed and Stephen J. Shellhammer
ISBN 978-1-4398-4879-1

**Wireless Sensor Networks: Current Status and Future Trends**
Shafiullah Khan, Al-Sakib Khan Pathan, and Nabil Ali Alrajeh
ISBN 978-1-4665-0606-0

**Wireless Sensor Networks: Principles and Practice**
Fei Hu and Xiaojun Cao
ISBN 978-1-4200-9215-8

## AUERBACH PUBLICATIONS
www.auerbach-publications.com
To Order Call: 1-800-272-7737 • Fax: 1-800-374-3401
E-mail: orders@crcpress.com

# Transmission Techniques for 4G Systems

Mário Marques da Silva, Américo M. C. Correia,
Rui Dinis, Nuno Souto, and João Carlos Silva

CRC Press
Taylor & Francis Group
Boca Raton   London   New York

CRC Press is an imprint of the
Taylor & Francis Group, an **Informa** business

CRC Press
Taylor & Francis Group
6000 Broken Sound Parkway NW, Suite 300
Boca Raton, FL 33487-2742

First issued in paperback 2016

Version Date: 20121015

ISBN 13: 978-1-138-19995-8 (pbk)
ISBN 13: 978-1-4665-1233-7 (hbk)

Visit the Taylor & Francis Web site at
http://www.taylorandfrancis.com

and the CRC Press Web site at
http://www.crcpress.com

# Contents

# Preface

Fourth Generation (4G) wireless communication systems aim to allow peak data rates in the range of 1 Gbps for nomadic access and 100 Mbps for vehicular mobility. 4G aims to support current and emergent multimedia services, such as mobile TV, social networks and gaming, high-definition television and video teleconference, multimedia messaging service, using the All-over IP concept and with improved quality of service.

This book describes transmission schemes suitable for future broadband wireless systems and proposes and studies several advances in transmission techniques and receiver design to support emergent wireless needs for 4G requirements. New requirements include increasing throughputs and bandwidths, increased spectrum efficiency and network capacity, and lower delays and round-trip times.

4G services require extensive exploitation of advanced schemes such as Multiple-Input Multiple-Output (MIMO), base station cooperation, macro-diversity, inter-cell interference cancellation, multihop relay techniques, and hierarchical constellations, as well as multi-resolution techniques. All of these principles are studied in this book, and advances are proposed for different propagation and multi-user environments, using block transmission techniques.

The purpose of this book is to concentrate in a single place several important R&D activities currently under way in the field of wireless communications for 4G systems including evolved Multimedia Broadcast and Multicast Service (E-MBMS). These aspects are normally covered in different books, and thus, the book reduces the time and cost required to learn and improve skills and knowledge in the field. Moreover, this book presents a compilation of the latest developments in the area, which is the outcome of several years of research and participation in many international activities and projects. The focus is on the key requirements of emergent services, with special emphasis on 4G systems. The purpose is to cover the several subjects including such key requirements, and providing the corresponding description of fundamentals and theory.

The transmission and detection techniques and schemes presented in this book are transversal to many digital communication systems (wireless, cellular, satellite, etc.). Nevertheless, we especially emphasize the transmission techniques for 4G systems, which is the main subject of this book. With such an approach, this book appeals to a wide range of potential readers. It can be used by either an engineer with a BSc degree to learn more about the latest R&D wireless activities for the purpose of an MSc or PhD program, or for business activities. This book can also be used by academic, institutional, or industrial researchers to support the study, planning, design, and development of prototypes and systems.

Although the subjects covered in this book are broad and generic, the final chapter focuses specifically on system-level evaluation of 4G using different transmission techniques. The combination of the enhancements is accomplished by adaptive transmission techniques. These techniques and enhancements will both meet and exceed the 4G requirements.

# About the Authors

**Mario Marques da Silva** is a professor at Universidade Autónoma de Lisboa (CESITI), and Lisbon, Portugal, Portuguese Naval Academy (CINAV). He is a researcher at the Portuguese Instituto de Telecomunicações. He received his BSc in electrical engineering in 1992, and his MSc and PhD in telecommunications/electrical engineering, in 1999 and 2005, respectively, from the Universidade Técnica de Lisboa.

He has been involved in several telecommunications projects. Between 2005 and 2008, he was with the NATO Air Command Control & Management Agency (NACMA) in Brussels, Belgium, where he managed the deployable communications of the new Air Command and Control System Program. His research interests include networking and mobile communications, including block transmission techniques (OFDM, SC-FDE), interference cancellation, space-time coding, MIMO systems, smart and adaptive antennas, channel estimation, and software defined radio.

Dr. Marques da Silva is the author of *Multimedia Communications and Networking* and *Transmission Techniques for Emergent Multicast and Broadcast Systems* (both CRC Press), as well as the author of numerous journal and conference papers. He is a member of the Institute of Electrical and Electronics Engineers (IEEE) and member of the Armed Forces Communications and Electronics Association (AFCEA), and serves as a reviewer for many international scientific journals and conferences.

**Américo Correia** is a full professor at Instituto Superior de Ciências do Trabalho e da Empresa–Institute University of Lisbon (ISCTE-IUL), Lisbon, Portugal, where he is the head of the Science and Technology Department. He is also a senior researcher at the Portuguese Instituto de Telecomunicações (Institute for Telecomunicacions). Dr. Correia received his BSc in electrical engineering from the University of Angola in 1983, his MSc and PhD from Instituto Superior Técnico (IST), Lisbon, Portugal, in 1990 and 1994, respectively. He was a visiting scientist at the Nokia Research Center from September to December 1998. From September 2000 to August 2001 he worked with Ericsson Eurolab Netherlands. His research interests include several areas of mobile communications, including, frequency-domain techniques (such as OFDM (orthogonal frequency division multiplexing) and SC-FDE (single carrier frequency domain equalization)), multi-user detection, interference cancellation, space-time coding, MIMO systems (multiple input, multiple output), smart and adaptive antennas, software defined radio, radio resource management, and multimedia broadcast/multicast services. He has been involved in several telecommunications projects, either as a researcher or as project leader, including research, architecture, development, analysis, simulation, and testing. He is the author of numerous international research papers. He is a member of IEEE communications society and serves as a reviewer of several international publications. Dr. Correia is one of the authors of the serves as a *Transmission Techniques for Emergent Multicast and Broadcast Systems.*

**Rui Dinis** received his PhD from the Instituto Superior Técnico (IST), Technical University of Lisbon, Portugal, in 2001. From 2001 to 2008, he was a professor at IST. Since 2008, he has taught at Faculdade de Ciências e Tecnologia da Universidade Nova de Lisboa FCT-UNL. He was a researcher at CAPS/IST (Centro de Análise e Processamento de Sinais) from 1992 to 2005; from 2005 to 2008, he was a researcher at ISR/IST (Instituto de Sistemas e Robótica); and in 2009, he joined the research center IT (Instituto de Telecomunicações).

Dr. Dinis currently serves as editor of *IEEE Transactions on Communications* in the transmission systems area, subarea frequency-domain processing and equalization. He has been involved in several research projects in the broadband wireless communications area. His main research interests include modulation, equalization, channel estimation, and synchronization.

**Nuno M. B. Souto** graduated in aerospace engineering, avionics branch, from the Instituto Superior Técnico, Lisbon, Portugal in 2000. From November 2000 to January 2002, he worked as a researcher in the field of automatic speech recognition at the Instituto de Engenharia e Sistemas de Computadores, Lisbon, Portugal. He received his PhD from Instituto Superior Técnico in 2006. He is an assistant professor at Instituto Universitário de Lisboa (ISCTE-IUL) and a researcher at Instituto de Telecomunicações (IT), Lisbon, Portugal. Dr. Souto has participated in several European research projects and has published in several international journals. His research interests include wideband CDMA systems, OFDM, channel coding, modulation, channel estimation, and MIMO systems.

**João Carlos M. Silva** received his BS in aerospace engineering from the Instituto Superior Técnico (IST) Technical University Lisbon, Portugal, in 2000. From 2000 to 2002, he worked as a business consultant for McKinsey & Company. He joined IT (Instituto de Telecomunicações) in 2002 as an investigator, and was promoted to senior investigator after he completed his PhD at IST, focusing on spread spectrum techniques, multi-user detection schemes, and MIMO systems. From 2006 to 2007, he was a professor at Escola Superior Tecnologia e Gestão (ESTG), in Leiria where he taught physics and serves as a reviewer, and has been a professor since 2007 in Instituto Superior de Ciencias do Trabalho e da Empresa (ISCTE), in Lisbon (teaching computer networks). In 2011, he moved from the IT-IST branch to the IT-ISCTE branch. He is the author of more than 60 papers and reviewer of several international publications. His research interests include computer networks and mobile communications, block transmission techniques (OFDM, SC-FDE), WCDMA, multi-user detection, interference cancellation, space-time coding, MIMO systems, smart and adaptive antennas, channel estimation and software defined radio.

# Chapter 1

# Requirements for Current and Emergent Multimedia Services

In order to implement the concept of "anywhere" and "anytime," as well as to support new and emergent services, users are demanding more and more from cellular communication systems. New requirements include increasing throughputs and bandwidths, enhanced spectrum efficiency, shorter delays and greater network capacity, made available by the air interface.* These are the key issues necessary to deliver new and emergent broadband data services.

The long term evolution (LTE) air interface was specified in Third Generation Partnership Project (3GPP)† Release 8, and enhanced in Release 9. Its initial deployment was in 2010 (see Table 1.1). The LTE comprises a completely new air interface based on orthogonal frequency division multiple access (OFDMA)‡ in the downlink and single carrier–frequency division multiple access (SC-FDMA) in the uplink. It allows a spectral efficiency improvement by a factor of 2 to 4, as compared to high speed packet access§ (HSPA), making a new spectrum, alongside multiple input multiple output (MIMO) systems and the All-over-IP¶ architecture.

---

* Where the bottleneck is typically located.
† 3GPP is responsible for specifying and defining the architecture of the European 3G and 4G evolution.
‡ As opposed to the wideband code division multiple access (WCDMA) transmission technique utilized in Universal Mobile Telecommunication System (UMTS).
§ Standardized in 3GPP Releases 5, 6, and 7 (see Table 1.1).
¶ Internet Protocol.

**Table 1.1 Comparison between Several Different 3GPP Releases**

| FDD | WCDMA | HSPA | HSPA+ | LTE | LTE/IMT |
| --- | --- | --- | --- | --- | --- |
| TDD | TD-SCDMA | TD-HSDPA | TD-HSUPA | TD-LTE | Advanced |
| Deployment | 2003 | 2006/8 | 2008/9 | 2010 | 2014 |
| 3GPP Release | 99 | 5/6 | 7 | 8 | 10 |
| Downlink data rate | 384 kbps | 14.4 Mbps(*) | 28 Mbps(*) | >160 Mbps(+) | 1 Gbps |
| Uplink data rate | 128 kbps | 5.76 Mbps(*) | 11 Mbps(*) | >60 Mbps(+) | 500 Mbps |
| Switching | Circuit + Packet Switching | Circuit + Packet Switching | Circuit + Packet Switching | IP based | IP based (Packet Switching) |
| Transmission Technique | WCDMA/ TD-SCDMA | WCDMA/ TD-SCDMA | WCDMA/ TD-SCDMA | Downlink: OFDMA | Downlink: OFDMA/ Uplink:SC-FDMA |
| MIMO | No | No | Yes | Yes | Yes |
| Multihop Relay | No | No | No | No | Yes |
| AMC | No | Yes | Yes | Yes | Yes |
| Cooperative Systems | No | No | No | No | Yes |
| Carrier Aggregation | No | No | No | No | Yes |

*Note:* (*) Peak data rates; (+) Assuming 20 MHz bandwidth and 2 × 2 MIMO.

Moreover, the fourth generation cellular system (4G) is expected to be fully implemented in 2014. It aims to support peak data rates in the range of 100 Mbps for vehicular mobility to 1 Gbps for nomadic access (in both indoor and outdoor environments). 4G aims to support current and emergent multimedia services, such as social networks and gaming, mobile TV, high-definition television (HDTV), digital video broadcast (DVB), multimedia messaging service (MMS), or video chat, using the All-over-IP concept and with improved quality of service[*] (QoS).

---

[*] Important QoS parameters include exchanged data rates, bit error rate, end to end packet loss, delay, and jitter for different traffic types.

The international mobile telecommunications-advanced (IMT-Advanced[*]) defined by the International Telecommunications Union–Radiocommunications (ITU-R) in [ITU-R 2008]. IMT-Advanced is intended to be an international standard of the next generation cellular systems.[†]

Standardization of LTE-Advanced[‡] is part of 3GPP Release 10 and was enhanced in Release 11. New topological approaches such as multihop relay, MIMO systems, cooperative systems, and carrier aggregation, as well as block transmission techniques allow an improved coverage of high data rate transmission and improved system performance and capabilities to be included in LTE-Advanced.

## 1.1 Multimedia Paradigm

There is still a great deal of ongoing investigation on ways to improve the delivery of multimedia information. The multimedia broadcast and multicast service (MBMS), already introduced in 3GPP Release 6, aims to use spectrum-efficient multimedia services, by transmitting data over a common radio channel. MBMS is a system which allows multiple mobile network users to efficiently receive data from a single content provider source by sharing radio and transport network resources. While conventional mobile communications are performed in unicast[§] mode, multimedia services are normally delivered in either broadcast or multicast mode. In broadcast mode, data is transmitted in a specific area (MBMS service area), and all users in the specific MBMS service area are able to receive the transmitted MBMS data. Very often, broadcast communications are established in a single direction (that is, there is no feedback from the receiver into the transmitter). In multicast mode, data is transmitted in a specific area, but only registered users in the specific MBMS service area are able to receive the transmitted MBMS data.

Before describing the MBMS requirements of the next generation mobile systems, we define two basic but important concepts: media and traffic. The media is generated by humans or machines and, as shown in Figure 1.1, can be classified as text, visuals, or sounds. While the text corresponds to digital data, the visuals and sounds are typically analog, requiring analog-to-digital conversion before transmission through modern digital networks. Examples of visuals include images, videos, and graphs. Speech and music are two examples of sounds. Multimedia is viewed as a mixture of these different types of media, exchanged in a synchronized manner [Marques da Silva 2012].

---

[*] IMT-Advanced is commonly referred to as IMT-A.
[†] Similarly, IMT2000 corresponds to a set of third generation cellular system standards, namely, IEEE 802.16e, CDMA2000, WCDMA, etc.
[‡] LTE-Advanced is commonly referred to as LTE-A.
[§] Unicast stands for a communication whose data destination is a single station.

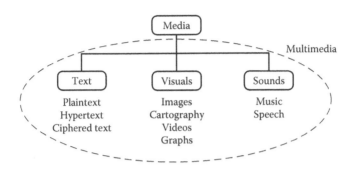

**Figure 1.1   Elementary media sources.**

When media sources are being exchanged through a network, it is generically referred to as traffic. In order to provide QoS, the networks should respond according to the traffic requirements. The telephony and video streaming traffic correspond to real-time traffic, while the file transfer and web-browsing traffic are non-real-time traffic. Different services present various bandwidth requirements. Figure 1.2 illustrates the bandwidth requirements for different services.

In order to provide QoS, a network should give priority to real-time traffic. Furthermore, traffic such as file transfer is sensitivity to errors. In this case, the provision of QoS is achieved by implementing a set of mechanisms that prevent data from being lost or corrupted.

Until recently, the aim of the telecommunications sector was to allow the implementation of the convergence. The convergence has different meanings, namely:

- A network infrastructure that supports multiple services (telephony, e-mail, web browsing, file transfer, etc.).
- A mixture of fixed and mobile communications.
- A mixture of telecommunications, information systems, and multimedia.

Finally, the convergence can be viewed as the convergence of all different convergences. This goal has been mainly achieved with the latest technological evolutions, which resulted in additional user features and improved QoS capabilities.

Currently, networks are viewed as enablers of knowledge sharing. This is the new paradigm of modern society and characterizes the collaborative era, where a deep collaboration and interaction of different entities (humans or machines) is permanently established. Knowledge consists of the ability of the right entity to have access to the right information, at the right time. This goal requires a high level of interaction, high bandwidth made available with the concept of "anywhere" and "anytime,"* along with business intelligence platforms. Business intelligence

---

* The concept of "anywhere" and "anytime" is mainly supported by mobile communications.

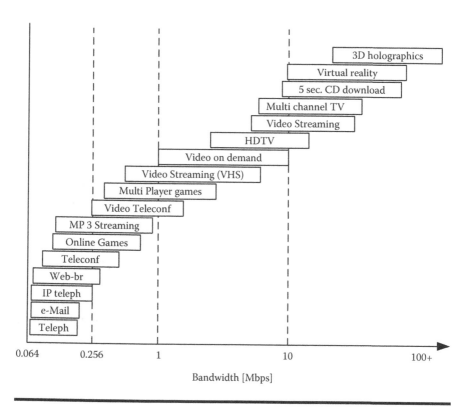

**Figure 1.2    Bandwidth requirements of different services.**

platforms allow entities to receive the necessary filtered information that enables decisions to be made at a specific moment. Naturally, all of these requirements should be implemented at the minimum cost.

In this collaborative era, the traffic tends to be multimedia based, and collaborative applications are mainly ad hoc applications* (e.g., social networks,† peer-to-peer applications), video, or even data applications [Marques da Silva 2012].

Figure 1.3 shows the evolution of network usage. Initially, this was employed merely for data applications. Afterward, convergence was an important issue to allow a better usage of the network. Increase of the interactivity level by Internet users became the Internet world, a space for deep collaboration between entities but with a high level of danger as well.

---

* Ad hoc applications allow users to inject nonstructured information into the Internet world.
† Social networks allow the exchange of unmanaged multimedia data by groups of people or interest.

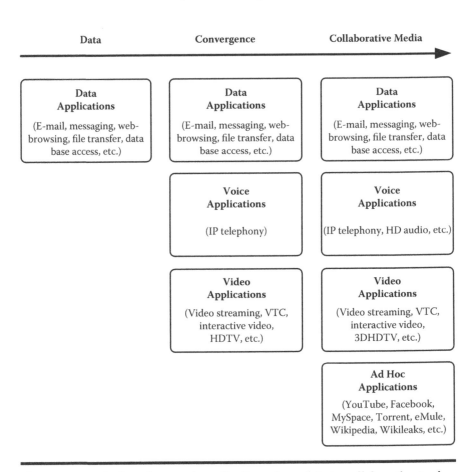

**Figure 1.3 Evolution of network applications: from data to collaborative tool.**

## 1.2 Evolution from 3G Systems to Long-Term Evolution

The third generation of cellular systems (3G) comprises different evolutions. The initial version, specified by 3GPP release 99, marked a sudden change in the multiple access technique (see Table 1.1). While the global system for mobile communications[*] (GSM) was based on time division multiple access (TDMA), 3G mainly uses code division multiple access (CDMA). This evolution allowed a rate improvement from tens of kbps up to 384 kbps for the downlink and 128 kbps for the uplink. These rates were improved in the following updates, achieving 28 Mbps in the downlink of HSPA+ (3GPP release 7). In order to respond to the increased speed demands of the emergent services, higher speeds became possible with the already deployed LTE, supporting 160 Mbps in the downlink (as defined by 3GPP Release 8), and even

---

[*] GSM corresponds to the second generation of mobile communications (2G).

higher speeds with some additional improvements to the LTE baseline introduced in 3GPP Release 9 (e.g., advanced MIMO systems). The LTE air interface was the result of a study item launched by 3GPP called Evolved UTRAN (E-UTRAN). The goal was to face the latest demands for voice, data, and multimedia services, improving spectral efficiency by a factor of 2–4, as compared to HSPA Release 7.

The LTE air interface is a completely new system based on OFDMA in the downlink and SC-FDMA in the uplink. Depending on the purpose, different types of MIMO systems are considered in 3GPP Release 8. The modulation employed in LTE comprises quadrature phase shift keying (QPSK), 16-QAM or 64-QAM (quadrature amplitude modulation), using adaptive modulation and coding (AMC). When in the presence of noisy channels, the modulation order is reduced and the code rate is increased. The opposite occurs when the channel presents better conditions.

The LTE comprises high spectrum flexibility, with different spectrum allocations of 1.4, 3, 5, 10, 15, and 20 MHz. This allows a more efficient spectrum usage and a dynamic spectrum allocation based on the bandwidths/data rates required by the users [Astely et al. 2009].

Intra-cell interference is avoided in LTE by allocating the proper orthogonal time slots and carrier frequencies between users in both uplink and downlink. However, inter-cell interference is a bigger problem than in the case of UMTS,* especially for users at the cell edge. Inter-cell interference can be mitigated by implementing mechanisms such as interference cancellation schemes, reuse partitioning, and advanced base station cooperation.

Another important modification of the LTE, as compared to UMTS, is the All-IP architecture (that is, all services are carried out on top of IP), instead of the circuit[†] plus packet[‡] switching network adopted by UMTS.

A significant improvement of the LTE, compared to the UMTS, relies on its improved capability to support multimedia services. The evolved MBMS (eMBMS) is implemented in LTE in two types of transmission scenarios [Astely et al. 2009]:

- Multi-cell transmission: Multimedia broadcast over a single frequency network (MBSFN)[§] on a dedicated frequency layer or on a shared frequency layer. The group of cells that receive the same MBSFN multicast data service is referred to as the MBSFN area.[¶]
- Single-cell transmission: Single cell–point to multipoint (SC-PTM) on a shared frequency layer.

---

* Due to the lower power spectral density of WCDMA signals, the level of interferences generated in UMTS tends to be lower.
† Circuit switching is employed in UMTS for voice service.
‡ Packet switching is employed in UMTS for data service.
§ The MBSFN allows delivering services such as mobile television.
¶ Within an MBSFN area, if one or more cells are not required to broadcast the multimedia data service, the transmission can be switched off, and the corresponding resources can be released to regular unicast or other services.

Multi-cell transmission in a single frequency network (SFN) area is a way to improve the overall network spectral efficiency. In MBSFN, when different cells transmit the same eMBMS multimedia data service, the signals are combined in order to provide diversity for user equipment (UE) located at a cell boundary. This results in improved performance and better service quality.

## 1.3 WiMAX—IEEE802.16

WiMAX, standardized by the Institute of Electrical and Electronics Engineers (IEEE) as IEEE 802.16, was created in 2001 and updated by several newer versions. It consists of a technology that implements a wireless metropolitan area network (WMAN) [Eklund et al. 2002; Andrews et al. 2007; Peters and Heath 2009]. WiMAX stands for Worldwide Interoperability for Microwave Access and allows fixed and mobile access. The basic idea is to provide wireless Internet access to the last mile, with a range of up to 50 km. Therefore, it can be viewed as a complement or competitor of the existing asynchronous digital subscriber line (ADSL) or cable modem, providing the service with the minimum effort in terms of required infrastructures. On the other hand, fixed WiMAX can also be viewed as a backhaul for Wi-Fi (IEEE 802.11), cellular base station (BS), or mobile WiMAX. As the standard only defines the physical layer and medium access control (MAC) sublayer, it can be used in association with either IP version 4 (IPv4) or IP version 6 (IPv6).

In order to allow the operation of WiMAX in different regulatory spectrum constraints faced by operators in different geographies, this standard specifies channel sizes ranging from 1.75 MHz up to 20 MHz, using either time division duplexing (TDD) or frequency division duplexing (FDD), with many options in between [Yarali and Rahman 2008].

The initial version of WiMAX was upgraded by several newer versions:

■ IEEE 802.16-2004, also referred to as IEEE 802.16d. This version only specified the fixed interface of WiMAX, without providing any support for mobility [IEEE 802.16-2004]. This version of the standard was adopted by the European Telecommunications Standards Institute (ETSI) as a base for the HiperMAN.*

■ IEEE 802.16-2005, also referred to as IEEE 802.16e. This is of an amendment to the previous version. It introduced support for mobility, handover, and roaming, among other new capabilities [IEEE 802.16e]. In addition, in order to achieve better performances, MIMO schemes were introduced.

■ Relay specifications are included in the IEEE 802.16j amendment. The incorporation of multihop relay capability in the foundation of mobile IEEE 802.16-2005 is a way to increase both the available throughput by a factor

---

* High-performance metropolitan area network.

of 3 to 10 and coverage (and higher channel reuse factor), or even to fill the "coverage hole" of indoor coverage [Oyman et al. 2007; IEEE 802.16; Peters and Heath 2009]. Multihop relay capability was included in IMT-Advanced [Astely et al. 2009].

In addition to these versions, requirements for the next version Mobile WiMAX called IEEE 802.16m were completed. The goal of IEEE 802.16m is to fulfill all the IMT-Advanced requirements as proposed in ITU-R 2008, making this standard a candidate for the IMT-A. Advances in IEEE 802.16m include wider bandwidths (up to 100 MHz, shared between uplink and downlink), adaptive and advanced TDMA/OFDMA access schemes, advanced relaying techniques (already incorporated in IEEE 802.16j), advanced multiple-antenna systems, adaptive modulation schemes such as hierarchical constellations and AMC, and frequency adaptive scheduling, among other advanced techniques [ITU-R 2008].

The original version of the standard specified a physical layer operating in the range of 10 to 66 GHz, based on OFDM and TDMA technology. IEEE 802.16-2004 added specifications for the 2 to 11 GHz range (licensed and unlicensed), whereas IEEE 802.16-2005 introduced the scalable OFDMA (SOFDMA) with MIMO (either space time coding based, spatial multiplexing based, or beamforming) or advanced antenna systems (AAS) [IEEE 802.16e], instead of the simple OFDM with 256 subcarriers considered by the previous version.

In terms of throughputs and coverage, these two parameters are subject to a trade-off [IEEE 802.16e]: typically, mobile WiMAX provides up to 10 Mbps per channel (symmetric) over a range of 10 km in rural areas (LOS environment) or over a range of 2 km in urban areas (NLOS environment) [Ohrtman 2008]. With the fixed WiMAX, this range can usually be extended. The mobile version considers an omnidirectional antenna, whereas fixed WiMAX uses a high-gain antenna (directional). Throughput and ranges may always change. Nevertheless, by enlarging one parameter the other has to reduce, otherwise the bit error rate (BER) would suffer a degradation. In the limit, WiMAX may deliver up to 70 Mbps per channel (in LOS, short distance, and fixed access), and may cover up to 50 km (in LOS for fixed access), with a high-gain antenna [Ohrtman 2008], but not both parameters simultaneously. In this case, a single SS is supported by the BS. Unlike UMTS, where handover is specified in detail, mobile WiMAX has three possibilities, but only the first one is mandatory: hard handover (HHO), fast base station switching (FBSS), and macro diversity handover (MDHO). FBSS and MDHO are optional because it is up to the manufacturers to decide their implementation specifications. Therefore, there is the risk that handover is not possible between two BSs from different manufacturers. Another drawback of the use of WiMAX is the maximum speed allowed in mobility, which is limited to 60 km/h. For higher speeds, the user experiences a great degradation in performance.

The WiMAX version currently available (IEEE 802.16-2005) incorporates most of the techniques also adopted by LTE (from 3GPP). Examples of such techniques are OFDMA, MIMO, advanced turbo coding, All-over-IP architecture, etc. In addition, the inclusion of multihop relay capabilities (IEEE 802.16j) aims to improve the speed of service delivery and coverage by a factor of 3 to 10. Moreover, IEEE 802.16m is expected to integrate and incorporate several advancements in transmission techniques in order to meet all the IMT-Advanced requirements, including 100 Mbps mobile and 1 Gbps nomadic access, as defined by ITU-R [ITU-R 2008].

## 1.4 LTE-Advanced and IMT-Advanced

4G aims to support the emergent multimedia and collaborative services, with the concept of "anywhere" and "anytime," meeting the latest bandwidth demands [Bhat 2012]. The LTE-Advanced (standardized by 3GPP) consists of a 4G system. Based on LTE, the LTE-Advanced presents an architecture using the All-over IP concept. The support for 100 Mbps in vehicular and 1 Gbps for nomadic access* is envisaged with the implementation of the following mechanisms:

- Carrier aggregation composed of multiple bandwidth components (up to 20 MHz) in order to support transmission bandwidths of up to 100 MHz
- Advanced antenna systems, increasing the number of downlink transmission layers to eight and uplink transmission layers to four (including downlink and uplink multi-user MIMO)
- Multihop relay (adaptive relay, fixed relay stations, configurable cell sizes, hierarchical cell structures, etc.) in order to achieve coverage improvement and/or increased data rate
- Advanced inter-cell interference cancellation (ICIC) schemes
- Multi-resolution techniques (hierarchical constellations, MIMO systems, OFDMA multiple access technique, etc.)
- Base station (BS) cooperation and macro-diversity

Standardization of LTE-Advanced is part of 3GPP Release 10 (completed in June 2011), and enhanced in Release 11 (slated for December 2012).

The IMT-Advanced refers to the international 4G system, as defined by the ITU-R [ITU-R 2008]. Moreover, the LTE-Advanced was ratified by the ITU as an IMT-Advanced technology in October 2010 [ITU 2010].

It is expected that, in future mobile radio networks, multihop relaying will be introduced [Sydir and Taori 2009]. Therefore, new topological approaches such as multihop or distributed antennas solutions and relaying allow an increased coverage of high rate data transmission, as well as improved system performance and capabilities.

Within 4G, voice, data, and streamed multimedia will be delivered to the user based on an All-over-IP packet switched platform, using IPv6. The goal is to attain

---

* 1 Gbps as a peak data rate in the downlink, whereas 500 Mbps is required for the uplink.

the necessary QoS and data rates in order to accommodate the emergent services. The potential transmission mechanisms listed above for LTE-Advanced are also included in the IMT-Advanced requirements.

Due to the improvements in address spacing with the 128 bits made available by IPv6, multicast and broadcast applications will be easily improved, as well as the additional security, reliability, intersystem mobility, and interoperability capabilities. In addition, the 4G concept consists of a pool of wireless standards that can be efficiently implemented using the software-defined radio (SDR) platform, which is currently an interesting R&D* area being explored by many industries worldwide.

## 1.5 Evolved Multimedia Broadcast and Multicast Service

The evolved MBMS (MBMS) framework [3GPP 2008b] constitutes the evolutionary successor of MBMS, and is envisaged to play an essential role in LTE-A proliferation in mobile environments. MBMS was introduced in 3GPP Release 8 (LTE), with further improvements to be implemented in LTE-A. The basic role of these techniques/mechanisms relies on power and resource optimization during MBMS transmissions. The objective of eMBMS is to provide services with different QoS requirements depending on the channel conditions experienced by different users. For services such as mobile TV, this means that users with better channel conditions will be able to receive video with the best possible quality, whereas users with poorer channel conditions will receive lower-quality video. Another example is given by Chen and Tsai [2009], where a cellular phone may only use a 1.4 MHz bandwidth to acquire audio service only, while a 10 MHz bandwidth can be employed by a laptop device to obtain both audio and video services. This impacts positively both the spectral efficiency of eMBMS and the service coverage achieved. UEs with good channel quality can receive higher data rates using the same number of subcarriers. UEs with poor channel conditions are guaranteed a minimum quality of service instead of no service availability.

### 1.5.1 Requirements for eMBMS

One of the most important properties of eMBMS is resource sharing among many UEs. This means that many users should be able to listen to the same eMBMS channel simultaneously. Therefore, power should be allocated to these eMBMS channels for arbitrary UEs in the cell to receive this service. This approach goes against the traditional power control concept used in unicast, where the power

---

* Research and development.

should be the minimum that provides the required quality of service (typically a minimum signal-to-noise ratio (SNR)) to a user located at the edge of a cell, while avoiding high interference levels to users in the same cell or in adjacent cells.

Services may be classified by the type of cast, namely, unicast, multicast, or broadcast. In the wireless medium, the broadcast is the basic mode, as the data is sent to all nodes. MBMS is a system that enables mobile networks to efficiently deliver data from a single content provider source to multiple users by sharing radio and transport network resources. Multimedia services are normally delivered either in multicast or broadcast mode. In multicast mode, the multimedia data is transmitted to a specific group of users (MBMS user groups) within a specific area (MBMS service area).

In broadcast mode, data is transmitted in a specific area (MBMS service area), but all users in the specific MBMS service area are able to receive the transmitted MBMS data.

For broadcast and multicast transmissions in a mobile cellular network, depending on the communication link conditions, some receivers will experience better SNR than others, and thus the capacity of the communication link for these users is higher. In broadcast transmissions, it is possible to transfer some of the capacity of the good communication links to the poor ones, and the trade-off can be worthwhile.

The specific feature for multi-resolution can be implemented using fast link adaptation: instead of compensating the variations of downlink radio conditions by means of power control, for point-to-point links the transmitted power is kept constant and the modulation and coding of the transport block is chosen every transmission time interval. This is possible with the AMC technique. To users experiencing good channel conditions, 16-QAM or even 64-QAM can be allocated to maximize throughput, whereas users experiencing poor channel conditions are penalized on throughput.

## 1.6 Introduction to 4G Transmission Techniques

The challenge facing the mobile telecommunications industry today is how to continually improve the end-user experience, to offer appealing services through a delivery mechanism that includes improved speed, service attractiveness, and service interaction. In order to deliver the required services to the users at the minimum cost, the technology should allow better and better performances, higher throughputs, improved capacities, and higher spectral efficiencies. The following sections describe several mechanisms that can be implemented in order to fulfill these goals.

The bandwidth requirements for eMBMS present a considerable challenge since multipath propagation leads to severe time-dispersion effects. In this case, conventional time-domain equalization schemes are not practical. Block transmission techniques, with appropriate cyclic prefixes and employing frequency-domain equalization (FDE) techniques, have been shown to be suitable for high data rate transmission over severely time-dispersive channels [Falconer et al. 2002], and therefore have advantages for use with emergent MBMS. The OFDM technique is the most popular modulation based on this technique. Moreover, MIMO schemes enhanced with state-of-the art receivers are also typically associated with MBMS in order to improve the overall system performance in terms of capacity, spectral efficiency, and coverage.

Since typical FDE receivers are coherent, we need accurate channel estimates at the receiver. These estimates can be obtained with the help of pilots multiplexed with data, either in the time or in the frequency domains [Hoher et al. 1997]. As an alternative, we can employ implicit pilots (also known as pilot embedding or superimposed pilots) [Lugo et al. 2004], where pilots or training signals are added to data so as to save bandwidth. The major problem associated with superimposed pilots is the interference between data and training signals: on the one hand the channel estimates are corrupted by the data signal, and on the other hand the detection performance will be degraded due to the interference from the training block. With advanced receivers it is possible to jointly perform detection and channel estimation, leading to a good trade-off with improved performance and improved spectral efficiency, while keeping complexity at a low level. This is especially advantageous when combined with multi-antenna systems, BS cooperation, and when hierarchical constellations are employed, situations where channel estimation can present a significant challenge.

For MBMS services, it makes sense to have two or more classes of bits with different error protection, to which different streams of information can be mapped. Depending on the propagation conditions, a given user can attempt to demodulate only the more protected bits or also the other bits that carry the additional information. By using nonuniformly spaced signal points in hierarchical modulations, it is possible to modify the different error protection levels [Vitthaladevuni and Alouini 2001, 2003]. These techniques are interesting for applications where the data being transmitted is scalable; that is, it can be split into classes of different importance. This technique is known as multi-resolution. For example, in the case of video transmission, the data from the video source encoders may not be equally important. This also happens in the transmission of coded voice. The nonuniform QAM constellation concept has already been incorporated in the DVB-T (digital video broadcast–terrestrial) standard.

Other aspects, such as synchronization, along with advanced equalizers and turbo decoders also come into play to obtain a correct estimate of the medium and sort all interferences between the different links.

## 1.6.1 Orthogonal Frequency Division Multiplexing

Orthogonal frequency division multiplexing is a transmission technique adopted by many high-data-rate communication systems such as Wi-Fi, WiMAX, and

LTE, because it is well suited to frequency selective fading channels [Cimini 1985; Liu and Li 2005]. The OFDM transmission technique splits the symbol streams into several lower rate streams, which are then transmitted in parallel subcarriers. As a consequence, the symbol period is increased, making the signal less sensitive to intersymbol interference (ISI). In the frequency domain, the frequency response of each subcarrier tends to be flat (that is, with a very low level of distortion), and an equalizer is normally utilized to mitigate the remaining effects of channel distortion (that is, the remaining effects of the frequency selectivity).

ISI occurs in digital transmission of symbols when the channel is characterized by the existence of several multipaths, some of which reach the receiver's antennas with delay higher than the root mean square (rms) delay spread of the channel (relating to the first path). In this case, this effect can be viewed in the frequency domain as having a signal composed of two or more sinusoids with frequency separation greater than the channel coherence bandwidth, being affected differently by the channel (in terms of attenuation and delay/phase shift). The ISI tends to increase with the increase of the signal bandwidth (increase of data rates, according to the Nyquist theorem). This is why the transmission of a signal with a data rate $N$ times higher cannot be achieved by simply using a signal with a transmission bandwidth $N$ times higher. In this case, in order to mitigate the increased ISI effects, a good solution is the OFDM transmission technique, which, as can be seen from Figure 1.4, splits a signal with a certain data rate into $N$ parallel lower rate subcarriers (symbols S0, S1, S2, and S3 are transmitted simultaneously but in different subcarriers) [Marques da Silva et al. 2010].

Although most of the processing of an OFDM transmitter is implicitly performed using discrete Fourier transform (DFT) and inverse DFT (IDFT) functions,

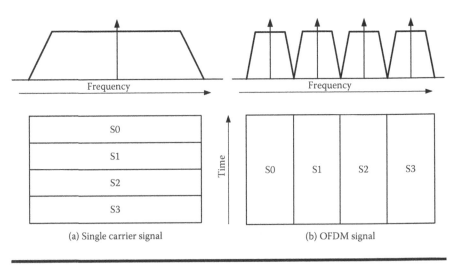

(a) Single carrier signal

(b) OFDM signal

**Figure 1.4   Spectrum of (a) single carrier signal versus (b) OFDM signal.**

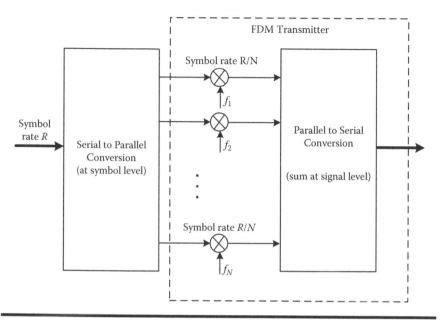

**Figure 1.5    Implicit baseband processing of an OFDM transmitter.**

the elementary processing of an OFDM transmitter is similar to that of a frequency domain multiplexing (FDM) transmitter. Figure 1.5 shows the implicit processing of an OFDM transmitter. As can be seen, such processing includes FDM processing, which consists of modulating different channels with different subcarriers, followed by an adder module (that is, sum at signal level). On the other hand, in OFDM processing, the source is a single channel (instead of multiple channels, as considered by a FDM transmitter), which is split into lower data rate channels (Figure 1.6).

The basic idea of the OFDM transmission technique consists of splitting a higher rate into a group of parallel lower rate streams, and modulating each lower rate stream with a subcarrier ( $f_1, f_2, ..., f_N$ in Figure 1.5) in such a way that the resulting parallel signals are ideally uncorrelated. Another difference between FDM and OFDM signals relies on how these subcarriers are uncorrelated. While FDM signals use guard bands to ensure uncorrelation between adjacent subcarriers, in OFDM, the uncorrelation is implemented using the IDFT (at the transmitter) and DFT (at the receiver) processing. This results in much greater efficiency, which typically improves spectrum efficiency. OFDM signals from adjacent subcarriers present some level of overlapping in the frequency domain. Nevertheless, the mathematical OFDM implementation using the IDFT/DFT ensures that the signals are uncorrelated at the receiver. Although these uncorrelated and parallel subcarrier signals are frequency separated, they are summed in time (second box in Figure 1.5) before being transmitted together. In practice, the DFT and IDFT

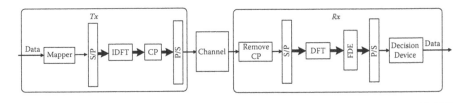

**Figure 1.6  Generic transmission chain for OFDM.**

operations are normally implemented through an efficient technique called Fast Fourier Transform (FFT) and inverse FFT (IFFT).*

A typical limitation of OFDM signals results from the fact that the dynamic range of an OFDM amplifier is very demanding, in order to respond to the amplitude variations which result from the instantaneous sum of $N$ subcarrier signals (parallel to serial conversion [S/P]). This is more visible for a higher number of subcarriers.

Besides the increased resistance against ISI that results from the lower symbol rate of each substream, another great advantage of OFDM signals results from the fact that even if a subcarrier experiences a deep fading or other type of interference, since the symbols are interleaved (serial-to-parallel conversion), these errors can easily be recovered using error correction techniques.

Moreover, in order to mitigate the residual effects of ISI, each block of $N$ IDFT coefficients is typically preceded by a cyclic prefix (CP) or a guard interval consisting of $N_g$ samples, such that the length of the CP is at least equal to the time span of the channel (channel length). The CP is simply a repetition of the last $N_g$ time-domain symbols.

The conventional equalization process of time domain signals consists of a series of convolution operations whose length is proportional to the time span of the channel. This means that for severely time-dispersive channels the receiver can be very complex and the level of processing can be high. Moreover, the computation of the channel impulsive response tends to be inaccurate, which results in an ineffective time domain equalization process. On the other hand, the equalization process of OFDM signals is simply performed as a multiplication of the OFDM signal spectrum (signal) by the frequency response of the channel. Note that the IDFT operation performed in the transmitter places the signal to be transmitted in the frequency domain. This represents a great advantage in terms of processing requirements and effectiveness, in comparison with the equalization process normally employed in time domain signals.

Due to its inherent immunity to ISI, the OFDM transmission technique is well suited for high-data-rate transmissions and is known for high performance in frequency-selective channels. It enables further frequency-domain adaptation, provides benefits in broadcast scenarios, and is well suited for MIMO processing.

---

* It is implemented computing the N-point FFT and IFFT.

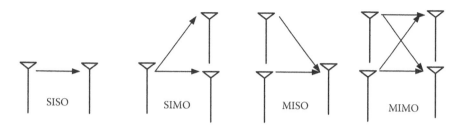

**Figure 1.7   Multiple Antenna configurations.**

The OFDMA can be viewed as a specific type of OFDM transmission where multiple access is achieved by assigning subsets of OFDM subcarriers to individual users.

The high peak to average power ratio (PAPR) of OFDM signals makes it unsuitable for mobile transmission (uplink). Consequently, due to its lower PAPR, SC-FDMA is the selected transmission technique for LTE uplink. Block transmission techniques are detailed in Chapter 2.

## 1.6.2 Multiple Input Multiple Output

The use of multiple antennas at both the transmitter and receiver aims to improve performance or to increase symbol rate of systems, but it usually involves higher implementation complexity. The antenna spacing must be larger than the coherence distance to ensure independent fading across different antennas elements[*] [Foschini 1996; Foschini and Gans 1998; Rooyen et al. 2000].

The various configurations, shown in Figure 1.7, are referred to as single-input single-output (SISO), multiple-input single-output (MISO), single-input multiple-output (SIMO), or multiple-input multiple-output (MIMO). The SIMO and MISO architectures are forms of receive and transmit diversity schemes, respectively. MIMO architectures can be used for combined transmit and receive diversity, for the parallel transmission of data or spatial multiplexing. When used for spatial multiplexing, MIMO technology promises high bit rates in a narrow bandwidth. Therefore, it is of great significance to spectrum users. In this case, the MIMO system involves the transmission of different signals from each transmit antenna element so that the receiving antenna array receives a superposition of all transmitted signals.

MIMO schemes are used to push the performance or capacity/throughput limits as high as possible without an increase of the spectrum bandwidth, although at the cost of an obvious increase of complexity [Rooyen et al. 2000; Marques

---

[*] Alternatively, different antennas should use orthogonal polarizations to ensure independent fading across different antennas.

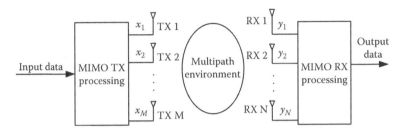

**Figure 1.8    Generic diagram of a MIMO scheme.**

da Silva and Correia 2001, 2002a, 2002b, 2003; Hottinen et al. 2003]. Figure 1.8 is a generic diagram of a MIMO scheme.

MIMO schemes are implemented based on multiple-antenna techniques. These multiple-antenna techniques may present the following configurations:

- Space-time block coding (STBC)
- Multi-layer transmission
- Space division multiple access (SDMA)
- Beamforming

Although STBC is essentially a MISO system, the use of receive diversity makes it a MIMO, which corresponds to the most common configuration for this type of diversity. STBC-based schemes focus on achieving a performance improvement through the exploitation of additional diversity, while keeping the symbol rate unchanged [Alamouti 1998; Tarokh et al. 1999]. Symbols are transmitted using an orthogonal block structure, which enables a simple decoding algorithm at the receiver [Alamouti 1998; Marques da Silva et al. 2004].

Multi-layer transmission and SDMA belong to the same group called spatial multiplexing (SM), whose principles are similar but whose purposes are quite different. The goal of the MIMO based on multi-layer transmission scheme is to achieve higher data rates in a given bandwidth, whose rate of increase corresponds to the number of transmit antennas [Foschini 1996; Foschini and Gans 1998]. An example of the multi-layer transmission scheme is Vertical–Bell Laboratories Layered Space-Time (V-BLAST). In this case, the number of receive antennas must be equal to or higher than the number of transmit antennas. The increase of symbol rate is achieved by "steering" the receive antennas to each one (separately) of the transmit antennas in order to receive the corresponding data stream. This is achieved through the use of the nulling algorithm. With a sufficient number of receive antennas, it is possible to resolve all data streams, as long as the antennas are sufficiently spaced so as to minimize the correlation [Marques da Silva et al. 2005].

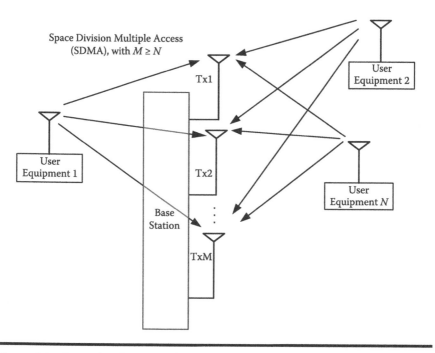

**Figure 1.9    Example of SDMA scheme applied to the uplink.**

The goal of the SDMA scheme is to improve the cell capacity (more users per cell) while keeping the spectrum allocation unchanged. It is usually considered in the uplink, where the transmitter (UE) has a single antenna while the receiver (BS) has several antennas. Figure 1.9 depicts an SDMA configuration applied to the uplink. SDMA assumes that the number of antennas at the receiver is higher than the number of users that share the same spectrum. With such an approach, the receiver can decode the signal from each transmitter while avoiding the signal from the other transmitters. Similar to the decoding performed in multi-layer transmission, this can be achieved through the use of the nulling algorithm.

In SM, the symbol with the highest SNR is first detected using a linear nulling algorithm such as zero forcing (ZF) or minimum mean square error (MMSE) [Foschini 1996]. The detected symbol is regenerated, and the corresponding signal portion is subtracted from the received signal vector, typically using a successive interference cancellation (SIC) algorithm. This cancellation process results in a modified received signal vector with fewer interfering signal components left. This process is repeated until all symbols from different transmit antennas (or users) are detected. According to the detection-ordering scheme in [Foschini 1996], the detection process is organized so that the symbol with the highest SNR is detected at each detection stage.

Unlike the STBC and SM MIMO schemes, where those antenna elements that form an array are usually widely separated in order to form a transmit diversity array with low correlation among them, beamforming is implemented by an antenna array with certain closely located array elements. Commonly, the antenna array forms a beam with an antenna element spacing of typically half a wavelength. This scheme is an effective solution to maximize the SNR, as it steers the transmit (or receive) beam toward the receive (or transmit) antenna [Marques da Silva et al. 2009a]. As a result, an improved performance or coverage is achieved with beamforming.

MIMO schemes require normally some additional processing and complexity from the receiver. Different receivers were analyzed in Silva et al. [2005a, 2005b], in order to establish a trade-off between performance and complexity for such systems. An alternative approach that minimizes the level of computation from the receiver, transferring it to the transmitter while retaining the ability to exploit diversity, has been proposed and studied in Marques da Silva et al. [2008, 2009a]. Such an approach is embodied in pre-coding schemes for MIMO systems. MIMO systems are detailed in Chapter 4.

### 1.6.3 Single-User MIMO and Multi-User MIMO

The MIMO techniques previously described are typically employed in the concept of single-user MIMO (SU-MIMO). SU-MIMO considers data being transmitted from a single user to another individual user. An alternative concept is multi-user MIMO (MU-MIMO), where multiple streams of data are simultaneously allocated to different users, using the same frequency bands. This increases spectral efficiency relating to the conventional SISO system, which also corresponds to a higher throughput delivered to different cell UEs using the same frequency bands.

The approach behind MU-MIMO is similar to SDMA. Nevertheless, while SDMA is typically employed in the uplink,* MU-MIMO is normally implemented in the downlink. This allows sending different data streams to different UEs. In this case, instead of performing the nulling algorithm at the receiver side, the nulling algorithm is performed using a pre-coding approach at the transmitter side (BS). This is possible because the BS can accommodate a high number of transmit antennas and the UE can only accommodate a single or reduced number (lower) of receive antennas. In the downlink of a MU-MIMO configuration, the number of transmit antennas is higher than the number of multiple data streams that are sent to multiple users at the same time and occupying the same frequency bands (the opposite of the SDMA approach). In this configuration, the nulling algorithm is implemented at the transmitter side using a pre-processing algorithm such as zero forcing, MMSE, dirty paper coding, etc. Alternatively, instead of implementing

---

* Because the BS typically has the ability to accommodate a higher number of receive antennas to perform the nulling algorithm.

the foregoing spatial multiplexing principle, MU-MIMO can be performed using the beamforming algorithm. In any case, MU-MIMO requires accurate downlink Channel State Information (CSI) at the transmitter side. This normally results in lower gains than those obtained under the assumption of ideal CSI. Obtaining CSI is trivial using TDD mode, being more difficult to be achieved when FDD is employed. In FDD mode, CSI is normally obtained using a feedback link in the opposite direction.

It is worth noting that when the goal relies on achieving a performance improvement, then SU-MIMO is normally employed using an algorithm such as STBC. On the other hand, when the aim is to achieve higher throughputs using a constrained spectrum, then MU-MIMO is typically the solution, using one of the approaches described above. Users located at the cell edge served by MU-MIMO may experience a degradation of the SNR due to inter-cell interference, inter-user interference,* additional path loss, or limited BS transmit power (which results from the use of a pre-coding). A mechanism that can be implemented to mitigate this limitation is to employ a dynamic MIMO system, where MU-MIMO is employed everywhere except at the cell edge. In this location, the BS switches into SU-MIMO, which improves performance [Liu 2012]. Alternatively, BS cooperation is known to be an effective mechanism that improves the performance at the cell edge, resulting in a more homogenous service quality, regardless of the users' positions.

## 1.6.4 Base Station Cooperation

An important requirement of 4G systems is the ability to deliver a homogenous service, regardless of the users' location. Users at the cell edge may experience a degradation of the SNR due to inter-cell interference, additional path loss, or limited Evolved NodeB[†] (eNodeB) transmit power. In the event MU-MIMO is employed, the power constraints become more important,[‡] and the SNR degradation of UEs located at the cell edge can be even deeper than in SISO environments. In these scenarios, BS cooperation plays an important role, as it allows the exploitation of additional diversity or the delivery of a high and constant throughput (MU-MIMO), regardless of the users' positions.

BS cooperation stands for the ability to send or receive data from/to multiple adjacent eNodeBs to/from UEs located at cell edges. Using BS cooperation, independent antenna elements of different eNodeBs are grouped, and the UEs can

---

* Users that share the spectrum and that are separated by MU-MIMO spatial multiplexing.
† eNodeB refers to the base station of LTE and 4G systems. This corresponds to the evolution of the NodeB, employed in previous UMTS.
‡ In MU-MIMO, the power at the transmitter (NodeB, in case of downlink) becomes very demanding because the pre-coding may require a high level of power and because the available power is split over different streams of data.

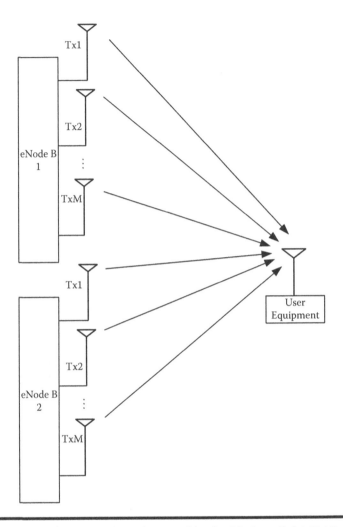

**Figure 1.10    Base station cooperation combined with downlink MIMO.**

experience a throughput increase (MU-MIMO) or performance improvement.[*]
The resulting MIMO can be viewed as a "giant MIMO," consisting of a combination of the independent antenna elements from different eNodeBs (see Figure 1.10).

If the aim is to achieve an increase of throughput, BS cooperation can be viewed as a special type of MU-MIMO. Similar to MU-MIMO, a preprocessing such as beamforming, zero forcing, MMSE, or dirty-paper is typically employed in order to ensure that the UE receives a signal coming from multiple eNodeBs.

---

[*] A performance improvement can be achieved through the exploitation of additional diversity using, for example, space-time coding.

This allows spectrum efficiency improvement, even at the cell edge. As for the downlink MU-MIMO case, accurate downlink CSI is also required at eNodeBs, which consists of an implementation difficulty. Moreover, BS cooperation may also be employed to minimize the level of interference experienced by served UEs.

BS cooperation can also be employed in the uplink. In this case, independent parallel streams of data can be sent out by different UE transmit antennas (typically, a different data stream is sent by each transmit antenna element), while different eNodeBs decode these different streams using the foregoing SDMA algorithm. Alternatively, traditional receive diversity can be employed. In both cases, the network needs to include a centralized post-processing, as opposed to traditional individual eNodeB processing.

## 1.6.5 Macro-Diversity

Macro-diversity refers to the transmission of the same information by different eNode-Bs to the UEs in the downlink. Macrodiversity aims at supplying additional diversity* in situations where the terminal is far from the BSs. This allows compensating the path loss affecting the transmission to a UE located at the edge of the cell. Consequently, it allows a reduction of the amount of transmit power needed to reach a distant receiver, thus increasing network capacity. Therefore, macrodiversity can be viewed as a special type of BS cooperation.

There are two types of networks to be considered: the multi-frequency networks (MFNs) and the single frequency networks. The eNodeB to which a terminal is linked to is referred to as the active set.

In broadcast, the global channel impulse response (CIR) is longer due to the longer distances between the transmitter (BS) and the farther of the different receivers (UEs). Nevertheless, if the cyclic extension is long enough, the global CIR will be the sum of the independent CIRs. This is shown in Figure 1.11. This enables the SFN concept by exploiting the macro-diversity effect.

Macro-diversity is used during soft handover in order to ensure smooth transitions between two cells or two sectors of the same cell, increasing the multipath diversity and reducing the risk of call drop.

Using the same transmitting frequencies (from different eNodeBs), a diversity is obtained and deep fadings tend to be avoided. In case of a single-path channel profile coming from two eNodeBs, the resulting signal is viewed as a two-path channel profile, and diversity is exploited. Nevertheless, since a single receiver is employed, although diversity is exploited, the frequency selectivity increases. In fact, this effect can be viewed as one signal received from one eNodeB being interfered with by the signal received from the other eNodeB. Nevertheless, since OFDM signals make use of equalizers at the subcarrier level and using the appropriate cyclic prefix,

---

\* Since it provides transmit diversity, macro-diversity can be viewed as a MISO system.

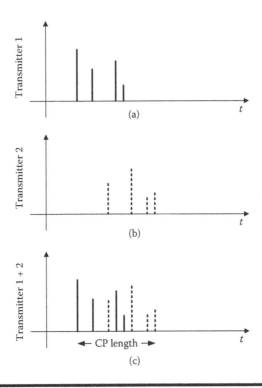

**Figure 1.11  Global CIR (c) is composed of the sum of the CIR of the several transmitters (a + b).**

this additional frequency selectivity is mitigated. Consequently, macro-diversity presents overall benefits in terms of diversity.

The performance gain brought about by macro-diversity depends on the diversity order of the channel. A two-path channel benefits more from macro-diversity than a six-path channel because the latter already exhibits a high multipath diversity order. Unlike the dedicated physical channel, macro-diversity for MBMS in SFN mode does not consume network resources, because MBMS is broadcasted simultaneously in several cells.

In the case of MFN, the UE is required to estimate the carrier of each eNodeB to which it is linked. This increases its power consumption. In this case, the signals received from different eNodeBs (especially distant ones) may be significantly delayed with regard to those received from near eNodeBs. This requires extra memory at the terminal in order to store the received signals for further combining. Alternatively, an additional synchronization procedure among the eNodeB transmitters is required. In the downlink, the combining takes place at the mobile, which has to demodulate and then combine the signals received from the different eNodeBs in the active set. The extra complexity added by macro-diversity then depends on the receiver type. In the case of an equalizer, one has to be set up and operate for each eNodeB to which the UE is linked. Moreover, the UE must estimate one transmission channel per eNodeB.

In the special case of OFDM, two main cases for macro-diversity can be distinguished:

■ eNodeBs are synchronized, at least to allow UEs to receive signals from two or more eNodeBs with a time difference smaller than the cyclic prefix.
■ eNodeBs are not synchronized.

In the first case, the eNodeBs can transmit identical signals to the terminal on the same time-frequency resource. This is possible because the signals will superpose within the cyclic prefix: no ISI occurs as long as the sum of the time differences plus the maximum delay of the channel impulse responses is shorter than the cyclic prefix. In this case, the terminal can employ a single receiver to demodulate the superimposed signals. This means that it will perform a unique DFT. In this case, macro-diversity behaves just like transmit diversity (from a unique transmitter with multiple spaced antennas). When different eNodeBs transmit the same data over the same subcarriers, the resulting propagation channel is equivalent to the cumulative sum of all propagation channels, which increases the diversity gain.

When eNodeBs send the same data over different subcarriers (MFN), the maximum diversity can be achieved since each data symbol benefits from the summation of the propagation channel powers (that is, frequency diversity). It could also be possible to form a MIMO scheme, but in that case the data and distinct pilot signals have to be sent on orthogonal time-frequency resources and several propagation channels have to be estimated.

If the eNodeBs are not synchronized, the terminal will need separate receiver chains to demodulate the signals from the distinct eNodeBs. Moreover, to avoid interference, orthogonal time-frequency resources have to be allocated to different eNodeBs. This is still very complex to achieve, and thus in the general case, interference will occur.

Fast cell selection is one option for macro-diversity in unicast mode (selective diversity). Intra-eNodeB selection should be able to operate on a subframe basis. An alternative Intra-eNodeB macro-diversity scheme for unicast consists of a simultaneous multi-cell transmission with soft combining. The basic idea of multi-cell transmission is that, instead of avoiding interference at the cell border by means of inter-cell-interference coordination, both cells are used for the transmission of the same information to a UE, thus reducing inter-cell interference as well as improving the overall available transmit power. Another possibility of intra-NodeB multi-cell transmission relies on the exploitation of diversity between the cells with space-time processing (that is, by employing STBC through two cells). Assuming eNodeB-controlled scheduling and that fast/tight coordination between different eNodeB is not feasible, multi-cell transmission should be limited to cells belonging to the same eNodeB. For multi-cell broadcast, soft combining of radio links should be supported, assuming a sufficient degree of inter-eNodeB synchronization, at least among a subset of eNode-Bs.

## 1.6.6 Multihop Relays

One important key to improving the coverage and capacity for high-quality multimedia broadcast and multicast transmissions in mobile networks is to provide a homogenous service, regardless of the users' location, that is, to allow high data rates for UEs even at the cell edge. UEs at the cell edge suffer from high propagation loss and high inter-cell interference from neighbor cells. Other UEs reside in areas that suffer from strong shadowing effects or require indoor coverage from outdoor BS. Thus, the overall goal of multihop relay is to bring more power to the cell edge and into shadowed areas while inducing minimal additional interference for neighboring cells. There are several methods of implementing multihop relays [Sydir and Taori 2009]. The next paragraphs describe some of these methods.

*Adaptive Relaying*: As stated before, UEs at the cell edge suffer from high propagation loss and high inter-cell interference from the neighboring cells. Other UEs reside in areas that suffer from strong shadowing effects. The obvious solution to this would be to decrease cell sizes by installing additional BSs, which would, of course, represent an increase in network infrastructure costs. On the other hand, adaptive relay nodes can provide temporary network deployment and service outage in the area. Their positioning is selected so that an improvement of service quality or cell coverage is achieved.

In opposition to conventional repeaters working with the amplify-and-forward strategy, adaptive relays are understood to work in a decode-and-forward style. By performing this, relays amplify and retransmit only the required component of the signal they receive and suppress the unwanted portions (that is, regeneration is performed). The disadvantages of relays compared to simple repeaters are the additional delay that they introduce into the transmission path between BS and UE and, depending on the algorithms, a possible signaling overhead. The gain is based on the fact that the transmission path is split up into smaller parts, which can reduce propagation loss.

Fixed relays stations positioned at a specified distance from the BS (Figure 1.12) could help increase the probability that a UE receives enough power from several BSs. This deployment concept would sectorize the cell in an inner region where the UEs (e.g., UE 1 in the figure) can receive their signals from the BS plus some relay stations and an outer region where only signals from relay stations are strong enough (e.g., UE 2 in the figure).

Relay nodes (RNs) were introduced in 3GPP Release 9 [3GPP 2010b] as a special type of eNB* that is not directly connected to the core network. An RN

---

* eNB stands for evolved NodeB, that is, a NodeB employed in LTE.

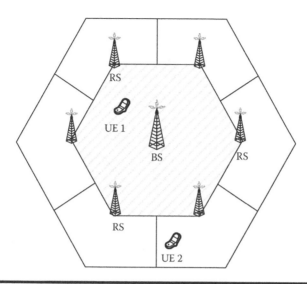

**Figure 1.12    Two-hop relaying architecture.**

receives data that was forwarded by an eNB* connected to the evolved packet core[†] (EPC). Upon receiving such data, the RN sends it to the UEs under its area of coverage. This technique is depicted in Figure 1.13. This is a very interesting option for operators, as RNs usually imply structures less expensive to deploy and to maintain, in comparison with eNBs. They can provide temporary network deployment and service outage in an area (e.g., during a sporadic event that concentrates a lot of people in the same geographical area, such as a summer festival). The use of an RN allows fast deployment, is an inexpensive way to solve the problem, and can also provide coverage in small areas not covered by eNBs.

*Configurable Virtual Cell Sizes.* Kudoh and Adachi [2003] proposed a wireless multihop virtual cellular network. It consists of a so-called central port that corresponds to a BS acting as a gateway to the core network. The so-called wireless ports correspond to relay stations that communicate with the UEs and relay the signal from and to the central port. The wireless ports that communicate directly with the UEs are called end wireless ports. The wireless ports are stationary and can act together with the central port as one virtual BS. The central port and the end wireless ports introduce additional diversity into the cell, so that the transmit power may be reduced. This also means a reduced interference for other virtual cells. The differences between present cellular networks and virtual cellular networks are illustrated in Figures 1.14 and 1.15.

---

* This eNB is denoted as Donor eNB or DeNB.
† Evolved packet core corresponds to the core of the LTE (All-IP) network.

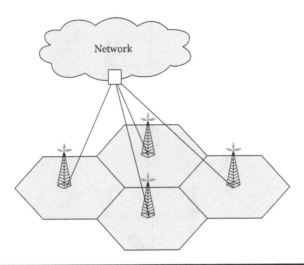

**Figure 1.13   Present cellular network.**

From the perspective of multimedia broadcasting and multicasting, configurable virtual cell sizes could be employed to adapt the cell size to the user distribution and service demands. Service needs can present spatial and temporal variations.

In 3GPP [2010c], four relay architectures are proposed and studied. These architectures differ from one another in terms of expected behavior of the RN/DeNB and how the data is sent within the EPC until it reaches the UE. This study

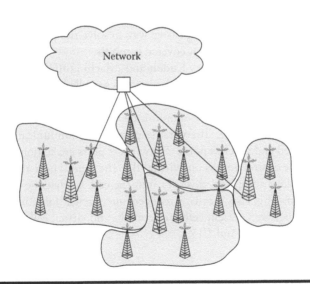

**Figure 1.14   Virtual cellular network.**

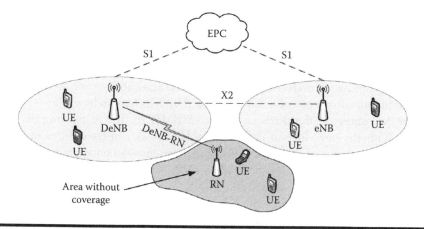

**Figure 1.15   Signal constellation for 16-QAM nonuniform modulation.**

concludes that an architecture where RN acts as a proxy for S1/X2 has the most overall benefits, having been incorporated in 3GPP Release 10 (LTE-Advanced).

The detailed RN architecture chosen for LTE-Advanced can be seen from Figure 1.16. In this architecture, we can see that the RN simultaneously plays two roles:

■ From the network point of view (particularly for the DeNB), the RN acts as an ordinary UE (denoted as Relay-UE).
■ From the UE point of view, the RN acts as a normal eNB.

In this way, the network can abstract itself from establishing a point-to-point connection with each and every UE. In fact, it only has to establish a connection to the

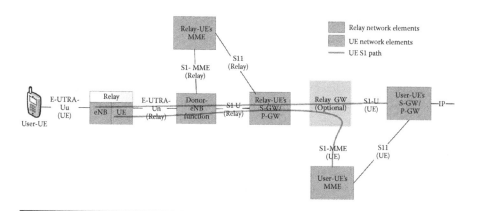

**Figure 1.16   Example of the type of demodulation used inside a cell for transmission of a 64-QAM hierarchical constellation.**

RN as it would normally with an ordinary UE, and then forward all data that was destined for UEs to the RN. Then, the RN, on its own, will forward all the data received from the network to the respective UEs. The process of relaying is completely transparent; that is, a UE always assumes it is connected to an eNB because RNs are viewed as an eNB from the UE side, and as a consequence no changes to the communication protocol or interfaces used by UEs have to be made.

In the context of multi-resolution for eMBMS, the use of relaying fits perfectly. With the introduction of RNs in eMBMS, we can have different zones with different grades of service. For instance, if we consider once again the example in Figure 1.15, in the areas covered by eNBs, UEs are expected to have high signal-to-interference plus noise ratio (SINR) when they are closer to the center of the cell, diminishing as the UEs approach the cell edges. If a hierarchic 64-QAM modulation with coding rate 3/4 is being used, UEs at the cell edge will experience low levels of QoS or even service outage due to low SINR. If an RN is strategically positioned at the cell edges of existing eNBs, we can expect that area to have improved coverage and the RN can adapt its transmitting conditions to provide a different resolution of the service in the area.* With such an approach, the use of an RN means that, in E-UTRAN, when providing eMBMS, UEs will experience higher coverage and different service resolutions for the same eMBMS contents depending on their location and their receiving conditions.

According to 3GPP [2010b], RNs can operate in two different modes:

- Inband mode: The link between the DeNB and the RN uses the same carrier frequency as the link between the RN-UE and the eNB-UE.
- Outband mode: The link between DeNB and the RN uses a different carrier frequency than that of the RN-UE link.

It is worth noting that the transmit power of RNs is typically lower than that of the eNBs. The objective is to diminish inter-cell interference caused by the introduction of RNs operating in the same band and carrier frequency as eNBs (inband mode), and to limit the area of coverage of a given RN (since RNs are used to cover small areas that cannot be properly covered by existing eNBs).

In order to reduce inter-cell interference, RNs can also be employed in eMBMS subframes to create "transmission gaps." These transmissions gaps are illustrated in Figure 1.17. As can be seen, in the first instant where $Time = t$, the DeNB transmits a segment of data labeled "Data 1" to all the UEs inside its area of coverage and also to the RN. At this instant, the RN stores "Data 1" in memory and is not transmitting any data to the users inside its area of coverage, thus not creating any inter-cell interference. At the next instant $Time = t + 1$, a second set of data labeled "Data 2" is transmitted

---

* Instead of transmitting video with high frame rate and low resolution, we can halve the frame rate and transmit with higher resolution.

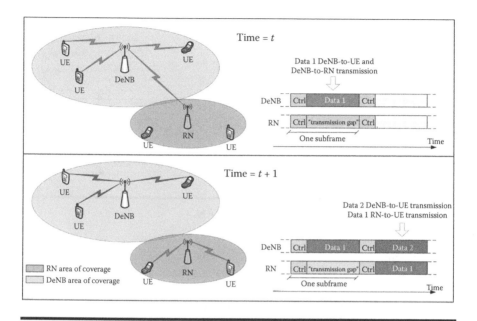

**Figure 1.17    Example of H-64QAM modulation.**

from DeNB to all the UEs within its area of coverage. In its turn, the RN is not receiving "Data 2" but, instead, is transmitting to the UEs within its area of coverage the "Data 1" that was stored in memory at *Time* = *t*, generating inter-cell interference.

As a result of this inter-cell interference pattern produced by RNs, the total inter-cell interference is greatly reduced (RNs are only causing interference half of the total transmission times) with a trade-off between throughput and inter-cell interference.

## 1.6.7  Hierarchical Constellations

Availabilty of limited spectrum resources is among the major constraints to achieving high-bit-rate transmissions in wireless communication networks. Multilevel quadrature amplitude modulation (M-QAM) is considered an attractive technique to achieve this objective due to its high spectral efficiency and it has been studied and proposed for wireless systems by several authors [Webb and Hanzo 1994; Webb and Steele 1995; Goldsmith and Chua 1997].

As discussed in Cover [1972], M-QAM constellations can be constructed in a hierarchical way so as to provide multi-resolution and improve the efficiency of the network in broadcast/multicast transmissions. In this case, the constellations can be referred to as hierarchical, embedded, or multi-resolution M-QAM (we will denote them as M-HQAM [Hierarchical QAM]).

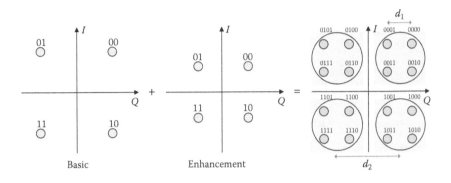

**Figure 1.18  Example of relay node in E-UTRAN.**

For example, 16-QAM hierarchical constellations can be constructed using a main QPSK constellation where each symbol is, in fact, another QPSK constellation. This is depicted in Figure 1.18. Such a construction procedure results in two classes of bits with different error protection that can be modified by using nonuniformly spaced signal points.

The bits used for selecting the symbols inside the small inner constellations are called weak bits, and the bits corresponding to the selection of the large outer QPSK constellation are called stronger bits. The idea is that the constellation can be viewed as a 16-QAM constellation if the channel conditions are good or as a QPSK constellation otherwise. In the latter situation, the received bit rate is reduced to half, which results in a service with lower resolution (e.g., video). Some alterations to the physical layer of the UMTS system to incorporate these modulations have already been proposed in Souto et al. [2005a, 2005b, 2007a].

Figure 1.19 shows an example of the usage of a 64-HQAM constellation in a cellular system. Depending on the position in the cell, the users will demodulate the received signal as 64-QAM, 16-QAM, or QPSK.

This type of multi-resolution technique is interesting for applications where the data being transmitted is scalable, that is, when it can be split into classes of different importance. For example, in the case of video transmission, the data from the video source encoders may not be equally important. The same happens in the transmission of coded voice. Several authors have studied the use of hierarchical constellations for this purpose. In Ramchandran et al. [1993] and Wei [1993], hierarchical QAM constellations were employed for the transmission of digital high definition television signals. Moreover, Engels and Roohling [1998] compare the performance of 64-QAM and 64-DAPSK hierarchical constellations, while Pursley and Shea [1999] study the application of M-PSK hierarchical constellations in multimedia transmissions. Hierarchical 16-QAM and 64-QAM constellations have been incorporated into DVB-T standards [ETSI 2004b]. Hierarchical constellations are detailed in Chapter 3.

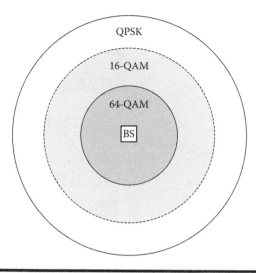

**Figure 1.19** **Relay architecture chosen for LTE-Advanced. (Adapted from 3GPP, Evolved Universal Terrestrial Radio Access (E-UTRA); Relay architectures for E-UTRA (LTE-Advanced), TR 36.806 v9.0.0, May 2010.)**

## *1.6.8 Multi-Resolution Techniques*

Multi-resolution techniques allow access to different service grades as a function of the channel conditions or mechanisms employed. Multi-resolution considers different layers of bits, corresponding to different resolutions: the most protected bits are available to every user and the ones carrying additional resolution are only available to users with higher SNR. Depending on the user's position inside a cell, more or fewer layers with additional resolution will be correctly received by the mobile.

Hierarchical modulation is considered a multi-resolution technique that can be employed in eMBMS because users with different levels of channel quality will have access to different types of service quality. This can be seen from Figure 1.20. An H-64QAM modulation is used at the cell, and depending on the channel conditions of each user,* the mobile user will receive:

■ The strong bits, corresponding to basic quality video (QPSK)
■ Both the strong and medium bits, corresponding to refined quality video (16-QAM)
■ All the strong, medium, and weak bits together, corresponding to a full-quality video (64-QAM)

---

* The distance to the center of the cell tends to be inversely proportional to the received SNR.

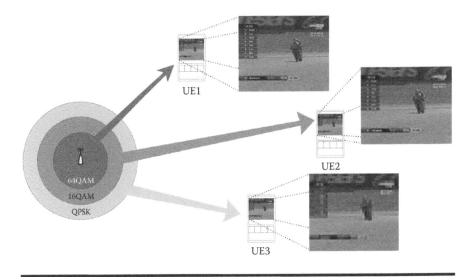

**Figure 1.20    Example of transmission gaps in eMBMS.**

The use of relay nodes is another possible multi-resolution technique that can be applied in eMBMS, and was introduced in 3GPP Release 9 specifications as a feature capable of improving cell-edge throughput and coverage [3GPP 2010b]. A UE served by a relay node and eNB can experience full service, whereas a UE only served by a far eNB may be limited in terms of service capabilities or resolution.

In LTE, there are several different classes of UEs, with differences among them regarding the combination of antennas, and consequently the maximum data rates that each one can achieve. No fewer than two receiving antennas at the UE are expected. Consequently, a SIMO antenna configuration is the most basic configuration that can be used in LTE UEs. The usage of two receiving antennas at the UE represents a diversity gain that boosts the performance that can be achieved with techniques such as soft combining.

According to 3GPP specifications for LTE, a MIMO system of up to $4 \times 4$ configurations can be adopted for UEs. A spatial multiplexing MIMO system is a method capable of offering multi-resolution and taking advantage of the different link capacities. In MIMO-based networks, the transmission of different sequences of data from different antennas ensures multi-resolution. In OFDMA-based networks, the transmission from each set of antennas of different fractions of the total set of subcarriers (physical resource blocks) is another way to offer multi-resolution.

Any of the previous methods is able to provide unequal bit error protection. In any case, there are two or more classes of bits with different error protection levels, to which different streams of information can be mapped.

Regardless of the channel conditions, a given user always attempts to demodulate every type of bit, the most protected and those carrying additional resolution.

Depending on the UE's position inside the cell, more or fewer blocks with additional resolution will be correctly received. The basic quality will always be correctly received independent of the position of any user, within the 95% target of coverage.

## 1.7 Energy Efficiency in Wireless Communications

The tremendous expansion of mobile network terminals has contributed to the increase of the environmental footprint. Currently, more than 4 billion subscribers are using mobile phones around the world. The operation of both mobile phones and network infrastructure require huge quantities of electrical energy, which currently represents up to 50% of the operational costs.

Energy-efficient wireless transmission techniques alongside energy-efficient network architecture and protocols, hardware optimization, and renewable energy sources contribute to the implementation of the green cell networks concept because they help reduce the carbon emission footprint. Summarizing, the carbon footprint reduction can be achieved through advancements in three different main dimensions:

- Architecture and protocols: Improvements in the transmission techniques and network protocols and architectures
- Components: More efficient hardware implementation in terms of energy dependency
- Energy provisioning: Including a new strategy based on the generalized use of renewable energy sources

The architecture and protocols dimension can be implemented on two different fronts: the network level and the link level.

The network level comprises the dimensioning and adjustment of the network as a whole such that a reduction of energy consumption is possible without radically compromising network performance and spectral efficiency [Correia et al. 2010b; Ferling et al. 2010]. This is achieved through the implementation of network protocols such as BS cooperation, multihop relay (see Figure 1.21), MIMO systems, hierarchical cellular structure,[*] or interference mitigation schemes (including pre-coding). These techniques contribute to a reduction in the transmit power by both eNodeBs and UE [Ericsson 2007].

Mobile ad hoc networking (MANET) is also a concept that allows the implementation of networks without use of eNodeBs, and thus achieving the same capability with less energy and with less visual impact and investment. Energy efficiency is also achieved by dynamically allowing an adaption of the cellular network architecture to traffic load fluctuations. In this case, the network architecture should

---

[*] Using macrocells, microcells, picocells, femtocells, etc.

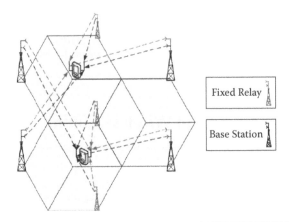

**Figure 1.21    Cooperative MIMO system with fixed relay station.**

be sufficiently dynamic such that it adapts to variations in the number of UEs, variations in the throughputs required by different UEs, or to various geographic densities of UES. Cognitive radio brings flexibility to the network as it constitutes a mechanism that allows a dynamic and more efficient use of the spectrum. This results in higher throughput per user or higher number of users per cell by exploiting spectrum opportunities. Cognitive radio constitutes an efficient alternative to dimensioning the network to peak traffic scenarios.

The link level front is associated with the individual energy utilization in the interface between a UE and an eNodeB. This includes synchronization and channel estimation techniques. Due to multiplexing of pilot/training and data symbols, some of the available bandwidth and energy has to be consumed for accomplishing the transmission of the pilot symbols. Since the channel impulse response is usually very long, especially for block transmission schemes, the required channel estimation overheads can be high, namely for fast-varying scenarios. This implies a reduction of the useful bit rate, decreasing the spectral efficiency and increasing the energy needs. A promising method for overcoming this problem relies on the employment of implicit pilots, also known as pilot embedding or superimposed pilots, which are added to the data block, instead of being multiplexed with it (see Chapter 4). This means that we can significantly increase the pilots' density, while retaining the system capacity. Note that the MIMO implementation constitutes a mechanism to improve the spectral efficiency, throughput, or number of users per cell. Nevertheless, it requires additional energy spent in pilots, as the channel estimation is independently performed for each antenna pair, using a different pilot stream. This results in an energy efficiency degradation of the MIMO relating to the SISO. The implementation of embedded pilots enables to become the MIMO system because an energy-efficient technique as the energy spent in pilot symbols is very much reduced, and thus the MIMO can improve the SNR, coverage, or throughput, without additional transmit power (or even with a power reduction).

From a component perspective, the basic concept relies on the development of signal processing techniques for smart components in order to reduce energy consumption. Moreover, the relationship between the output power and the consumed power of apiece of equipment or component (e.g., a transmitter) should be maximized. This can be viewed as power efficiency, and can also be improved by implementing a careful design and advanced signal processing. It is known that the energy consumed in a power amplifier represents between 50% and 80% of the electrical energy consumed in an eNodeB [Correia et al. 2010b]. The OFDM transmission technique is characterized by high PAPR[*] levels, whose implementation requires a power amplifier (PA) operating well below the saturation point. This results in poor power efficiency. A signal processing technique commonly employed to improve the energy efficiency of a PA relies on decreasing the PAPR through the use of a pre-coding technique that reduces the dynamic range of a block transmission signal. This technique allows a PA operation closer to the saturation threshold, which leads to an energy efficiency gain. Another signal processing technique currently in development relies on the implementation of smart cooling.[†] It is also worth noting that cooling systems tend to be more demanding in high-power transmitters. Reducing the required transmit power allows a simpler and less powerful cooling system.

A new energy provisioning strategy based on the use of renewable energy sources is another dimension that can be exploited in order to achieve reduction of the carbon emission footprint. This includes the implementation of photovoltaic cells and wind generators in BSs. Moreover, photovoltaic cells can also be employed in mobile phones or in human clothing to charge it.

---

[*] The PAPR level increases with the increase of the number of OFDM subcarriers.
[†] Cooling systems represent up to 30% of the consumed BS energy.

# Chapter 2

# Block Transmission Techniques

This chapter describes block transmission techniques combined with frequency-domain implementations that are suitable for high rate transmission over severely time-dispersive channels. These techniques include OFDM, selected for DVB-T, Wi-Fi, WiMAX, and 3GPP LTE, as well as SC-FDE, also considered for 3GPP LTE.

This chapter is organized as follows: Section 2.1 describes OFDM signals and FFT-based transmitter and receiver implementations. Section 2.2 is concerned with PAPR issues for OFDM signals, as well as appropriate PAPR-reducing techniques. Section 2.3 describes SC-FDE schemes and its relationship with OFDM schemes. We describe linear and iterative FDE receivers for SC-FDE. Section 2.4 presents some performance comparisons of OFDM and SC-FDE.

## 2.1 OFDM Schemes

### 2.1.1 Concept

To understand the OFDM concept, we will start with a more general multicarrier signal whose complex envelope can be written as

$$s(t) = \sum_{k=-\frac{N}{2}}^{\frac{N}{2}-1} S_k r(t) \exp(j2\pi tkF), \tag{2.1}$$

where $r(t)$ is an appropriate pulse shape, $F = 1/T$ is the subcarriers' separation, $N$ is the number of subcarriers ($N$ even), and $S_k$ is the $k^{th}$ frequency-domain symbol resulting from a direct mapping rule of the original data bits into a selected signal constellation (e.g., a PSK [phase shift keying] or a QAM (quadrature amplitude modulation) constellation).

Applying the Fourier transform to both sides of (2.1) yields

$$S(f) = F\{s(t)\} = \sum_{k=-\frac{N}{2}}^{\frac{N}{2}-1} S_k R(f - kF).$$ (2.2)

Clearly, multicarrier singles can be regarded as a dual version of conventional single-carrier signals. In fact, (2.2) can be regarded as the frequency-domain pulse amplitude modulation (PAM) signal. For uncorrelated symbols the power spectral density (PSD) of the transmitted signals is proportional to

$$\sum_{k=-\frac{N}{2}}^{\frac{N}{2}-1} E\left[|S_k|^2\right] |R(f - kF)|^2.$$ (2.3)

The simplest case of a multicarrier modulation is a conventional frequency division multiplexing (FDM) scheme where the spectra associated with different symbols do not overlap. If we assume that the bandwidth associated with $R(f)$ is smaller than $F/2$,* then each symbol $S_k$ occupies a fraction $1/N$ of the total transmission band, as shown in Figure 2.1.

It is well known that for conventional single carrier modulations we have an ISI-free transmission at the receiver's matched filter output if

$$\int_{-\infty}^{\infty} r(t - kT_s) r^*(t - k'T_s) dt = 0, \ k \neq k'.$$ (2.4)

From the duality relation just referred, it is easy to see that the orthogonality condition between the subcarriers for a multicarrier modulation is given by

$$\int_{-\infty}^{\infty} R(f - kF) R^*(f - k'F) df = 0, \ k \neq k',$$ (2.5)

which, from Parseval's theorem, is equivalent to

$$\int_{-\infty}^{\infty} |r(t)|^2 \exp(-2j\pi(k - k')Ft) dt = 0, \ k \neq k'.$$ (2.6)

---

* Clearly, $F$ is the bilateral bandwidth, and $F/2$ is the unilateral bandwidth.

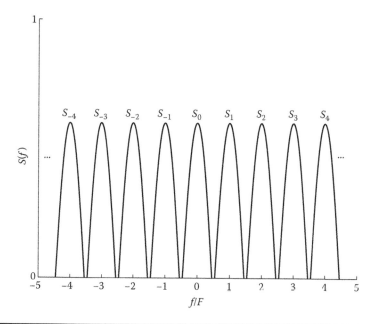

**Figure 2.1   Conventional FDM.**

This means that it is possible to verify the orthogonality between subcarriers, expressed by (2.5) (or (2.6)), even when $\{R(f-kF), k=0,1,...,N-1\}$ overlap in the frequency domain. Therefore, we do not need to restrict ourselves to the conventional FDM case of Figure 2.1, where the spectra associated with different frequency channels do not overlap. In fact, if we have

$$R(f) = \sin c\left(\frac{f}{F}\right), \tag{2.7}$$

with $\sin c(x) = \sin(\pi x)/(\pi x)$, then the corresponding time-domain impulse $r(t)$ is rectangular with duration $T = 1/F$ (for instance, a rectangular impulse from $t_0$ to $t_0 + T$). In this case, (2.6) reduces to

$$\int_{t_0}^{t_0+T} \exp(-2 j\pi(k-k')Ft)dt = 0, \quad k \neq k' \tag{2.8}$$

and, for a subcarrier separation $F$, the $N$ subcarriers are orthogonal when $T = 1/F$ (see Figure 2.2).

For conventional OFDM modulations, the adopted impulses $r(t)$ are rectangular:

$$r(t) = \begin{cases} 1, & [-T_G, T] \\ 0, & \text{elsewhere} \end{cases}, \tag{2.9}$$

where $T = 1/F$ and $T_G \geq 0$ represents the guard period. It will be shown in the following that this guard period can be used to cope with time-dispersive channels. Although (2.6) is not verified with impulses given by (2.9), we can say that the different subcarriers are orthogonal for the interval [0, *T*], which will be the effective detection interval. In fact,

$$\int_0^T |r(t)|^2 \exp(-j2\pi(k-k')Ft)dt = \int_0^T \exp(-j2\pi(k-k')Ft)dt = 0, \quad k \neq k'. \quad (2.10)$$

In this case, since

$$s^{(P)}(t) = \sum_{k=-\frac{N}{2}}^{\frac{N}{2}-1} S_k \exp\left(j2\pi\frac{kt}{T}\right) \quad (2.11)$$

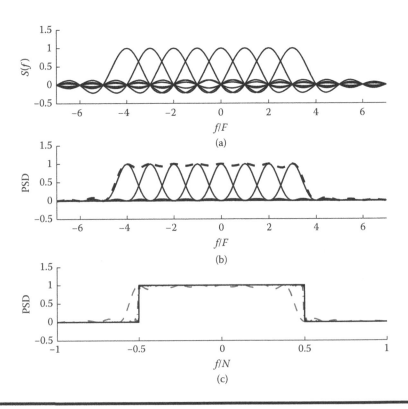

Figure 2.2  **Spectrum of each OFDM subcarrier (a), PSD of an OFDM signal with N = 8 (orthogonal) subcarriers (- - -) and PSD of each subcarrier (- - -) (b) and PSD of an OFDM signal with N = 8 (- - -), N = 64 (- -), and N = 512 (- - -) subcarriers (c).**

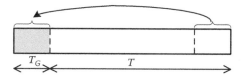

**Figure 2.3 Repetitions of the MC burst's final part in the guard interval.**

is a time-domain periodic function with period $T$, the complex envelope associated with the guard period is a repetition of the MC burst's final part as illustrated in Figure 2.3, that is,

$$s(t) = s(t+T), \quad -T_G \leq t \leq 0. \tag{2.12}$$

Unlike classical FDM systems, where the total frequency band is subdivided into $N$ nonoverlapping frequency subchannels, each one modulated with a separate symbol and subsequently frequency multiplexed, OFDM modulations are multicarrier modulations that make much more efficient use of bandwidth since they verify the orthogonality conditions between different subcarriers although spectra of individual subchannels overlap.

### 2.1.2 Transmitter Implementation

The complex envelope of OFDM signals can be described as a sum of bursts of duration $T_B \geq T$ (i.e., the bursts are transmitted at a rate $1/T_B \leq F$), with the separation between subcarriers denoted by $F$ and the duration of the useful part $T = 1/F$, that is,

$$s^{Tx}(t) = \sum_m s^{(m)}(t - mT_B). \tag{2.13}$$

Given (2.1), the $m^{th}$ OFDM burst can be written as

$$s^{(m)}(t) = \sum_{k=-\frac{N}{2}}^{\frac{N}{2}-1} S_k^{(m)} r(t) \exp(j2\pi kFt) = s^{(m,P)}(t) r(t), \tag{2.14}$$

where, according with (2.11),

$$s^{(m,P)}(t) = \sum_{k=-\frac{N}{2}}^{\frac{N}{2}-1} S_k^{(m)} \exp\left( j2\pi \frac{kt}{T} \right). \tag{2.15}$$

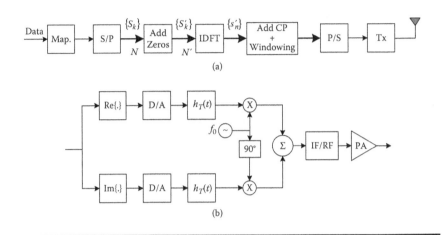

**Figure 2.4 OFDM transmitter structure (a) and detail of Tx block (b).**

The coefficient $S_k^{(m)}$ denotes the $k^{\text{th}}$ symbol of the $m^{\text{th}}$ burst and $r(t)$, a rectangular impulse whose duration should be greater than $1/F$ ($T_B = T + T_G \geq T = 1/F$) and appropriately adjusted so as to deal with the time-dispersive conditions of the channels as described in the following.

The characterization of the complex envelope of OFDM signals given by (2.13) and (2.14) suggests a conceptual FDM transmitter structure with a bank of $N$ parallel single carrier modulators with frequencies $f_k = f_c + kF$, $k = 0,1,...,N-1$, with $f_c$ denoting the frequency of the first subcarrier. However, this is only feasible for a small number of subcarriers $N$.

Fortunately, the OFDM transmitter can be implemented using a single modulator. The basic OFDM transmitter structure is shown in Figure 2.4, where the "in-phase" and "quadrature" components of each OFDM burst are obtained from a sequence of samples corresponding to the IDFT of the block to be transmitted.

The motivation behind this transmitter implementation is the following. Let us consider the signal

$$s^{(P)}(t) = \sum_{k=0}^{N-1} S_k \exp\left( j2\pi \frac{kt}{T} \right). \tag{2.16}$$

It should be pointed out that the subcarrier index in $s^{(P)}(t)$ runs from 0 to $N-1$, unlike in (2.11), where the subcarrier index runs from $-N/2$ to $N/2-1$. The reason for this slight convention change is that $s^{(P)}(t)$ is directly related to the conventional DFT definition. Clearly, $s^{(P)}(t)$ is periodic with period $T$ and occupies a two-sided band $N/T = NF$. This means that $s^{(P)}(t)$ can be completely recovered from its samples taken in the interval $[0,T]$ with a sample rate $1/T_s = N/T$, that is,[*]

$$s_n^{(P)} \triangleq s^{(P)}\left(\frac{nT}{N}\right) = \sum_{k=0}^{N-1} S_k \exp\left(j2\pi\frac{kn}{N}\right) = Ns_n, \tag{2.17}$$

where the block of time-domain samples $\{s_n; n = 0,1,...,N-1\}$ is the IDFT of $\{S_k; k = 0,1,...,N-1\}$, with the IDFT defined as $\{x_n; n = 0,1,...,N-1\} = IDFT\{X_k; k = 0,1,...,N-1\}$ with

$$x_n = \frac{1}{N}\sum_{k=0}^{N-1} X_k \exp\left(j2\pi\frac{kn}{N}\right). \tag{2.18}$$

This means that, apart from a scalar factor $N$, the sampled version of $s^{(P)}(t)$ in the interval $[0,T[$ corresponds to the IDFT of the block $\{S_k; k = 0,...,N-1\}$ in the frequency domain,[†] which can be efficiently implemented with the well-known FFT algorithm [Cooley and Tukey 1965].

Since the power spectrum of an OFDM signal with bursts given by (2.14) has an approximately rectangular shape with bandwidth $N/T = NF$, the required sample rate is approximately $NF$. Actually, there are slight aliasing effects with this sampling rate, but they become negligible for moderate and high values of $N$ (see Figure 2.2).

After generating the samples of $s^{(P)}(t)$, the wave shape associated with a given burst is obtained by multiplying those samples with the samples of the "time window" $r(t)$ whose duration is higher than $T$ (see (2.14)). This means that OFDM burst samples are given by $s_n r_n$, with $r_n \triangleq r(nT/N)$ (note that, unlike the samples $s_n$, the samples $s_n r_n$ are not periodic).

Finally, the analog signal associated with a given OFDM burst is generated with the samples $s_n r_n$ by digital-to-analog conversion (D/A) followed by reconstruction filtering (see Figure 2.4b). The complex envelope can be written as

$$s(t) = \sum_{n=-\infty}^{+\infty} s_n r_n h_T\left(t - n\frac{T}{N}\right), \tag{2.19}$$

where $h_T(t)$ is the impulsive response of the reconstruction filter.

To simplify the reconstruction filter $h_T(t)$, usually the samples of the OFDM burst given by (2.14) are taken with a sample rate $M_{Tx}N/T > N/T$, that is, with an oversampling factor $M_{Tx} > 1$, not necessarily an integer (the adoption of $M_{Tx} > 1$

---

[*] With our DFT definition, the first sample corresponds to instant (or frequency) 0.
[†] We also assumed that the time-domain and frequency-domain inherent to the DFT are periodic.

also helps reduce the aliasing effects, especially when the number of subcarriers is small). Usually, the original block $\{S_k; k = -N/2, -N/2+1, \ldots, N/2-1\}$ already includes $2N_I$ "idle" subcarriers (i.e., with $S_k = 0$), half of them at the beginning and the other half at the end of the burst. This is, in fact, equivalent to oversampling an OFDM burst by a factor of

$$M_{Tx} = \frac{N}{N - 2N_I}, \qquad (2.20)$$

with $N - 2N_I$ useful subcarriers.

When we have an oversampling factor $M_{Tx}$ for a reference burst with $N$ subcarriers, the samples of $s^{(P)}(t)$ in the interval $[0,T]$ are given by

$$s_n^{(M_{Tx})} \triangleq s^{(P)}\left(n\frac{T}{N'}\right) = \sum_{k=-\frac{N}{2}}^{\frac{N}{2}-1} S_k \exp\left(j2\pi\frac{nk}{N'}\right), \quad n = 0,1,\ldots,N'-1, \qquad (2.21)$$

where $N' = NM_{Tx}$. In this case,

$$s_n^{(M_{Tx})} = N'\left(\frac{1}{N'}\sum_{k=0}^{N'-1} S_k' \exp\left(j2\pi\frac{nk}{N'}\right)\right) = N's_n'', \qquad (2.22)$$

$n = 0,1,\ldots,N'-1$, with

$$s_n' = \frac{1}{N'}\sum_{k=0}^{N'-1} S_k' \exp\left(j2\pi\frac{nk}{N'}\right), \qquad (2.23)$$

Clearly, the time-domain block $\{s_n'; n = 0,1,\ldots,N'-1\}$ is the IDFT of the extended block $\{S_k'; k = 0,1,\ldots,N'-1\}$ obtained by adding $N'-N$ zeros to the original block in the frequency-domain, $\{S_k; k = -N/2, -N/2+1, \ldots, N/2-1\}$, in the following way:

$$S_k' = \begin{cases} S_k, & 0 \leq k \leq \dfrac{N}{2}-1 \\[2mm] 0, & \dfrac{N}{2} \leq k \leq N' - \dfrac{N}{2} - 1 \\[2mm] S_{k-N'}, & N' - \dfrac{N}{2} \leq k \leq N' - 1. \end{cases} \qquad (2.24)$$

Once again, apart from a scalar factor $N'$, the sampled version of (2.16) with an oversampling factor $M_{Tx}$ corresponds to the IDFT of the block $\{S_k'; k = 0,\ldots,N'-1\}$.

The complex envelope of the analog signal associated with a given OFDM burst with an oversampling factor $M_{Tx}$ is given by

$$s^{(M_{Tx})}(t) = \sum_{n=-\infty}^{+\infty} s'_n r_n h_T \left( t - n \frac{T}{N'} \right). \tag{2.25}$$

Although signal $s^{(M_{Tx})}(t)$ does not exactly equal the reference representation of the OFDM burst given by (2.14), the difference is small, especially for a large number of subcarriers or when the oversampling factor is high, with differences manifesting mainly at the extremes vicinities of the interval occupied by $r(t)$; obviously, if $M_{Tx} \rightarrow +\infty$, the signal given by (2.25) converges to the OFDM reference burst.

The time window $r(t)$ does not need to have a rectangular shape. In fact, it can take different shapes, and it is common to employ a square-root raised-cosine window, which allows us to reduce the out-of-band radiation levels on the spectrum of OFDM bursts. This means that the signal associated with a given burst still has a complex envelope given by (2.25), but with

$$r(t) = r'(t) * h_W(t), \tag{2.26}$$

where

$$h_W(t) = \frac{\pi}{2T_W} \cos \left( \frac{\pi t}{T_W} \right) rect \left( \frac{t}{T_W} \right), \tag{2.27}$$

and $r'(t)$ is a rectangular impulse with duration $T_B = T + T_G + T_W$. As shown in Figure 2.5, the duration of the time window $r(t)$ is $T_B + T_W = T + T_G + 2T_W$, resulting in an overlap of $T_W$ between adjacent bursts. This means that this raised-cosine window has a roll-off factor of $T_W / (T + T_G + T_W)$. In the following, we assume that $T_W = 0$.

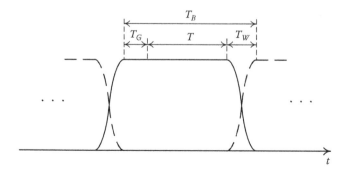

**Figure 2.5  Raised-cosine time window.**

## 2.1.3 Receiver Implementation

Figure 2.6a shows the OFDM receiver structure, and Figure 2.6b shows the input section equivalent scheme of the receiver, corresponding to RF (radio frequency) and IF (intermediate frequency) stages and the signal down-conversion and filtering circuitries to the orthogonal demodulator. The signal sampling is performed with a sample rate

$$\frac{1}{T_s} = \frac{N'}{T},$$
(2.28)

that is, the reception oversampling factor can be the same as the transmission oversampling factor.

Due to the multipath propagation, the received burst will overlap as illustrated in Figure 2.7a. Moreover, some interference between the different subcarriers of the same burst will occur. However, since detection of OFDM signal operates on signal samples associated with a useful period of duration $T$, this means that the use of CPs with duration $T_G$ longer than the overall channel impulse response (which includes the impact of the transmission and detection filters as well as radio channel itself) prevents the effects of bursts overlapping in the received samples associated with the useful interval (see Figure 2.7b); this is usually referred to as absence of intersymbol interference (ISI), although the term interblock interference (IBI) would probably be more appropriate.

Let us consider the received time-domain samples $y_n$ given by the discrete convolution

$$y_n = \sum_{l=0}^{N_h - 1} s_{n-l} h_l + v_n,$$
(2.29)

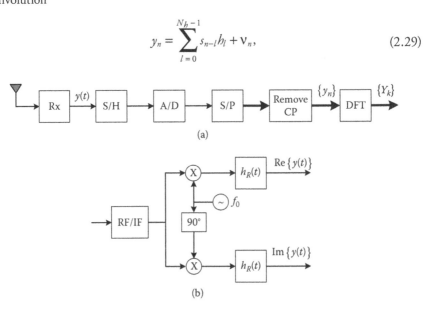

(a)

(b)

**Figure 2.6** OFDM receiver structure (a) and detail of *Rx* block (b).

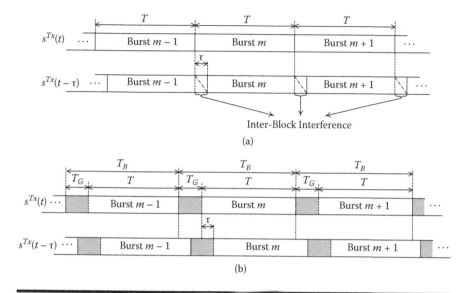

**Figure 2.7 Multipath channel impact leading to IBI (a) and elimination of IBI through guard periods (b).**

where $\{s_n; n = 0,\ldots,N-1\}$ is the block of samples associated with a transmitted burst, $\{h_n; n = 0,\ldots,N_h-1\}$ is the channel impulsive response, with $N_h < N$ denoting the channel length, and $\{v_n; n = 0,\ldots,N-1\}$ the corresponding Gaussian channel noise samples, assumed independent and identically distributed (i.i.d.) in each received burst.

Since the first $N_h$ samples associated with the guard period $T_G$ are a repetition of the $L$ final burst samples, it can easily be shown that the samples associated with the useful part of the received signal, $\mathbf{y} = \begin{bmatrix} y_0 & y_1 & \ldots & y_{N-1} \end{bmatrix}^T$, can be written as

$$\mathbf{y} = \mathbf{h}_{CP}\,\mathbf{s} + \mathbf{v} \qquad (2.30)$$

where $\mathbf{s} = [s_0\ s_1\ \ldots\ s_{N-1}]^T$, $\mathbf{v} = [v_0\ v_1\ \ldots\ v_{N-1}]^T$ and $\mathbf{h}_{CP}$ is a size-$N \times N$ circulant matrix given by

$$\mathbf{h}_{CP} = \begin{bmatrix} h_0 & 0 & \cdots & & 0 & h_{N_h-1} & \cdots & h_2 & h_1 \\ h_1 & h_0 & 0 & \cdots & & 0 & h_{N_h-1} & \cdots & h_2 \\ & \ddots & \ddots & & & & & \ddots & \ddots \\ 0 & \cdots & 0 & h_{N_h-1} & \cdots & h_0 & 0 & \cdots & h_{N_h-1} \\ & & & & \ddots & & \ddots & & \\ 0 & \cdots & & 0 & h_{N_h-1} & \cdots & h_1 & h_0 & 0 \\ 0 & \cdots & & & 0 & h_{N_h-1} & \cdots & h_1 & h_0 \end{bmatrix},$$

$$(2.31)$$

whose elements verify $[\mathbf{h}_{CP}]_{i,i'} = h_{(i-i') \bmod N}$ ($x \bmod y$ denotes the modulo operation, i.e., the remainder of division of $x$ by $y$).

It is well known that circulant matrices can be diagonalized using a Fourier matrix, that is,

$$\mathbf{h}_{CP} = \mathbf{F}^{-1} \Lambda \mathbf{F}, \tag{2.32}$$

where the size-$N \times N$ matrix $\mathbf{F}$, given by

$$\mathbf{F} = \frac{1}{\sqrt{N}} \begin{bmatrix} 1 & 1 & 1 & \cdots & 1 \\ 1 & \omega & \omega^2 & \cdots & \omega^{N-1} \\ 1 & \omega^2 & \omega^4 & \cdots & \omega^{2(N-1)} \\ \vdots & \vdots & \vdots & & \vdots \\ 1 & \omega^{N-1} & \omega^{2(N-1)} & \vdots & \omega^{(N-1)(N-1)} \end{bmatrix}, \tag{2.33}$$

with $\omega = \exp(-j2\pi/N)$, is the unitary (i.e., $\mathbf{F}^H = \mathbf{F}^{-1}$) DFT matrix whose columns are the eigenvectors of $h_{CP}$, and $\Lambda$ is a size-$N \times N$ diagonal matrix whose elements, the eigenvalues of $h_{CP}$, equal the DFT of the first column of $h_{CP}$, that is,

$$\Lambda = \begin{bmatrix} \lambda_0 & & & 0 \\ & \lambda_1 & & \\ & & \ddots & \\ 0 & & & \lambda_{N-1} \end{bmatrix} = \begin{bmatrix} H_0 & & & 0 \\ & H_1 & & \\ & & \ddots & \\ 0 & & & H_{N-1} \end{bmatrix} = \mathbf{H}, \tag{2.34}$$

where

$$\lambda_k = \sum_{n=0}^{N-1} h_n \exp\left(-j2\pi \frac{kn}{N}\right) = DFT\{h_n\} = H_k, \quad k = 0,1,\ldots,N-1. \tag{2.35}$$

Using (2.32) and (2.34) in (2.30), it follows that (for the sake of notation simplicity, we will drop the superscript $m$)

$$\mathbf{y} = \mathbf{F}^{-1}\mathbf{HFs} + \mathbf{v} \Leftrightarrow$$

$$\Leftrightarrow \mathbf{Fy} = \mathbf{HFs} + \mathbf{Fv} \Leftrightarrow \tag{2.36}$$

$$\Leftrightarrow \mathbf{Y} = \mathbf{HS} + \mathbf{N},$$

with the vectors

$$\mathbf{Y} = \mathbf{Fy} = \left[Y_0\ Y_1 \dots Y_{N-1}\right]^T,\qquad(2.37)$$

$$\mathbf{S} = \mathbf{Fs} = \left[S_0\ S_1 \dots S_{N-1}\right]^T\qquad(2.38)$$

and

$$\mathbf{N} = \mathbf{Fv} = \left[N_0\ N_1 \dots N_{N-1}\right]^T\qquad(2.39)$$

denoting the DFTs of **y**, **s**, and **v**, respectively. This means that the received sample at the $k^{\text{th}}$ subcarrier is given by

$$Y_k = H_k S_k + N_k,\qquad(2.40)$$

$k = 0,1,\dots,N-1$, where $H_k$ is the channel frequency response for the $k^{\text{th}}$ subcarrier and $N_k$ is the channel noise component for that subcarrier. Therefore, the channel acts like a simple multiplicative factor for each subcarrier, preserving the orthogonality between subcarriers in the useful interval; this is usually referred to as absence of *interchannel interference* (ICI). In fact, since the CP corresponds to a cyclic extension of each burst, this means that, for each useful interval $T$, the transmitted signal corresponds, not to a sequence of OFDM bursts expressed by (2.13), but to the periodic signal $s^{(P)}(t)$ given by (2.11) (see Figure 2.8). Therefore, the linear convolution associated with the channel is formally equivalent to a circular convolution with respect to the useful part of the OFDM block.

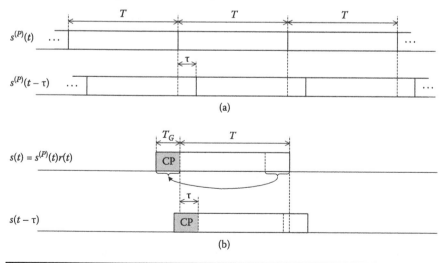

**Figure 2.8  Multipath channel impact on $S(P)$ (a) and on the corresponding cyclic prefix (CP) extended OFDM burst (b).**

It should be mentioned that the elimination of ISI can be achieved by extending each burst with any known fixed sequence, including an all-zero sequence (zero padding (ZP)) or a pseudo noise (PN) symbol sequence, denoted PN extension or unique word (UW). ZP schemes can be a good alternative to CP-assisted schemes [Muquet et al. 2000a]. However, complex receiver structures must be employed, involving the inversion or the multiplication of matrixes whose dimensions can grow with the block length, which are not suitable when large blocks are employed. By employing overlap-and-add techniques [Muquet et al. 2000b], the receiver complexity becomes similar to that of conventional CP-assisted schemes, but the performance is also identical. Efficient FFT-based receivers can also be designed for ZP OFDM [Araújo and Dinis 2006].

To avoid power and spectral degradation, the CP length should be a small fraction of the overall length of the blocks. In this book, we will consider only CP-assisted block transmission.

Since CP-assisted transmission can be seen as an $N$ parallel nonselective subchannel transmission expressed by (2.40), channel distortion effects for an uncoded OFDM transmission can easily be compensated for using the receiver depicted in Figure 2.9a, where the equalized frequency-domain sample at $k^{th}$ the subcarrier $\tilde{S}_k$ is obtained by

$$\tilde{S}_k = F_k Y_k, \tag{2.41}$$

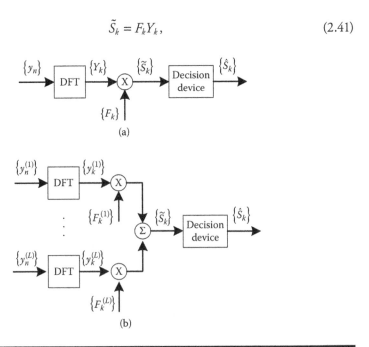

**Figure 2.9** **Channel distortion effects compensation for an uncoded OFDM transmission with no space diversity (a) and with an *L*-order space diversity receiver (b).**

with the set of coefficients $\{F_k; k = 0,1,...,N-1\}$ given by

$$F_k = \frac{1}{H_k} = \frac{H_k^*}{|H_k|^2}, \tag{2.42}$$

corresponding to an FDE under the zero-forcing (ZF) criterion.

In the case where we have $L$-order space diversity, the received sample at the $k$th subcarrier and $l$th diversity branch is given by

$$Y_k^{(l)} = S_k H_k^{(l)} + N_k^{(l)}, \tag{2.43}$$

$(l = 1,...,L)$, where $H_k^{(l)}$ and $N_k^{(l)}$ denote the channel frequency response and noise term for the $k$th subcarrier and the $l$th diversity branch, respectively. The corresponding equalized sample is

$$\tilde{S}_k = \sum_{l=1}^{L} F_k^{(l)} Y_k^{(l)}, \tag{2.44}$$

where the set $\{F_k^{(l)}; k = 0,1,...,N-1\}$ $(l = 1,...,L)$ denotes the FDE coefficients associated to the $l$th diversity branch, which can be set as

$$F_k^{(l)} = \frac{H_k^{(l)*}}{\displaystyle\sum_{l'=1}^{L} |H_k^{(l')}|^2}. \tag{2.45}$$

Therefore, the receiver can be the one depicted in Figure 2.9b and, from (2.44),

$$\tilde{S}_k = S_k + \frac{\displaystyle\sum_{l=1}^{L} H_k^{(l)*}}{\displaystyle\sum_{l'=1}^{L} |H_k^{(l')}|^2} N_k^{(l)}. \tag{2.46}$$

A decision on the symbol transmitted through subcarrier $k$ can be made based on $\tilde{S}_k$. Clearly, the receiver structure of Figure 2.9b implements a maximal-ratio combining (MRC) diversity scheme for each subcarrier $k$.

## 2.2 Amplification Issues

### 2.2.1 Envelope Fluctuations of OFDM Signals

One important drawback of conventional OFDM transmission schemes is their strong envelope fluctuation and high peak-to-mean envelope power ratio, high peak-to-mean envelope power ratio (PMEPR), as depicted in Figure 2.10, which lead to amplification difficulties. In fact, to avoid the out-of-band radiation levels that are inherent to nonlinear distortion, power amplifiers for OFDM transmission are required to have linear characteristics, and/or a significant input backoff has to be adopted.

In fact, when the number of subcarriers is high, the complex envelope of OFDM signals has a Gaussian-like nature and the distribution of its envelope is almost Rayleigh. Therefore, its PDF probability density function (PDF) is approximately given by

$$p(R) = \frac{R}{\sigma^2} \exp\left(-\frac{R^2}{2\sigma^2}\right), \qquad (2.47)$$

which means a strong envelope fluctuation.

Since the highest envelope values have a very small probability, it is reasonable to define a PMEPR in a statistical way, for example,

$$\text{PMEPR} \stackrel{\Delta}{=} \frac{X^2(P)}{2\sigma^2} \qquad (2.48)$$

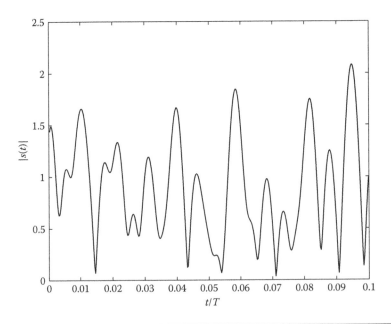

**Figure 2.10  Evolution of the envelope of an OFDM signal.**

where $X(P)$ is the envelope value that is exceeded with probability $P$. For a Rayleigh-distributed envelope,

$$P = \mathrm{Prob}(R > X) = \int_X^{+\infty} p(R)dR = \exp\left(-\frac{X^2}{2\sigma^2}\right) \tag{2.49}$$

and

$$X(P) = \sqrt{-2\sigma^2 \log(P)}. \tag{2.50}$$

A reasonable value for $P$ is $P = 10^{-3}$, which corresponds to PMEPR $\approx 8.4$ dB, regardless of the value of $N$ (as long as $N \gg 1$).

As an alternative, we could define the PMEPR as

$$\mathrm{PMEPR} \overset{\Delta}{=} \frac{X_B^2(P)}{2\sigma^2}, \tag{2.51}$$

where $X_B(P)$ is the maximum envelope per OFDM block, which is exceeded with probability $P$ (when $\max_{t \in [0,T[}|s^{(P,m)}(t)| = R_B,\, P = \mathrm{Prob}(R_B > X_B(P)))$ [Müller and Huber 1997].

When the number of subcarriers is high, $X(P)$ is almost independent of $N$. On the other hand, $X_B(P)$ increases with $N$ even when $N$ is high enough to allow the Gaussian approximation for the OFDM signals. It should be noted that, in the following, we will adopt the PMEPR definition given by (2.48), which means a PMEPR value practically independent of $N$ for $N \gg 1$. Several methods have been proposed to reduce these amplification difficulties by means of digital signal processing. Some of these methods operate in the frequency domain, appropriately using a reasonable amount of redundancy to avoid high-amplitude peaks when the number of subcarriers is low [Jones and Wilkinson 1996]. However, as the number of subcarriers increases, the code rate of the required PMEPR-reducing frequency-domain codes becomes lower and lower. The so-called techniques partial transmit sequences (PTS) [Müller et al. 1997; Müller and Huber 1997; Cimini and Sollenberger 1999], which also operate in the frequency domain, are able to achieve a strongly reduced PMEPR for a large number of subcarriers while using a very small redundancy. The main drawback of the PTS techniques is their very high computational complexity for large OFDM data blocks, mainly due to the optimization procedures that are required on a block-by-block basis.

Some other signal processing methods rely on a time-domain operation, such as a digital clipping of the high amplitude peaks, for PMEPR reduction purposes. Following the pioneering contributions in this direction (namely in O'Neill and Lopes [1995] and Li and Cimini [1998]), it has been recognized that the clipping operation should be performed on oversampled OFDM bursts and followed by a filtering procedure, so as to reduce the out-of-band radiation levels while alleviating the "peak regrowth problem."

The simplest and most flexible techniques for reducing the envelope fluctuations of OFDM signals involve a nonlinear operation [O'Neill and Lopes 1995; Li and Cimini 1998; Dinis and Gusmão 1997; May and Rohling 1998], eventually followed by a filtering procedure. A promising class of low-complexity signal processing schemes for reduced-PMEPR spectrally efficient OFDM transmission was originally proposed in Dinis and Gusmão [2000]. Such schemes combine a nonlinear operation in the time domain (possibly according to a clipping characteristic) with a linear, filtering operation in the frequency domain. This burst-by-burst frequency-domain filtering, besides not requiring an increased guard time to avoid ISI, can be very selective, for example, completely removing the out-of-band radiation effects of the preceding nonlinear time-domain operation. A similar technique was proposed in Dinis and Gusmão [2001a]. Its whose main difference lies in the type of nonlinear operation employed with each class: a nonlinearity that operates on the complex OFDM samples in Dinis and Gusmão [2000] and a nonlinearity that separately operates on the real (I) and imaginary (Q) parts of these samples in Dinis and Gusmão [2001a]. In the following, we present a wide class of signal processing schemes, including the schemes of both Dinis and Gusmão [2000] and Dinis and Gusmão [2001a], which can be regarded as belonging to subclasses of the broad class. We also include an appropriate statistical modeling for the blocks along the signal processing chain that is derived from well-known results on memoryless nonlinearities with Gaussian inputs.

## 2.2.2 A Broad Class of Signal Processing Schemes

The general signal processing schemes considered here were first proposed in Dinis and Gusmão [2000] and assume the transmitter structure depicted in Figure 2.11. A block of time-domain samples, concerning any OFDM burst to be transmitted, is generated as described below:

◼ An augmented frequency-domain block $\{S'_k; k = 0,1,\ldots,N'-1\}$, with $N' = NM_{Tx}$ for a selected $M_{Tx} \geq 1$, is formed by adding $N'-N$ zeros to the original frequency-domain block $\{S_k; k = -\frac{N}{2}, -\frac{N}{2}+1, \ldots, \frac{N}{2}-1\}$, directly related to data, in the following way:

$$S'_k = \begin{cases} S_k, & 0 \leq k \leq \dfrac{N}{2}-1 \\[2mm] 0, & \dfrac{N}{2} \leq k \leq N' - \dfrac{N}{2}-1 \\[2mm] S_{k-N'}, & N' - \dfrac{N}{2} \leq k \leq N'-1 \end{cases} \qquad (2.52)$$

(it is assumed that the selected $N'$ is a power of two).

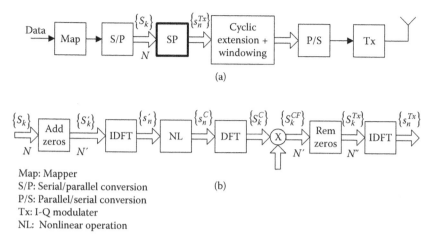

(a)

(b)

Map: Mapper
S/P: Serial/parallel conversion
P/S: Parallel/serial conversion
Tx: I-Q modulater
NL: Nonlinear operation

**Figure 2.11   Proposed transmitter structure (a) and detail of the signal processing scheme within the SP block (b).**

■ The IDFT of this frequency-domain block is computed, leading to the block $\{s'_n; n = 0,1,\ldots,N'-1\}$, with $s'_n$ given by

$$s'_n = \frac{1}{N'} \sum_{k=0}^{N'-1} S'_k \exp\left( j2\pi \frac{nk}{N'} \right).$$

■ Each time-domain sample, $s'_n$, is submitted to a nonlinear operation, leading to a modified sample $s^C_n$. We consider two types of nonlinearities:

1. Cartesian nonlinearities (represented in Figure 2.12a), which can be characterized as I-Q memoryless nonlinearities since they separately operate on the real and the imaginary parts of each complex sample $s'_n$. This is done in accordance with

$$s^C_n = g\left( \mathrm{Re}\{s'_n\} \right) + jg\left( \mathrm{Im}\{s'_n\} \right), \tag{2.53}$$

where $g(x)$ denotes an appropriately chosen odd function.
2. Polar nonlinearities (represented in Figure 2.12b), which can be regarded as bandpass memoryless nonlinearities [Saleh 1981] since they operate on each complex sample $s'_n$ according to

$$s^C_n = g_C\left( |s'_n| \right) \exp\left( j \arg(s'_n) \right), \tag{2.54}$$

with a selected function $g_C(\cdot)$ of the "envelope" $R = |s'_n|$. The underlying AM-to-AM and AM-to-PM nonlinear characteristics are $A(R) = |g_C(R)|$ and $\Theta(R) = \arg(g_C(R))$, respectively.

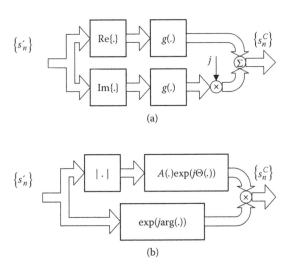

**(a)**

**(b)**

**Figure 2.12    Models for Cartesian (a) and polar (b) memoryless nonlinearities.**

■ A DFT operation brings the nonlinearly modified block back to the frequency domain, where a linear shaping operation is performed by a multiplier bank with selected coefficients $G_k, k = 0,1,\ldots,N'-1$. A frequency-domain-filtered block $\{S_k^{CF} = G_k S_k^C ; k = 0,1,\ldots,N'-1\}$ is achieved with

$$S_k^C = \sum_{n=0}^{N'-1} s_n^C \exp\left(-j2\pi \frac{kn}{N'}\right).$$

■ Since $S_k^{CF} = 0$ for a range of values of $k$ (namely due to $G_k = 0$), a reduced block $\{S_k^{Tx}; k = 0,1,\ldots,N''-1\}$, with $N \leq N'' \leq N'$, can be derived from $\{S_k^{CF} = G_k S_k^C ; k = 0,1,\ldots,N'-1\}$, with

$$S_k^{Tx} = \begin{cases} S_k^{CF}, & 0 \leq k \leq \dfrac{N''}{2}-1 \\[2mm] S_{k+N'-N''}^{CF}, & \dfrac{N''}{2} \leq k \leq N''-1 \end{cases} \qquad (2.55)$$

(it is assumed that $N''$ is a power of two).

■ A second IDFT converts this modified frequency-domain block into the time domain, leading to the block $\{s_n^{Tx}; n = 0,1,\ldots,N''-1\}$.

■ The OFDM burst to be transmitted is then obtained, within the transmitter structure of Figure 2.11, by appending the required cyclic extension to $\{s_n^{Tx}; n = 0,1,\ldots,N''-1\}$, possibly modifying the amplitude of the samples at the burst tails according to a windowing procedure, and generating the analog I and Q signal components through D/A conversion and selected low-pass reconstruction filtering.

Appending $N'-N$ zeros to each initial frequency-domain block prior to computing the required IDFT is a well-known OFDM implementation technique, which is equivalent to oversampling, by a factor of $M_{Tx} = N'/N$, the "ideal" OFDM burst. The subsequent nonlinear operation proposed here is crucial for reducing the envelope fluctuations, whereas the frequency-domain filtering using the set $\{G_k; k = 0,1,\ldots,N'-1\}$ can provide a complementary filtering effect (of course, with some regrowth of the envelope fluctuations). The reduced size that is allowed for the last IDFT essentially means reduced computational effort; the removal of $N'-N''$ subcarriers with zero amplitude can be regarded as a decimation in the time domain.

For a given input block size $N$, a careful selection of $M_{Tx}$, the nonlinear characteristic (for a given input level), and $\{G_k; k = 0,1,\ldots,N'-1\}$ ensures reduced envelope fluctuations while maintaining low out-of-band radiation levels.

The selection of either a Cartesian or a polar nonlinearity, for PMEPR reduction purposes, defines the choice of a subclass of signal processing schemes within the broad class of signal processing considered here.

It should be mentioned that this broad class of signal processing schemes (following and extending our original proposal in Dinis and Gusmão [2000]) includes, as specific cases, signal processing schemes proposed by other authors in the meantime: this is the case with Ochiai and Imai [2000] ($M_{Tx} = \frac{N'}{N} = 1$ and a polar nonlinearity that operates as an envelope clipper). This is also the case with Armstrong [2001] and Ochiai and Imai [2002], where the same envelope clipping is adopted when assuming $M_{Tx} \geq 1$, with $G_k = 1$ for the $N$ "in-band" subcarriers and $G_k = 0$ for the $N'-N$ "out-of-band" ones.

It should also be mentioned that a more sophisticated technique, allowing improved PMEPR-reducing results, could be simply developed on the basis of the signal processing approach analyzed here. Such a technique consists of repeatedly using, in an iterative way, the signal processing chain that leads from $\{S_k'\}$ to $\{S_k^{CF}\}$ in Figure 2.11b (of course, $\{S_k'^{(l)}\} = \{S_k^{CF(l-1)}\}$ for $l > 1$ and $\{S_k'^{(1)}\} = \{S_k'\}$, where each superscript indicates a given iteration). The iterative techniques of Armstrong [2002] and Dinis Gusmão [2003a] correspond to the particular case where the nonlinear operation is an envelope clipping and the frequency-domain filtering is characterized by $G_k = 0$ for the $N'-N$ out-of-band subcarriers, with $G_k = 1$ in-band.

## 2.2.2.1 Peak Cancellation Techniques

The wide range of signal processing schemes described above can be employed to implement appropriate peack cancellation schemes such as the ones proposed in May and Rohling [1998] and Dinis and Gusmão [2001b]. For the sake of simplicity, we will assume that $N'' = N'$, which means that the final frequency-domain block $\{S_k^{Tx}; k = 0,1,\ldots,N''-1\}$ is identical to the block $\{S_k^{CF}; k = 0,1,\ldots,N'-1\}$. Let us consider that the nonlinear operation is an ideal clipping. In the Cartesian case, this means a nonlinearity where $g(x)$ is given by

$$g(x) = \begin{cases} -x_M, & x < -x_M \\ x, & |x| \le x_M, \\ x_M, & x > x_M \end{cases} \tag{2.56}$$

with $x_M$ denoting the clipping level. In the polar case, this means a nonlinearity where $g_C(R)$ is given by

$$g_C(R) = \begin{cases} R, & R \le s_M \\ s_M, & x > s_M \end{cases} \tag{2.57}$$

with $s_M$ denoting the clipping level; this nonlinearity operates as an "envelope clipper" on the bandpass signal that corresponds to the I-Q pair.

In this case, the signal processing scheme shown in Figure 2.13a is equivalent to that described above, provided that

$$p_n = \begin{cases} g_p^I(Re\{s_n'\}) + jg_p^I(Im\{s_n'\}), & \text{for the Cartesian nonlinearity} \\ g_p^{II}(|s_n'|)\exp(j\arg(s_n')), & \text{for the polar nonlinearity} \end{cases} \tag{2.58}$$

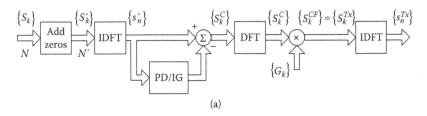

(a)

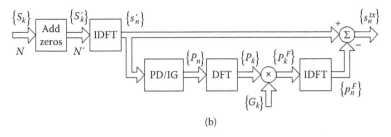

(b)

**Figure 2.13 Signal processing schemes for peak cancellation with OFDM transmission: general case (a); specific case, when $G_k = 1$ for the N "in-band" subcarrier (b).**

with

$$g_p^I(x) = x - g(x) \tag{2.59}$$

and

$$g_p^{II}(R) = R - g_C(R), \tag{2.60}$$

since

$$s_n' - p_n = s_n^C. \tag{2.61}$$

This means that

$$g_p^I(x) = \begin{cases} x - x_M, & x > x_M \\ x + x_M, & x < -x_M \\ 0, & \text{otherwise} \end{cases} \tag{2.62}$$

and

$$g_p^{II}(R) = \begin{cases} R - s_M, & R > s_M \\ 0, & \text{otherwise} \end{cases}. \tag{2.63}$$

The set of samples $\{p_n; n = 0,1,\dots,N'-1\}$ can be regarded as the result of a "peak detection/impulse generation" (PD/IG) procedure, which is followed by a "peak cancellation" one, for those values of $n$ where a peak has been detected.

Let us assume, additionally, that $G_k = 1$ for the $N$ "in-band" subcarriers ($0 \le k \le \frac{N}{2} - 1$ and $N' - \frac{N}{2} \le k \le N' - 1$). Therefore, $S_k' G_k = S_k'$ for $k = 0,1,\dots,N'-1$, and it is easy to see that the signal processing scheme shown in Figure 2.13b is equivalent to the scheme of Figure 2.13a: in fact, $S_k^{Tx} = S_k^{CF} = S_k' - P_k^F$ with $P_k^F = P_k G_k$ ($k = 0,1,\dots,N'-1$), which means that $s_n' - p_n^F = s_n^{CF}$ ($n = 0,1,\dots,N'-1$), when using $\{P_k; k = 0,1,\dots,N'-1\}$ and $\{P_k^F; k = 0,1,\dots,N'-1\}$, respectively, to denote the DFTs of $\{p_n; n = 0,1,\dots,N'-1\}$ and $\{p_n^F; n = 0,1,\dots,N'-1\}$. The block $\{p_n^F; n = 0,1,\dots,N'-1\}$ can be described as a circular convolution of $\{p_n = 0,1,\dots,N'-1\}$ and $\{g_n; n = 0,1,\dots,N'-1\} = $IDFT $\{G_k; k = 0,1,\dots,N'-1\}$.

Within the peak cancellation subclass, if $G_k$ equals zero "out-of-band" ($\frac{N}{2} \le k \le N' - \frac{N}{2} - 1$), besides being equal to one "in-band," then

$$g_n = \frac{1}{N'} \sum_{k=-\frac{N}{2}}^{\frac{N}{2}-1} \exp(j2\pi kn/N'), n = 0,1,\dots,N'-1 \tag{2.64}$$

($G_k = G_{k-N'}$ for any integer $k$). Therefore, $g_0 = N/N' = 1/M_{Tx}$, $|g_n| \approx |\sin c(n/M_{Tx})|$ $/M_{Tx}$ for $0 \le n \ll N'$ and

$$|g_n| \approx \left| \sin c \left( \frac{N'-n}{M_{Tx}} \right) \right| / M_{Tx}$$

for $1 \le N' - n \ll N'$; when $M_{Tx} = 1$ ($N' = N$), $g_n = 0$ for $n = 1, 2, \ldots, N' - 1$, and $g_0 = 1$. For the polar case, this is equivalent to the peak cancellation method reported by Nee and Prasad [2000] (Chapter 6), according to the ideas of May and Rohling [1998].

### 2.2.3 Signal Characterization along the Signal Processing Chain

This section shows how to obtain an appropriate statistical characterization for the modified block of frequency-domain samples in Figure 2.11b, $\{S_k^{Tx}; k = 0, 1, \ldots, N'' - 1\}$, that replaces the block $\{S_k'; k = 0, 1, \ldots, N' - 1\}$ of conventional OFDM schemes. For this purpose, we need to characterize statistically the blocks along the signal processing chain.

#### 2.2.3.1 Time-Domain Block at the Input to the Nonlinearity

It is assumed that $E[S_k] = 0$ and $E[S_k S_{k'}^*] = 2\sigma_S^2 \delta_{k,k'}$ ($\delta_{k,k'} = 1$ for $k = k'$ and 0 otherwise), with $\sigma_S^2 = \frac{1}{2} E[|S_k|^2]$ ($E[\cdot]$ denotes "ensemble average"), although this analytical approach can easily be extended to other cases, such as the ones where different powers are assigned to different subcarriers [Araújo and Dinis 2011; Araújo and Dinis 2012]. Therefore, it can be easily demonstrated that $E[s_n'] = 0$ and

$$E\left[ s_n' s_{n'}^{*} \right] = R_s(n - n') = 2\sigma^2 \frac{\sin c((n-n')N/N')}{\sin c((n-n')/N')} \exp\left( -\frac{j\pi(n-n')}{N'} \right) \quad (2.65)$$

($n = 0, 1, \ldots, N' - 1$; $n' = 0, 1, \ldots, N' - 1$), with $\sigma^2 = \frac{N}{(N')^2} \sigma_S^2$.

When the number of subcarriers is high ($N \gg 1$), the time-domain coefficients $s_n'$ can be approximately regarded as samples of a zero-mean complex Gaussian process [Dinur and Wulich 2001] with autocorrelation given by (2.65).

#### 2.2.3.2 Time-Domain Block at the Output of the Nonlinearity

In the following, we take advantage of the quasi-Gaussian nature of the samples $s_n'$ for obtaining the statistical characterization of the time-domain block at the output of the nonlinearity. In fact, it is well known that the output of a memoryless nonlinear device with a Gaussian input can be written as the sum of two uncorrelated components [Rowe 1982]: a useful one, proportional to the input; and a self-interference one. Clearly, this is the case with polar nonlinearities. This is also the case with cartesian nonlinearities, since the real and imaginary parts of the input complex envelope (i.e., the "in-phase" and "quadrature" components) are separately submitted to two identical memoryless nonlinearities. As a consequence, for both classes of nonlinearities, we can write

$$s_n^C = \alpha s_n' + d_n, \tag{2.66}$$

where $E[s_n' d_{n'}^*] = 0$. This means that the nonlinearly modified samples can be decomposed into uncorrelated useful and self-interference components. For cartesian nonlinearities, the $\alpha$ coefficient is given by

$$\alpha = \frac{E[xg(x)]}{E[x^2]} = \frac{1}{\sqrt{2\pi}\sigma^3} \int_{-\infty}^{+\infty} xg(x)\exp\left(-\frac{x^2}{2\sigma^2}\right) dx, \tag{2.67}$$

and for polar nonlinearities we have

$$\alpha = \frac{E[Rg_C(R)]}{E[R^2]} = \frac{1}{2\sigma^2} \int_0^{+\infty} RA(R)\exp(j\Theta(R))\frac{R}{\sigma^2}\exp\left(-\frac{R^2}{2\sigma^2}\right) dR. \tag{2.68}$$

The average power of the useful component is $S = |\alpha^2|\sigma^2$, and the average power of the self-interference component is given by $I = P_{out} - S$, where $P_{out}$ denotes the average power of the signal at the nonlinearity output. $P_{out}$ is given by

$$P_{out} = E[g^2(x)] = \frac{1}{\sqrt{2\pi}\sigma} \int_{-\infty}^{+\infty} g^2(x)\exp\left(-\frac{x^2}{2\sigma^2}\right) dx \tag{2.69}$$

for cartesian nonlinearities, and by

$$P_{out} = \frac{1}{2}E[A^2(R)] = \frac{1}{2} \int_0^{+\infty} A^2(R)\frac{R}{\sigma^2}\exp\left(-\frac{R^2}{2\sigma^2}\right) dR \tag{2.70}$$

for polar nonlinearities.

For a Cartesian nonlinearity, it can be shown [Dinis and Gusmão 1999; Dinis and Gusmão 2004] that the autocorrelation of the output samples can be expressed as a function of the autocorrelation of the input samples in the following way:

$$E[s_n^C s_{n'}^{C*}] = R_s^C(n-n') = \sum_{\gamma=0}^{+\infty} 2P_{2\gamma+1} \frac{\left(\text{Re}\{R_s(n-n')\}\right)^{2\gamma+1} + j\left(\text{Im}\{R_s(n-n')\}\right)^{2\gamma+1}}{\left(R_s(0)\right)^{2\gamma+1}},$$

$$\tag{2.71}$$

where $R_s(n-n')$ is given by (2.65) and the coefficient $P_{2\gamma+1}$ denotes the total power associated to the inter-modulation product (IMP) of order $2\gamma+1$, given by

$$P_{2\gamma+1} = \frac{1}{2^{2\gamma+1}(2\gamma+1)!}\left(\int_{-\infty}^{+\infty} g(x)p(x)H_{2\gamma+1}\left(\frac{x}{\sqrt{2}\sigma}\right)dx\right)^2, \tag{2.72}$$

where $H_n(x)$ denotes a Hermite polynomial of degree $n$, defined as [Abramowitz and Stegun 1972]

$$H_n(x) = (-1)^n \exp(x^2)\frac{d^n}{dx^n}\{\exp(-x^2)\}. \tag{2.73}$$

For a polar nonlinearity, it can be shown [Dinis and Gusmão 2004] that

$$E[s_n^C s_{n'}^{C*}] = R_s^C(n-n') = \sum_{\gamma=0}^{+\infty} 2P_{2\gamma+1}\frac{\left(R_s(n-n')\right)^{\gamma+1}\left(R_s^*(n-n')\right)^\gamma}{\left(R_s(0)\right)^{2\gamma+1}}, \tag{2.74}$$

where, once again, the coefficient $P_{2\gamma+1}$ denotes the total power associated to the IMP of order $2\gamma+1$. This coefficient is given by

$$P_{2\gamma+1} = \frac{|v_{2\gamma+1}|^2}{2\gamma!(\gamma+1)!}, \tag{2.75}$$

where

$$v_{2\gamma+1} = \int_0^{+\infty} Rg_C(R)W_{2\gamma+1}\left(\frac{R}{\sqrt{2\sigma^2}}\right)dR \tag{2.76}$$

with

$$W_{2\gamma+1}(x) = \int_0^{+\infty} t^{2\gamma+2}\exp(-t^2)J_1(2xt)dt = \frac{\gamma!}{2}\exp(-x^2)xL_\gamma^{(1)}(x^2) \tag{2.77}$$

(see [Stette 1974]); $L_\gamma^{(1)}(x)$ denotes a generalized Laguerre polynomial of order $\gamma$, defined as [Abramowitz and Stegun 1972]

$$L_\gamma^{(1)}(x) = \frac{1}{\gamma!x}\exp(x)\frac{d^\gamma}{dx^\gamma}\{\exp(-x)x^{\gamma+1}\}. \tag{2.78}$$

Since

$$R_s^C(n-n') = |\alpha|^2 R_s(n-n') + E[d_n d_{n'}^*], \tag{2.79}$$

it can be easily recognized that $P_1 = |\alpha|^2 \sigma^2$ and

$$E[d_n d_{n'}^*] = R_d(n-n')$$

$$= \sum_{\gamma=1}^{+\infty} 2 P_{2\gamma+1} \frac{\left(\mathrm{Re}\{R_s(n-n')\}\right)^{2\gamma+1} + j\left(\mathrm{Im}\{R_s(n-n')\}\right)^{2\gamma+1}}{\left(R_s(0)\right)^{2\gamma+1}} \qquad (2.80)$$

for cartesian nonlinearities, and

$$E[d_n d_{n'}^*] = R_d(n-n') = \sum_{\gamma=1}^{+\infty} 2 P_{2\gamma+1} \frac{\left(R_s(n-n')\right)^{\gamma+1}\left(R_s^*(n-n')\right)^{\gamma}}{\left(R_s(0)\right)^{2\gamma+1}}, \qquad (2.81)$$

for polar nonlinearities. The total power of the self-interference term is $I = \sum_{\gamma=1}^{+\infty} P_{2\gamma+1} = P_{out} - S$.

This method for statistical characterization of the transmitted blocks is quite appropriate whenever the power series in (2.80) and (2.81) can be reasonably truncated while ensuring an accurate computation. However, for strongly nonlinear conditions, the required number of terms becomes very high. In such cases, one can simplify the computation as explained below. When $\gamma \gg M_{Tx}$,

$$\mathrm{Im}\left\{\frac{R_s(n-n')}{R_s(0)}\right\}^{2\gamma+1} \approx 0$$

and

$$\left(\frac{R_s(n-n')}{R_s(0)}\right)^{2\gamma+1} \approx \mathrm{Re}\left\{\frac{R_s(n-n')}{R_s(0)}\right\}^{2\gamma+1} \approx \begin{cases} 1, & n = n' \\ 0, & n \neq n' \end{cases}, \qquad (2.82)$$

which means that the frequency-domain distribution of the power associated to a given IMP, $\{G_{s,2\gamma+1}(k),$

$$k = 0,1,\ldots,N'-1\} = DFT\left\{\left(\frac{R_s(n)}{R_s(0)}\right)^{2\gamma+1}; n = 0,1,\ldots,N'-1\right\},$$

is almost constant when $\gamma \gg M_{Tx}$, i.e., $G_{s,2\gamma+1}(k)/G_{s,2\gamma+1}(0) \approx 1$, $\gamma \gg M_{Tx}$ (see Figure 2.14).

As a consequence, the contribution of the $(2\gamma+1)$th IMP to the output auto-correlation can be approximated by $2 P_{2\gamma+1}\delta_{n,n'}$, leading to

$$R_s^C(n-n') \approx 2 \sum_{\gamma=0}^{\gamma_{max}} P_{2\gamma+1} \frac{\left(\mathrm{Re}\{R_s(n-n')\}\right)^{2\gamma+1} + j\left(\mathrm{Im}\{R_s(n-n')\}\right)^{2\gamma+1}}{\left(R_s(0)\right)^{2\gamma+1}} + 2 P_{2\gamma_{max}+1}^{+\infty}\delta_{n,n'}$$

$$(2.83)$$

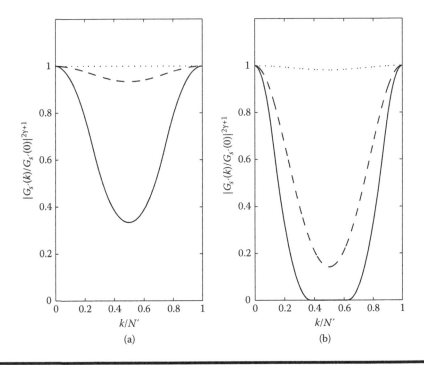

**Figure 2.14** Evolution of $|G_{s,2\gamma+1}(k)/G_{s,2\gamma+1}(0)|$ for $M_{Tx}=2$ (a) and $M_{Tx}=4$ (b) with $\gamma=1$ (solid line), $\gamma=4$ (dashed line), and $\gamma=25$ (dotted line).

($\gamma_{max} \gg M_{Tx}$) for a cartesian nonlinearity, with

$$P_{2\gamma_{max}+1}^{+\infty} = \sum_{\gamma=\gamma_{max}+1}^{+\infty} P_{2\gamma+1} = I - \sum_{\gamma=1}^{\gamma_{max}} P_{2\gamma+1}, \qquad (2.84)$$

where $I$ denotes the average power of the self-interference component. For polar nonlinearities,

$$R_s^C(n-n') \approx 2\sum_{\gamma=0}^{\gamma_{max}} P_{2\gamma+1} \frac{\left(R_s(n-n')\right)^{\gamma+1}\left(R_s^*(n-n')\right)^{\gamma}}{\left(R_s(0)\right)^{2\gamma+1}} + 2P_{2\gamma_{max}+1}^{+\infty}\delta_{n,n'}, \qquad (2.85)$$

with $P_{2\gamma_{max}+1}^{+\infty}$ also given by (2.84).

This means that, besides the computation of $I = P_{out} - S$, we just have to calculate the terms corresponding to the first $\gamma_{max}$ IMPs.

### 2.2.3.3 Analysis of Peak Cancelation Schemes

Let us consider now the peak cancellation schemes described above. From (2.58), it is clear that the coefficients $p_n$ can be regarded as the output of a Cartesian or polar (characterized by (2.62) or (2.63), respectively) memorylesss nonlinearity, with input $s'_n = p_n + s_n^C$ (see (2.15)). Since $s_n^C$ is the output of an ideal cartesian or polar clipping for the same input $s'_n$, it can be decomposed into useful and self-interference components as explained in (2.66). Therefore, it is clear that $p_n$ can also be decomposed into useful and self-interference components in the following way:

$$p_n = (1-\alpha)s'_n - d_n. \tag{2.86}$$

When $s_M/\sigma \gg 1$, $\alpha \approx 1$ and $p_n \approx -d_n$.

## 2.2.3.4 *Final Frequency-Domain and Time-Domain Blocks*

Keeping in mind (2.66) and the signal processing chain in Figure 2.11, the frequency-domain block $\{S_k^{CF} = S_k^C G_k; k = 0,1,\ldots,N'-1\}$ can be broken down into useful and self-interference components:

$$S_k^{CF} = \alpha S'_k G_k + D_k G_k, \tag{2.87}$$

where $\{D_k; k = 0,1,\ldots,N'-1\}$ denotes the DFT of $\{d_n; n = 0,1,\ldots,N'-1\}$.

Clearly, $E[D_k] = 0$ and

$$
E\left[D_k D_{k'}^*\right] = \sum_{n=0}^{N'-1}\sum_{n'=0}^{N'-1} E\left[d_n d_{n'}^*\right]\exp\left(-j2\pi\frac{kn-k'n'}{N'}\right)
$$
$$
= \sum_{n=0}^{N'-1}\sum_{n'=0}^{N'-1} R_d(n-n')\exp\left(-j2\pi\frac{kn-k'n'}{N'}\right) = \begin{cases} N'G_d(k), & k = k' \\ 0, & \text{otherwise} \end{cases}
\tag{2.88}
$$

$(k = 0,1,\ldots,N'-1; k' = 0,1,\ldots,N'-1)$, where $\{G_d(k); k = 0,1,\ldots,N'-1\}$ denotes the DFT of the block $\{R_d(n); n = 0,1,\ldots,N'-1\}$. Similarly, it can be shown that

$$
E[S_k^C S_{k'}^{C*}] = \begin{cases} N'G_s^C(k), & k = k' \\ 0, & \text{otherwise,} \end{cases}
\tag{2.89}
$$

where $\{G_s^C(k) = |\alpha|^2 G_s(k) + G_d(k); k = 0,1,\ldots,N'-1\}$ denotes the DFT of $\{R_s^C(n); k = 0,1,\ldots,N'-1\}$ (given by (2.71) or (2.74), according to the type of nonlinearity), with $\{G_s(k); k = 0,1,\ldots,N'-1\} = \text{DFT } \{R_s(n); n = 0,1,\ldots,N'-1\}$. Therefore, $E[S_k^{CF} S_{k'}^{CF*}] = 0$ for $k \neq k'$, and $E[|S_k^{CF}|^2] = |G_k|^2 E[|S_k^C|^2] = N'|G_k|^2 G_s^C(k)$.

By employing the statistical characterization of the frequency-domain block to be transmitted, one can calculate a signal-to-interference ratio (SIR) for each subcarrier, given by

$$SIR_k = \frac{E\left[|\,\alpha S_k'\,|^2\right]}{E\left[|\,D_k\,|^2\right]}. \tag{2.90}$$

For the situation without oversampling ($M_{Tx} = 1$), (2.65) leads to $R_s(n - n') = 2\sigma^2 \delta_{n,n'}$. From (2.71) and (2.74), we thus have

$$R_s^C(n - n') = 2\sum_{\gamma=0}^{+\infty} P_{2\gamma+1} = 2P_1 + 2\sum_{\gamma=1}^{+\infty} P_{2\gamma+1} \tag{2.91}$$

for $n = n'$ and $R_s^C(n - n') = 0$ for $n' \neq n$; as a consequence,

$$SIR_k = \frac{E[|\,\alpha S_k'\,|^2]}{E[|\,D_k\,|^2]} = \frac{P_1}{\displaystyle\sum_{\gamma=1}^{+\infty} P_{2\gamma+1}}, \tag{2.92}$$

which is independent of $k$. It should be noted that, for $M_{Tx} > 1$ (i.e., when $N' > N$), $SIR_k$ is a function of $k$, since $E[|\,D_k\,|^2]$ is also a function of $k$.

When $N$ is high enough to validate a Gaussian approximation for the time-domain samples at the nonlinearity input, $s_n'$, our modeling approach for the final frequency-domain and time-domain blocks is quite accurate and $D_k$ exhibits quasi-Gaussian characteristics for any $k$ [Araújo and Dinis 2010, 2012b].

### 2.2.3.5 Signal Characterization for Iterative Clipping and Filtering Techniques

Let us consider the repeated use of the signal processing chain leading from $S_k'$ to $S_k^{CF}$ in Figure 2.11b. After the first iteration, the Gaussian approximation for the samples at the input to the nonlinear device ceases to be reasonable. However, the $k$th component of the frequency-domain block, for the $l$th iteration, can still be broken down as a sum of two uncorrelated components,

$$S_k^{CF(l)} = \alpha_k^{(l)} S_k' G_k + D_k^{(l)} G_k, \tag{2.93}$$

with $\alpha_k^{(l)}$ depending on $k$ for a given $l > 1$. Therefore, the frequency-domain samples at the output of the nonlinearity can still be decomposed into a useful component and a self-interference component, but that useful component is no longer proportional to the input, after the first iteration (this means that a decomposition similar to (2.66) is not valid for $l > 1$). Of course, $\alpha_k^{(1)} = \alpha$ in (2.93), and this equation can be regarded as a generalization of (2.41). As with the first iteration, our simulations indicate that $D_k^{(l)} = S_k^{C(l)} - \alpha_k^{(l)} S_k'$ can exhibit quasi-Gaussian characteristics, provided that $N$ is large enough (see Section 5 for more details); moreover, $E[D_k^{(l)}D_{k'}^{(l)*}] \approx 0, k \neq k'$, which means that $E[S_k^{CF(l)}S_{k'}^{CF(l)*}] \approx 0, k \neq k'$. The values of $\alpha_k^{(l)}$ and $E[|D_k^{(l)}|^2]$ can be obtained by simulation in the following way:

$$\alpha_k^{(l)} = \frac{E[S_k^{C(l)}S_k'^*]}{E[|S_k'|^2]} \tag{2.94}$$

and

$$E[|D_k^{(l)}|^2] = E[|S_k^{C(l)} - \alpha_k^{(l)}S_k'|^2], \tag{2.95}$$

respectively. Therefore, we can use (2.94) and (2.95) to calculate

$$SIR_k^{(l)} = \frac{E\left[|\alpha_k^{(l)}S_k'|^2\right]}{E\left[|D_k^{(l)}|^2\right]} \tag{2.96}$$

for each subcarrier.

## 2.2.4 Numerical Results

In this section, we present a set of performance results concerning the signal processing schemes studied here. We consider OFDM schemes with $N = 256$ subcarriers, unless otherwise stated, and two types of nonlinearities, both leading to a maximum output envelope $s_M$:

■ A Cartesian nonlinearity, corresponding to an ideal I-Q clipping with a clipping level, $x_M = \frac{s_M}{\sqrt{2}}$, that is, where $g(x)$ is given by (2.56).
■ A polar nonlinearity, corresponding to an ideal envelope clipping with a clipping level $s_M$, that is, wher $g_C(R)$ is given by (2.57)

(In both cases, there is an equivalent family of peak cancellation schemes.)

For these clipping characteristics, it can easily be shown that $S$ and $I$ can be written in closed form. In fact, $S = |\alpha|^2 \sigma^2$, with

$$\alpha = 1 - 2Q\left(\frac{x_M}{\sigma}\right) \tag{2.97}$$

for Cartesian clipping, and

$$\alpha = 1 - \exp\left(-\frac{s_M^2}{2\sigma^2}\right) + \frac{\sqrt{2\pi}s_M}{2\sigma}Q\left(\frac{s_M}{\sigma}\right), \tag{2.98}$$

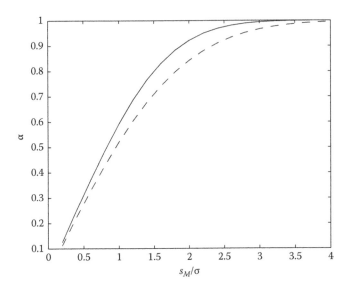

**Figure 2.15** **"Attenuation factor"** α for Cartesian (dashed line) and polar (solid line) clippings, as functions of $s_M/\sigma$.

for polar clipping. With respect to $I$, we have $I = P_{out} - S$, with

$$P_{out} = \frac{2\sigma^2}{\sqrt{2\pi}}\left(-\frac{x_M}{\sigma}\exp\left(-\frac{(x_M)^2}{2\sigma^2}\right) - \sqrt{2\pi}Q\left(\frac{x_M}{\sigma}\right) + \sqrt{\frac{\pi}{2}} + \frac{(x_M)^2\sqrt{2\pi}}{\sigma^2}Q\left(\frac{x_M}{\sigma}\right)\right)$$

(2.99)

for Cartesian clipping, and

$$P_{out} = \sigma^2\left(1 - \exp\left(-\frac{s_M^2}{2\sigma^2}\right)\right)$$

(2.100)

for polar clipping.

The α parameter is depicted in Figure 2.15, and the constant SIR when $M_{Tx} = 1$ is depicted in Figure 2.16. As expected, the values of α and SIR increase with $s_M/\sigma$. For very high values of $s_M/\sigma$, $\alpha \to 1$ and SIR $\to +\infty$.

By using some oversampling, that is, $M_{Tx} = 1$, the value of $SIR_k$ increases for all subcarriers and becomes no longer constant, with the subcarriers at the central region of the spectrum having the lower value, as shown in Figure 2.17. In all cases, the values of $SIR_k$ with $M_{Tx} = 2$ are very close to the corresponding values with $M_{Tx} = +\infty$. Of course, the set of multiplying coefficients $\{G_k; k = 0, 1, \ldots, N'-1\}$, used to shape the power spectral density of the transmitted signals (as will be shown in the following), does not change the $SIR_k$

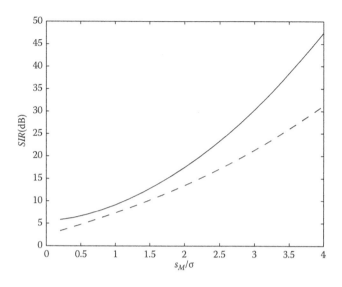

**Figure 2.16** SIR when $M_{Tx} = 1$, for Cartesian (dashed line) and polar (solid line) clippings, as functions of $s_M/\sigma$.

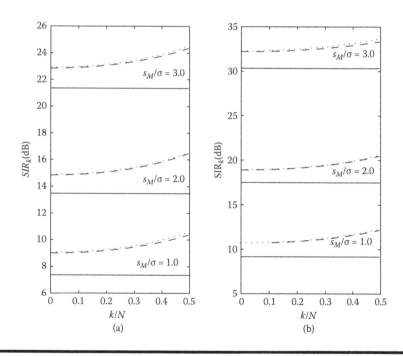

**Figure 2.17** Evolution of $SIR_k$ for Cartesian (a) and polar (b) clippings, when $M_{Tx} = 1$ (solid line), 2 (dashed line), or $+\infty$ (dotted line).

levels in all cases. The power distribution of the self-interference component within the subcarriers, required for the computation of the $SIR_k$ values, can be obtained from $R_d(n)$, given by (2.80) or (2.81) (see (2.90)). From Figures 2.16 and 2.17, it is clear that for a given $s_M/\sigma$ (and, inherently, a given maximum output envelope), the polar clipping has higher $SIR_k$ values.

Let us consider now the iterative clipping and filtering technique, where we repeatedly use the signal processing chain that leads from $\{S'_k\}$ to $\{S_k^{CF}\}$ in Figure 2.11b so as to provide an additional reduction of the envelope fluctuations while preserving a compact spectrum. We assume $L$ iterations and $s_M/\sigma = 2.0$, with an oversampling factor $M_{Tx} = 4$. It is also assumed that $G_k = 1$ for the $N$ in-band subcarriers and 0 for the remaining $N' - N$. Figure 2.18 shows the evolution of $\alpha_k^{(l)}$, obtained by simulation, as explained above. Except for the first iteration ($l = 1$), as expected, $\alpha_k^{(l)}$ significantly depends on $k$, taking lower values at the edge of the in-band region; by increasing the iteration order, $\alpha_k^{(l)}$ decreases for any $k$. On the other hand, the self-interference levels increase in the in-band region when we increase the number of iterations, leading to decreased $SIR_k$ values, as shown in Figure 2.19.

It should be noted that the statistical characterization of the transmitted blocks is independent of the number of subcarriers, provided that $N$ is large enough to

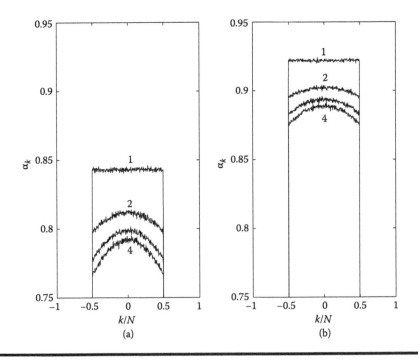

**Figure 2.18** Evolution of $\alpha_k$ for iterations 1 to 4, for Cartesian (a) and polar (b) clippings.

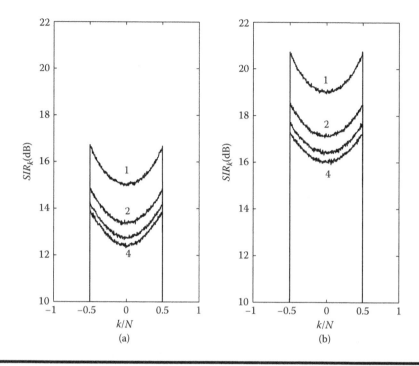

**Figure 2.19** $SIR_K$ **distribution with for iterations 1 to 4, for Cartesian (a) and polar (b) clippings.**

validate the Gaussian approximation. In this case, the statistical characterization is also independent of the adopted constellation on each subcarrier; however, since the self-interference noise behaves just as an additional Gaussian noise, the impact of a given SIR level changes with the adopted constellation. In the following, we will show that, when combined with a given OFDM transmission scheme, the signal processing schemes considered here can reduce the envelope fluctuation of the transmitted signals while retaining a high spectral efficiency.

As was already mentioned, the statistical characterization of the transmitted blocks presented here is based on the Gaussian behavior of the samples at the nonlinearity input, $s'_n$, which is valid for $N \rightarrow +\infty$. Since the exact analytical characterization with a small-to-moderate number of subcarriers is very complex (see [Araújo and Dinis 2010; Araújo and Dinis 2012b] for details), it is important to know how accurate our method is for low values of $N$.

Let us first consider the basic transmitter structure. It should be noted that, through available results of classical intermodulation analysis [Saleh 1981], each frequency-domain sample $S_k^C$ can always be broken down into two terms, one of them (the "self-interference" term) uncorrelated with the input sample $S'_k$; moreover, the correlated, "useful" term can be written as $\alpha_k^{eff} S'_k$, where $\alpha_k^{eff}$ denotes the "effective attenuation factor," given by $\alpha_k^{eff} = E[S_k^C S'^*_k] / E[|S'_k|^2]$. A set of simulation results on $\alpha_k^{eff}$ has been obtained for several values of $N$ and $s_M/\sigma$. These results show that $\alpha_k^{eff}$ is practically independent of $k$; moreover, we have the same attenuation factor for the time-domain samples and the frequency-domain samples; that is, $\alpha_k^{eff} \approx \alpha^{eff}$ as given by $\alpha^{eff} = E[s_n^C s_n^*]/E[|s_n'|^2]$), even for values of $N$ as low as 16. From these results, we could note that $\alpha^{eff}$ is very close to the $\alpha$ factor given by (2.67) or (2.68); with regard to the self-interference levels, we could note a very small decrease for lower values of $N$ (especially with high $s_M/\sigma$), as compared to the values computed under the Gaussian approach, leading to a slightly improved $SIR_k$. For $N = 64$ and a QPSK constellation on each subcarrier, the simulated $SIR_k$ values differ from the "theoretical" $SIR_k$ values by only about 0.1 to 0.3 dB, depending on $k$ and $s_M/\sigma$; for larger constellations, the differences are even lower. With $N = 256$, no difference could be observed between the theoretical and simulated results.

For the iterative transmitter structure, our simulations show that the frequency-domain samples $S_k^{C(l)}$ can still be broken down into uncorrelated "useful" and "self-interference" components as in (2.93), that is,

$$S_k^{C(l)} = \alpha_k^{(l)} S_k' + D_k^{(l)}, \tag{2.101}$$

with the frequency-dependent attenuation coefficient $\alpha_k^{(l)}$ given by (2.94). The evolution of $\alpha_k^{(l)}$ within the in-band region is almost independent of the number of subcarriers, provided that $N \geq 16$. As with the first iteration, there is a slight decrease on the in-band self-interference levels for low values of $N$, leading to higher $SIR_k$ values.

Another important issue is the accuracy of the Gaussian characterization for the frequency-domain, self-interference samples at the subcarrier level, $D_k$, which will be very helpful for deriving BER performances, as will be shown in the following. To evaluate the accuracy of this approach, we will consider the situation without oversampling (similar results could be reported with oversampling): Figure 2.20 shows the complementary cumulative distributions of $\text{Re}\{D_k\}$ (the same distributions were obtained for $\text{Im}\{D_k\}$), for several values of $s_M/\sigma$ and $N$. Clearly, the Gaussian approximation is accurate for small values of $s_M/\sigma$ (say, $s_M/\sigma \leq 2.0$), even with $N = 64$, which could be explained by the central limit theorem; however, for higher values of $s_M/\sigma$ this approximation is only appropriate for increased values of $s_M/\sigma$. The reason for this behavior is as follows: $\alpha \approx 1$ for moderate to high values of $N$ and, from (2.20), $d_n \approx s_n^C - s_n'$; this means that, for these high values of $s_M/\sigma$, the average number of significant samples $d_n$ can be low (unless $N$ is very high). As a consequence, the Gaussian approximation for the $D_k$ values, given by

$$D_k = \sum_{n=0}^{N'-1} d_n \exp\left(-j2\pi \frac{kn}{N'}\right), \tag{2.102}$$

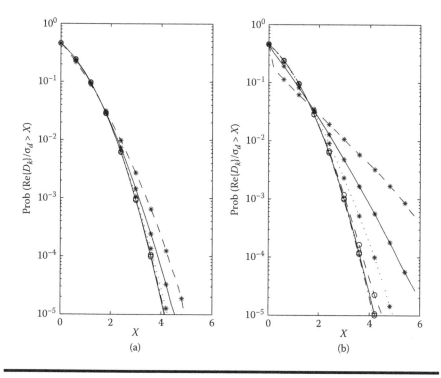

**Figure 2.20 Cumulative distribution function for (Re{$D_k$}/$\sigma_d$ $\sigma_d = \frac{1}{2}E[|D_k|^2]$) for Cartesian (a) and polar (b) clippings with $s_M/\sigma = 2.0$ (*) or 3.0 (O), when N = 64 (dashed line), 256 (solid line), or 1024 (dotted line) (dashed-dotted line) for the corresponding Gaussian distribution.**

is not appropriate since the central limit theorem ceases to apply to the situation. It should be noted that, for a cartesian clipping, the "true" clipping level is $\frac{s_M}{\sqrt{2}}$; therefore, the Gaussian approximation can be accurate for slightly higher values of $s_M/\sigma$.

For the iterative procedure, the self-interference levels slightly increase with the number of iterations, leading to an average number of significant samples $d_n^{(l)}$ ($\{d_n^{(l)}; n = 0,1,...,N'-1\} = $IDFT $\{D_k^{(l)}; k = 0,1,...,N'-1\}$) that is not lower than with the first iteration. As a consequence, we can employ the Gaussian approximation of $D_k^{(l)}$ with almost the same accuracy for the several iterations.

It will be shown later that we can use the Gaussian approach for accurate and efficient BER computations, regardless of the $s_M/\sigma$ values. In fact, when $s_M/\sigma$ is high enough to prevent a good Gaussian approximation for the $D_k$ coefficients, the channel noise levels became clearly dominant and a precise characterization of those coefficients is no longer relevant.

## 2.2.5 Characterization of the Analog Transmitted Signals

The complex envelope of the transmitted OFDM signal can be written as

$$s_{Tx}(t) = \sum_m s^{(m)}(t - mT_B),\qquad (2.103)$$

where $s^{(m)}(t)$ denotes the $m$th OFDM burst and $T_B$ is the burst interval. The OFDM samples for each burst, taken at a rate of $\frac{N''}{T}$ (i.e., with an oversampling factor $N''/N$ ), are given by $s_n^{Tx} r_n$, where $r_n \triangleq r(n\frac{T}{N''})$ with $r(t)$ denoting an appropriate window shape. The block $\{s_n^{Tx}; n = 0,1,\ldots,N''-1\}$ is the IDFT of $\{S_k^{Tx}; k = 0,1,\ldots,N''-1\}$ (for the sake of notation simplicity, we ignore the dependency with $m$).

When assuming that the time-domain samples $\{s_n^{Tx}; n = 0,1,\ldots,N''-1\}$ and the frequency-domain samples $\{S_k^{Tx}; n = 0,1,\ldots,N''-1$ (i.e., the DFT pairs) are periodic with period $N''$, the complex envelope of a given OFDM burst generated by the transmitter structure of Figure 2.11 can be written as

$$s(t) = \sum_{n=-\infty}^{+\infty} s_n^{Tx} r_n h_T\left(t - n\frac{T}{N'}\right).\qquad (2.104)$$

As with conventional OFDM signals with linear transmitters, $h_T(t)$ denotes the impulse response of the low pass reconstruction filter (the $\{s_n^{Tx}\}$ sequence is periodic but the $\{r_n\}$ sequence is not, of course).

For the spectral characterization of the transmitted signals, we will assume that the different OFDM bursts are uncorrelated and have the same statistical properties, with $E[S_k] = 0$ and

$$E[S_k S_{k'}^*] = 2\sigma_S^2 \delta_{k,k'}\qquad (2.105)$$

($\delta_{k,k'} = 1$ for $k = k'$ and 0 otherwise), where

$$\sigma_S^2 = \frac{1}{2}E\left[|S_k|^2\right]\qquad (2.106)$$

( $E[\cdot]$ denotes "ensemble average").

The PSD of the complex envelope $s_{Tx}(t)$ of the transmitted OFDM signals is given by

$$G_{Tx}(f) = \frac{1}{T_B}E\left[|S(f)|^2\right],\qquad (2.107)$$

where $S(f)$ denotes the Fourier transform of the OFDM burst $s(t)$. It can be easily shown that

$$S(f) = \frac{1}{T}\left(\sum_{k=-\frac{N''}{2}}^{\frac{N''}{2}-1} S_k^{Tx} R^{eq}\left(f - \frac{k}{T}\right)\right) \cdot H_T(f), \qquad (2.108)$$

with

$$R^{eq}(f) = \sum_{l=-\infty}^{+\infty} R\left(f - \frac{lN''}{T}\right), \qquad (2.109)$$

since

$$F\left\{\sum_{n=-\infty}^{+\infty} s_n^{Tx}\delta\left(t - n\frac{T}{N''}\right)\right\} = \frac{1}{T}\sum_{k=-\infty}^{+\infty} S_k^{Tx}\delta\left(f - \frac{k}{T}\right) \qquad (2.110)$$

($F(\cdot)$ denotes "Fourier transform").

Therefore, by using (2.108) in (2.107), we get

$$G_{Tx}(f) = \frac{1}{T_B} E\left[|S(f)|^2\right]$$

$$= \frac{1}{T^2 T_B} E\left[\sum_{k=-\frac{N''}{2}}^{\frac{N''}{2}-1} S_k^{Tx} R^{eq}\left(f - \frac{k}{T}\right) \sum_{k'=-\frac{N''}{2}}^{\frac{N''}{2}-1} S_{k'}^{Tx*} R^{eq*}\left(f - \frac{k}{T}\right)\right]|H_T(f)|^2$$

$$= \frac{|H_T(f)|^2}{T^2 T_B} \sum_{k=-\frac{N''}{2}}^{\frac{N''}{2}-1} E[|S_k^{Tx}|^2]\left|R^{eq}\left(f - \frac{k}{T}\right)\right|^2. \qquad (2.111)$$

since $E[S_k^{Tx} S_{k'}^{Tx*}] = 0$. Clearly, there are three filtering effects associated with the generation of the OFDM signals: the $\{S_k^{Tx}\}$ block depends on the frequency-domain filtering through the $\{G_k\}$ block; $R^{eq}(f)$ depends on $r(t)$, and, consequently, on $h_w(t)$; and $H_T(f)$ is the frequency response of the reconstruction filter.

## 2.2.6 Analytical BER Performance

It should be mentioned that the proposed signal processing schemes only involve modifications at the transmitter side, being compatible with conventional OFDM receivers.

The received signal is submitted to the receive filter, with impulse response $h_R(t)$, and then sampled at a rate $F_S = \frac{N''}{T}$, leading to blocks $\{y_n; n = 0,1,\ldots,N''-1\}$ after removal of the guard period. A square-root raised-cosine receive filtering is assumed (as well as for the transmit filtering), with roll-off $\rho \in [0,1]$; it is also assumed that $|H_R(f)| = |H_T(f)| = 1$ in the frequency band where $G_k \neq 0$. Next, a DFT operation leads to $\{Y_k; k = 0,1,\ldots,N''-1\} = \text{DFT}\{y_n; n = 0,1,\ldots,N''-1\}$, where $Y_k$ is concerned with the $k$th subchannel. It is well known that, thanks to the cyclic prefix,

$$Y_k = S_k^{Tx} H_k + N_k \tag{2.112}$$

(when the guard interval is longer than the overall channel impulse response), where $N_k$ and $H_k$ denote the noise component and the overall CFR (channel frequency response), respectively, at the $k$th subchannel. It can also be shown that these frequency-domain noise samples, uncorrelated and Gaussian, are zero-mean, and the variance of their real and imaginary parts is given by $N_0 \frac{(N'')^2}{T}$, when $\frac{N_0}{2}$ denotes the PSD of the white Gaussian noise at the receiver input.

Next, we will show how one can obtain the BER performance with the signal processing schemes considered here for a frequency-selective channel, characterized by the set of coefficients $\{H_k; k = 0,1,\ldots,N''-1\}$, when conventional OFDM receivers are employed.

## 2.2.6.1 Basic (Single-Iteration) Signal Processing Schemes

It was already shown that the frequency-domain block at the nonlinearity output can be broken down into two uncorrelated components: a useful component, $\{\alpha S_k' G_k, k = 0,1,\ldots,N'-1\}$; and a self-interference component $\{D_k G_k, k = 0,1,\ldots,N'-1\}$, where the $D_k$ coefficients have an approximately Gaussian distribution. This means that, besides the channel noise component, the received frequency-domain block can also be broken down into uncorrelated useful and self-interference components, since the frequency-selective channel identically affects the complex symbol to be transmitted and the corresponding self-interference $(H_k S_k^{Tx} = H_k \alpha S_k' G_k + H_k D_k G_k)$.

This broken down of the frequency-domain blocks into useful and self-interference components involves two complementary degradation effects: on the one hand, just part of the transmitted power is useful; on the other hand, the self-interference component is added to the channel noise, leading to an additional degradation.

For an ideal, coherent receiver with perfect synchronization and channel estimation, the BER for the $k$th subchannel can be expressed as

$$P_{b,k} \approx \alpha_M Q\left(\sqrt{\beta_M \cdot SNR_k}\right), \tag{2.113}$$

for $0 \geq k \geq \frac{N}{2}-1$ and $N''-\frac{N}{2} \geq k \geq N''-1$. In the equation above, $\alpha_M$ and $\beta_M$ are functions of the adopted constellation. For M-QAM constellations,

$$\alpha_M = \left(1-\frac{1}{\sqrt{M}}\right)\frac{4}{\log_2(M)} \qquad (2.114)$$

and

$$\beta_M = \frac{3}{M-1}. \qquad (2.115)$$

$Q(\cdot)$ denotes the well-known error function, given by

$$Q(x) = \frac{1}{\sqrt{2\pi}} \int_x^{+\infty} \exp\left(-\frac{t^2}{2}\right) dt, \qquad (2.116)$$

and $SNR_k$ denotes an equivalent SNR. This ratio is given by

$$SNR_k = \frac{E[|\alpha G_k S_k' H_k|^2]}{E[|D_k G_k H_k|^2] + E[|N_k|^2]}, \qquad (2.117)$$

where

$$E[|N_k|^2] = 2N_0 \frac{(N')^2}{T} \qquad (2.118)$$

and the other expectations can be analytically computed as described above. Of course, the overall BER is then the average of the BER associated with each of the $N$ "in-band" subchannels.

Asymptotically, when the channel noise effects become negligible, (2.113) takes the form

$$P_{b,k} \approx \alpha_M Q\left(\sqrt{\beta_M \cdot SIR_k}\right), \qquad (2.119)$$

where

$$SIR_k = \frac{E[|\alpha S_k'|^2]}{E[|D_k|^2]]}, \qquad (2.120)$$

which means that the nonlinear distortion should lead to an irreducible BER, depending on $SIR_k$ only, for any given constellation.

As described above, when $M_{Tx} = 1$ $SIR_k$ is independent of $k$ and given by $S/I$, with $S$ denoting the overall useful power and $I$ denoting the overall self-interference power. This allows very simple BER computations, since we do not need to evaluate separately the powers of the different IMPs (inter-modulation products), $P_{2\gamma+1}$, required for the computation of $E[|D_k|^2]$.

## 2.2.6.2 Iterative Clipping and Filtering Schemes

Let us consider now the iterative clipping and filtering technique. The frequency-domain block for the last iteration $\{S_k^{CF(L)} = S_k^{C(L)}G_k; k = 0,1,\ldots,N'-1\}$ can still be broken down into uncorrelated useful and self-interference components (see (2.93)), where $\alpha_k^{(L)}$ is variable across the in-band region, and the self-interference coefficients, $D_k^{(L)} = S_k^{C(L)} - \alpha_k^{(L)}S_k'$, exhibit quasi-Gaussian characteristics with zero mean and $E[D_k^{(L)}D_{k'}^{(L)*}] \approx 0, k \neq k'$. This means that we can still use the BER performance approach proposed for the basic, single-iteration case. We just have to replace $\alpha$ by

$$\alpha_k^{(L)} = \frac{E[S_k^{C(L)}S_k'^*]}{E[|S_k'|^2]}, \tag{2.121}$$

which has to be obtained by simulation for the $N$ "in-band" values of $k$. We also need to resort to simulations so as to obtain

$$E[|D_k^{(L)}|^2] = E[|S_k^{C(L)} - \alpha_k^{(L)}S_k'|^2]. \tag{2.122}$$

## 2.2.7 Performance Results

In this section, we present a set of performance results concerning the basic signal processing schemes studied here. For the sake of comparisons, we also include numerical results concerning two alternative techniques of higher complexity: one of them is the partial transmit sequence (PTS) technique; the other technique involves the repeated use of a signal processing scheme of the proposed class, within an iterative process.

We consider an OFDM modulation scheme with $N = 256$ subcarriers and an M-QAM constellation, with a Gray mapping rule, on each subcarrier (in fact, the performance of the signal processing schemes proposed here is almost independent of $N$, provided that $N$ is high enough to allow the Gaussian approximation of the OFDM signals). The set of multiplying coefficients $\{G_k; k = 0,1,\ldots,N'-1\}$ has a trapezoidal shape, with $G_k = 1$ for the $N$ data subcarriers (in-band region), dropping linearly to 0 along the first $(N_1 - N)/2$ out-of-band subcarriers at both sides of the in-band region (this means that we have $N_1$ non-zero subcarriers). We consider the two clipping characteristics previously described: an ideal I-Q clipping, and an ideal envelope clipping. We assume in all cases a square-root raised-cosine shape with $\rho = 0.25$ and a one-sided bandwidth $N''/T$. It is also assumed that the transmitter employs a power amplifier that is quasi-linear within the range of variations of the input envelope.

## 2.2.7.1 Performance Results for the Basic Signal Processing Scheme

Let us first consider the basic, single iteration, signal processing schemes proposed here (Figure 2.11b).

Figures 2.21 and 2.22 are concerned with the bandwidth efficiency issues with, respectively, "conventional OFDM" schemes and "modified OFDM" schemes, by using a signal processing of the proposed class. A well-known PSD-related function was adopted in both cases: the so-called fractional out-of-band power (FOBP). The FOBP results in Figure 2.21 are helpful to evaluate the impact of the windowing choice within a conventional OFDM scheme: even when $T_W$ is just a very small fraction of $T$, the spectrum becomes much more compact than with $T_W = 0$; on the other hand, the impact of $T_G/T \neq 0$ is rather small.

Figure 2.22 shows the FOBP when using the modified OFDM schemes proposed here, with $s_M/\sigma = 2.0$, for Cartesian and polar clippings and $T_G/T = 0.2$. Clearly, this clipping can lead to high out-of-band radiation levels. By using a frequency-domain filtering as reported above, with $N_1/N = 1$ (i.e., $G_k = 1$ for the $N$ data subcarriers and 0 for the remaining $N' - N$ ones) we can reduce the

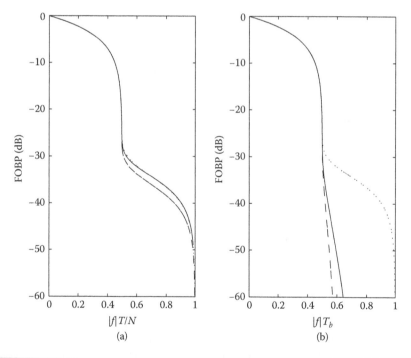

**Figure 2.21**  **FOBP with conventional OFDM: (a) for $T_W = 0$ and $T_G/T = 0$ (solid line), 0.2 (dotted line) or 0.5 (dashed line); (b) for $T_G/T = 2.0$ and $T_W/T = 0$ (dotted line), 0.025 (solid line), and 0.05 (dashed line).**

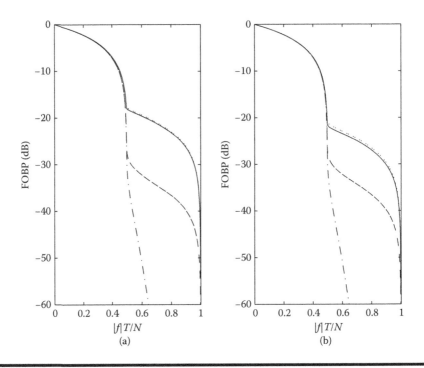

**Figure 2.22** FOBP for Cartesian (a) and polar (b) clippings, when $N_1 = N$, $s_M/\sigma = 2.0$ and $T_G/T = 2.0$, for $T_W = 0$ (dashed line) or $T_W/T = 0.025$ (dash-dotted line). For the sake of comparison, we include the FOBP when $G_k = 1$ for any $k$ and $T_W/T = 0$ (dotted line) or 0.025 (solid line), respectively.

out-of-band radiation to the levels of conventional OFDM; a further reduction in the out-of-band radiation levels is achieved if this frequency-domain filtering is combined with $T_W/T = 0.025$ (The use of $T_W/T = 0.025$ with $G_k = 1$ for any $k$ practically does not provide any improvement in the FOBP when $s_M/\sigma = 2.0$.) For other values of $s_M/\sigma$, we can still have out-of-band radiation levels similar to those of conventional OFDM schemes, provided that $N_1/N$ is low enough.

In Figure 2.23 we show the envelope distribution of the samples $s_n^{CF}$ for a polar clipping (similar results are obtained for cartesian clippings). As expected, an over-all reduction of several dB can be observed on the resulting PMEPR.

Figure 2.24 shows the PMEPR, defined according to (2.2) with $P = 10^{-3}$, when $N_1 = N$, $N_1 = 1.5N$ or, for the sake of comparison, when we do not employ any frequency-domain filtering ($G_k = 1$ for any $k$), for different values of $M_{Tx}$. From this figure, we can see that the PMEPR increases with $s_M/\sigma$ and is slightly higher for $N_1 = N$ (i.e., when lower out-of-band radiation levels are intended). We can also see that the higher the oversampling factor, the lower the PMEPR; on the other hand, a high filtering effort leads to a regrowth of the PMEPR. Without frequency-domain

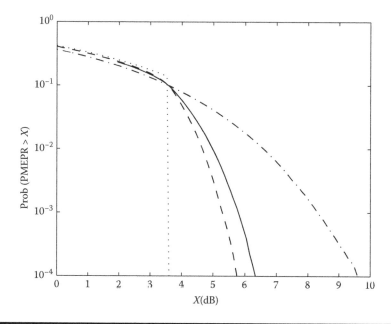

**Figure 2.23** Envelope distribution for a polar clipping with $s_M/\sigma = 2.0$: solid line with $N_1 = N$, dashed line with $N_1 = 1.5N$, dotted line with $G_k = 1$ for any $k$ and dash-dotted line with conventional OFDM.

filtering, we just have with $M_{Tx} = 4$ a PMEPR close to the one for $M_{Tx} = +\infty$; however, under a strong filtering effort, the PMEPR for $M_{Tx} = 2$ is already close to the PMEPR for $M_{Tx} = +\infty$.

In Figure 2.25 we present the required $E_b/N_0$ to have BER = $10^{-4}$ on an ideal AWGN channel, when $T_G = T_W = 0$ and we employ cartesian and polar clippings (an ideal coherent receiver with perfect synchronization was assumed). As expected, the required $E_b/N_0$ decreases with $s_M/\sigma$ in both cases. Clearly, the BER performances with $M_{Tx} = 2$ are already close to the corresponding performances with $M_{Tx} = +\infty$.

Figure 2.26 shows the required $E_b^{(p)}/N_0$ to have BER = $10^{-4}$ ($E_b^{(p)}/N_0(dB) = E_b/N_0(dB) + PMEPR\ (dB)$). This "peak $E_b/N_0$" is appropriate for an overall comparison of the "power efficiency" with the different transmission alternatives, since it combines the "detection efficiency" issue (through the required $E_b/N_0$) and the requirements on power amplification back-off (through the PMEPR). Clearly, the envelope clipping (polar clipping) has a better power efficiency than the I-Q clipping (Cartesian clipping). From Figure 2.26, we can see that the optimum values of $s_M/\sigma$ for a Cartesian clipping are 2.6 for $M = 4$, 3.6 for $M = 16$, and 4.2 for $M = 64$; for a polar clipping the optimum values of $s_M/\sigma$ are 2.0 for $M = 4$, 2.6 for $M = 16$, and 3.2 for $M = 64$. In both cases, the optimum values of $s_M/\sigma$ are almost independent of the filtering effort.

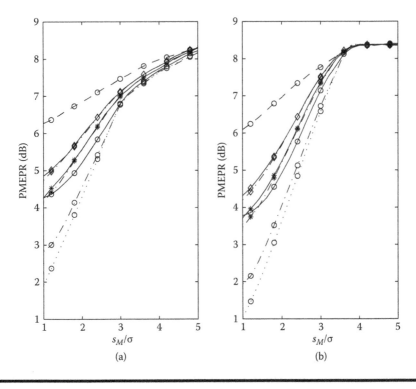

**Figure 2.24   PMEPR for Cartesian (a) and polar (b) clippings with $N_1 = N$ (◊), $N_1 = 1.5N$ (*), and for $G_k = 1$ for any $k$ (o), when $M_{Tx} = 1$ (dashed line), 2 (solid line), 4 (dash-dotted line), or $+\infty$ (dotted line).**

An important issue is the accuracy of the analytical approach for the performance evaluation presented here, which relies on both the quasi-Gaussian nature of OFDM signals with a high number of subcarriers, and the quasi-Gaussian characterization of the self-interference term at the subcarrier level. Our simulation results show that, whenever $N \geq 64$, the analytical approach for computation of the PSD of the transmitted signals is very accurate, regardless of the type of clipping, the clipping level, the oversampling factor, and the constellation size.

However, the accuracy of the analytical approach for computation of the BER performance depends not only on the number of subcarriers but also on the clipping level and the constellation size. For example, the theoretical and simulated BER performances of Figure 2.27, for a polar clipping when $M_{Tx} = 2$ and $N = 256$, are in close agreement (similar conclusions could be taken for other oversampling factors and for a Cartesian clipping). For $N = 64$, the analytical approach is not accurate when we are close to the irreducible error floor associated to the self-interference: for small values of $s_M/\sigma$ the theoretical performance is slightly pessimistic, the analytical $SIR_k$ is slightly lower than the "true" $SIR_k$.

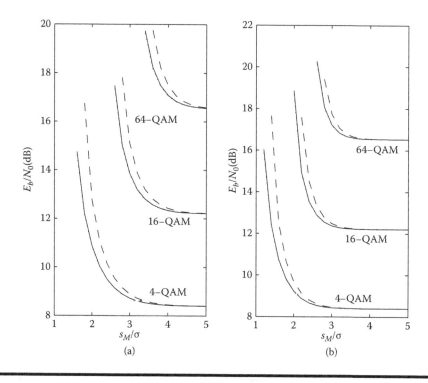

**Figure 2.25** Required $E_b/N_0$ for BER = $10^{-4}$, with Cartesian (a) and polar (b) clippings and $G_k = 1$ for any $k$, when $M_{Tx} = 1$ (dashed line), 2 (solid line), and $+\infty$ (dotted line).

For higher values of $s_M/\sigma$, the self-interference term at the subcarrier level can no longer be assumed to be Gaussian; however, we can still employ our analytical approach, since the corresponding self-interference levels are usually significantly below the channel noise levels.

### 2.2.7.2 Iterative Clipping and Filtering Schemes

Let us consider now the more sophisticated technique where we repeatedly use, in an iterative way, the signal processing chain that leads from $\{S'_k\}$ to $\{S_k^{CF}\}$ in Figure 2.11b. As stated previously, the purpose of this iterative technique is to further reduce the PMEPR values while keeping a compact spectrum.

For the following results, we consider $L$ iterations, and a polar (envelope) clipper, although a Cartesian clipping could also have been employed. It is assumed that $s_M/\sigma = 2.0$, the frequency-domain filtering is characterized by $N_1 = N$, and a 4-QAM constellation is employed on each subcarrier. The adopted oversampling factor is $M_{Tx} = 4$.

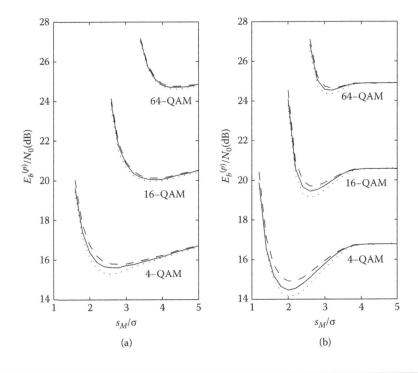

**Figure 2.26 Required $E_b^{(p)}/N_0$ for BER=$10^{-4}$, with Cartesian (a) and polar (b) clippings and $M_{Tx} = 2$, for the following cases: $G_k = 1$ for any $k$ (dotted line); $N_1 = N$ (dashed line); $N_1 = 1.5\ N$ (solid line).**

Figure 2.28 shows the impact of the number of iterations on the distribution of the signal envelope. From this figure, it is clear that we can greatly reduce the envelope fluctuations by using the iterative technique: the maximum envelope can be already close to $s_M$ with just three or four iterations. The corresponding PMEPR values, defined as in (2.48), are approximately 5.8dB, 4.9dB, 4.7dB, and 4.5dB for iterations 1, 2, 3, and 4, respectively. Since $G_k = 0$ for the $N'-N$ out-of-band subcarriers, and taking into account the PSD of the transmitted signals (given by (2.111)), it is clear that these PMEPR reductions are still achieved with out-of-band radiation levels similar to those of conventional OFDM.

Figure 2.29 shows the evolution of the self-interference levels before each frequency-domain filtering operation. These levels decrease in the out-of-band region and increase in the in-band region when the order of the iteration increases. Due to these lower out-of-band levels, the subsequent frequency-domain filtering operation produces a smaller PMEPR regrowth, which explains the results of Figure 2.28.

The higher in-band self-interference levels, which appear in Figure 2.29, combined with the corresponding decreased values of $\alpha_k^{(l)}$, lead to lower $SIR_k$ values.

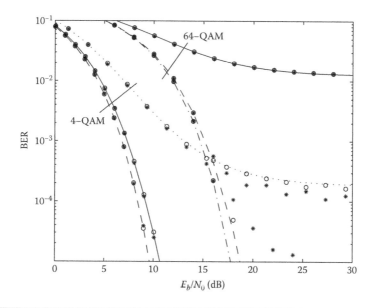

**Figure 2.27**  BER performance for 4-QAM and 64-QAM constellations, when a polar clipping is employed with $s_M/\sigma = 1.0$ (dotted line), 2.0 (solid line), 3.0 (dashed line) and 4.0 (dash-dotted line) ((o) for $N = 256$ and (*) for $N = 64$).

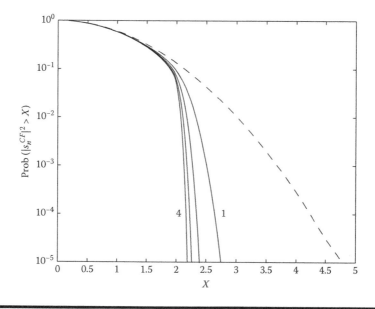

**Figure 2.28**  Envelope distribution for iterations 1 to 4 (solid lines), together with the envelope distribution for conventional OFDM (dashed line).

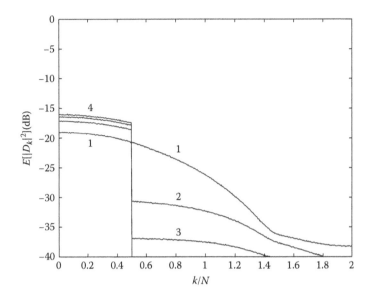

**Figure 2.29** Evolution of $E[|D_k^{(l)}|^2]$ for $l=1,2,3,4$.

These lower $SIR_k$ values give rise to an additional degradation on the BER performance when we increase the number of iterations.

Figure 2.30 shows the BER performances when an ideal AWGN channel is assumed, for different numbers of iterations. These results have been obtained by using the semi-analytical approach described above and are closely matched to the corresponding simulated results. As expected, the degradation increases with the number of iterations. With four iterations, leading to a PMEPR of 4.5 dB, the corresponding overall performance degradation is only about 1.9 dB for BER = $10^{-4}$, 1.2 dB for BER = $10^{-3}$, and 0.6 dB for BER = $10^{-2}$. This means that, for BER $\approx 10^{-2}$ to $10^{-3}$, the additional degradation when we increase the number of iterations is lower than the corresponding PMEPR gain (these "raw-BER" values are good enough for most systems, thanks to the use of channel coding).

### 2.2.7.3 Comparison with PTS Techniques

An alternative method for reducing the PMEPR of OFDM signals, while retaining a high spectral efficiency, is to employ partial transmit (PTS) techniques [Müller et al. 1997; Müller and Huber 1997; Cimini and Sollenberger 1999], which have been described as "distortionless methods," where the frequency-domain OFDM block is broken down into $V$ sub-blocks that are phase-rotated and combined to generate a "minimum-PMEPR" OFDM signal. It is assumed that the first sub-block is not phase-rotated; for each of the remaining blocks, there are $W$ possible phase

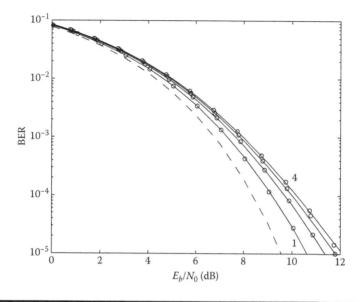

**Figure 2.30   BER for iterations 1 to 4 (solid line), together with BER for conventional OFDM (dashed line) (the circles denote simulated values).**

rotations, selected from the set $\{0, \frac{2\pi}{W}, \ldots, \frac{2\pi(W-1)}{W}\}$. With these PTS techniques, we obtain the same BER performance and about the same spectral efficiency as with the corresponding conventional OFDM schemes.

In Figure 2.31 we show the envelope distribution of the samples of the transmitted OFDM signal when the PTS technique is employed, and an M-QAM constellation with $M = 4$ is assumed on each subcarrier (similar results could be obtained for other values of $M$). The reduction in the PMEPR is as follows: for $N = 64$ the the PMEPR values are approximately 6.8 dB with $V = 4$ and $W = 2$, 6.4 dB with $V = W = 4$, and 5.8 dB with $V = 8$ and $W = 4$; for $N = 256$ these PMEPR values increase to 7.5, 7.2, and 6.9 dB, respectively. For the signal processing schemes considered here, with an envelope clipping operation according to $s_M/\sigma = 2.0$, we obtain a PMEPR of 5.7 dB for $L = 1$ iteration (corresponding to a degradation of 0.2 dB at BER $= 10^{-2}$ and 0.4 dB at BER $= 10^{-3}$) and a PMEPR of 4.5 dB for $L = 4$ iterations (corresponding to a degradation of 0.6 dB at BER$= 10^{-2}$ and 1.2 dB at BER $= 10^{-3}$). This means that the PMEPR values are worse with the PTS technique than with the basic signal processing scheme ($L = 1$) unless the number of subcarriers is very low; even when we take into account the BER performance degradations due to the nonlinear distortion, the proposed schemes are shown to provide better results, especially for high $N$. When used iteratively ($L > 1$), the signal processing schemes studied here exhibit an improved advantage. With very large constellations, however, the optimum values of $s_M/\sigma$ (and the corresponding PMEPR values) are a bit higher, therefore decreasing the difference between the two techniques.

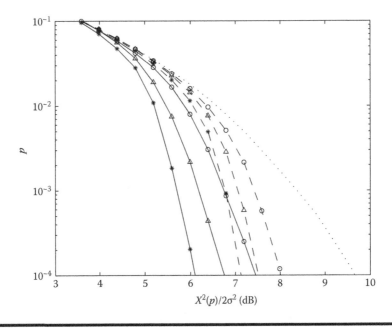

**Figure 2.31** **Envelope distribution with the PTS technique, when $N = 64$ (solid line) or 256 (dashed line), and, for the sake of comparison, with a conventional OFDM scheme (dotted line): ($\triangle$) for $V = 4$ and $W = 2$, ($\bigcirc$) for $V = W = 4$, and (\*) for $V = 8$ and $W = 2$.**

Moreover, as an additional advantage, the signal processing schemes considered here do not require any modification in the OFDM receiver, and the implementation complexity at the transmitter side is much lower than with the PTS techniques: essentially "three FFT operations" (for a single iteration) per OFDM transmitted block versus "$V$ FFTs plus the signal processing for the optimization of the phase rotations." The PTS techniques also require the transmission of an extra overhead for each block carrying the information on the selected set of phase rotations.

## 2.2.8 LINC-Type Amplification of OFDM Signals

The clipping techniques described above allow significant reductions in the envelope fluctuations of OFDM signals and, therefore, more efficient power amplification. However, we still need to employ quasi-linear amplifiers. However, it is known that grossly nonlinear amplifiers are simpler, cheaper, and have higher output power and higher amplification efficiency. Therefore, it would be desirable to employ grossly nonlinear power amplifiers with OFDM signals. A promising way of doing this is to use two-branch LINC (LInear amplification with Nonlinear Components) transmitter structures [Cox 1974].

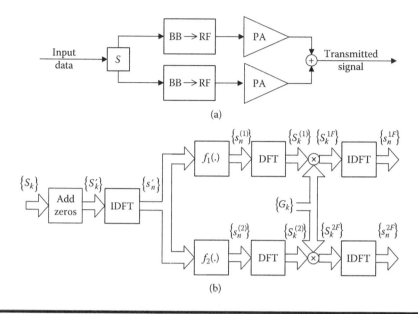

**Figure 2.32** **General LINC transmitter structure (a) and detail on the "signal separation" processor S (b).**

The two-branch LINC transmitter structure we are considering here is based on the techniques proposed in [Dinis and Gusmão 1996a, 2001c, 2008] and depicted in Figure 2.32, where S denotes the "signal separation" processor. In both transmitter branches, the analog I and Q signals are submitted to a conventional quadrature modulator, and the resulting signal is converted to RF and power amplified. Since the signals at the input of each amplifier have small envelope fluctuations, we can employ grossly nonlinear power amplifiers, which have higher amplification efficiency and higher output power, and are simpler to implement.

Figure 2.32b presents part of the signal processing structure proposed for the LINC decomposition. As with the signal processing schemes proposed for conventional transmitters, we form an augmented block, $\{S'_k; k = 0,1,\ldots,N'-1\}$, by adding $N'-N$ zeros to the original frequency-domain block and an IDFT operation leads to the corresponding time-domain block $\{s'_n; n = 0,1,\ldots,N'-1\}$. These samples are submitted, in parallel, to two nonlinear operations, to form the blocks $\{s_n^{(1)}; n = 0,1,\ldots,N'-1\}$ and $\{s_n^{(2)}; n = 0,1,\ldots,N'-1\}$, with

$$s_n^{(1)} = f_1(|s'_n|)\exp(j\arg(s'_n))$$

(2.123)

and

$$s_n^{(2)} = f_2(|s'_n|)\exp(j\arg(s'_n)).$$

(2.124)

The sets of samples $\{s_n^{(i)}; n = 0,1,\ldots,N'-1\}$ $(i = 1,2)$ can be regarded as the outputs of two bandpass memoryless nonlinearities, characterized by

$$f_1(R) = f_c(R) + jf_e(R) \tag{2.125}$$

and

$$f_2(R) = f_c(R) - jf_e(R), \tag{2.126}$$

$(|s_n'| = R)$, with $f_c(R)$ and $f_e(R)$ as follows:

$$f_c(R) = \begin{cases} \dfrac{1}{2}R, & R \leq s_M \\[2mm] \dfrac{1}{2}s_M, & R > s_M; \end{cases} \tag{2.127}$$

$$f_e(R) = \begin{cases} \dfrac{1}{2}\sqrt{s_M^2 - R^2}, & R \leq s_M \\[2mm] 0, & R > s_M. \end{cases} \tag{2.128}$$

The corresponding conversion functions are $|f_i(R)|$ (AM/AM) and $\arg(f_i(R))$ (AM/PM), $i = 1, 2$.

Next, a pair of DFTs brings the modified blocks $\{s_n^{(i)}; n = 0,1,\ldots,N'-1\}$ $(i = 1, 2)$ back to the frequency domain, where a spectral shaping is performed with the help of the set of multiplying coefficients $G_k, k = 0,1,\ldots,N'-1$. After these frequency-domain filtering operations, a pair of IDFTs leads to the final time-domain blocks for each branch:

$$\{s_n^{1F}; n = 0,1,\ldots,N'-1\} = IDFT\{S_k^{1F} = S_k^{(1)}G_k; k = 0,1,\ldots,N'-1\} \tag{2.129}$$

and

$$\{s_n^{2F}; n = 0,1,\ldots,N'-1\} = IDFT\{S_k^{2F} = S_k^{(2)}G_k; k = 0,1,\ldots,N'-1\}, \tag{2.130}$$

respectively.

Usually a cyclic extension is appended to each block and the corresponding analog in-phase and quadrature components are obtained by an appropriate D/A conversion, concluding the so-called "LINC decomposition." Once again, appending $N'-N$ zeros to each initial frequency-domain block is equivalent to oversampling, by a factor $M_{Tx} = N'/N$, the "ideal" OFDM burst. This oversampling leads to a reduction of the nonlinear in-band self-interference levels; moreover, a relatively high oversampling factor (i.e., $M_{Tx} \geq 4$), is required to ensure a quasi-constant envelope for the analog signal associated with each amplification branch, due to the increased signal bandwidth.

The generation of the blocks $\{s_n^{(i)}; n = 0,1,\ldots,N'-1\}$ $(i = 1, 2)$ from the block $\{s_n'; n = 0,1,\ldots,N'-1\}$ can be regarded as a LINC decomposition of an ideal envelope clipped version of the OFDM signal [Dinis and Gusmão 1996a]. In fact, from (2.123) and (2.124), we get

$$s_n^{(1)} + s_n^{(2)} = 2s_n^c = 2f_c(|s_n'|)\exp(j\arg(s_n')) = \begin{cases} s_n', & |s_n'| \leq s_M \\ s_M \exp(j\arg(s_n')), & |s_n'| > s_M. \end{cases} \quad (2.131)$$

On the other hand,

$$|s_n^{(1)}| = |s_n^{(2)}| = \frac{s_M}{2}, \quad (2.132)$$

which means analog signals with very low envelope fluctuations (for $M_{Tx} \geq 4$ an almost constant envelope is achieved if frequency-domain filtering is not employed); therefore, these signals can be amplified without distortion by two grossly nonlinear amplifiers.

It should be noted that $s_n^{(1)} = s_n^c + je_n$ and $s_n^{(2)} = s_n^c - je_n$, where

$$s_n^c = f_c(|s_n'|)\exp(j\arg(s_n')) = \begin{cases} \dfrac{1}{2}s_n', & |s_n'| \leq s_M \\ \dfrac{1}{2}s_M \exp(j\arg(s_n')), & |s_n'| > s_M \end{cases} \quad (2.133)$$

corresponds to the samples after an ideal envelope clipping, and the complementary terms required for (2.132) are $je_n$ and $-je_n$, with

$$e_n = f_e(|s_n'|)\exp(j\arg(s_n')) = \begin{cases} \dfrac{1}{2}s_n'\sqrt{\dfrac{s_M^2}{|s_n'|^2}-1}, & |s_n'| \leq s_M \\ 0, & |s_n'| > s_M. \end{cases} \quad (2.134)$$

It should also be noted that these complementary terms cancel in the LINC combination process (see (2.131)); however, when there are phase or gain imbalances between the two amplifying branches they can lead to a significant performance degradation, especially in terms of out-of-band radiation levels. The frequency-domain filtering operation, through the coefficients $G_k, k = 0,1,\ldots,N'-1$, can be very useful for reducing these levels, but it also leads to some envelope fluctuations at each amplifier input.

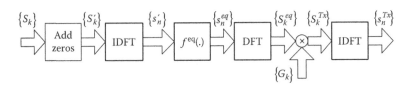

**Figure 2.33 Equivalent transmission model for LINC transmitters (possibly imbalanced).**

If we assume that the two amplifiers are linear in the region corresponding to the small input envelope fluctuations, then the impact of the two amplifiers can be described as two complex gains, $g_1$ and $g_2$. This takes into account both the phase ($\arg(g_1) \neq \arg(g_2)$) and gain ($|g_1| \neq |g_2|$) imbalances. Since the set of coefficients $\{G_k; k = 0,1,\ldots,N'-1\}$ is the same in the two branches, the equivalent transmission model represented in Figure 2.33 can also be employed, with the equivalent bandpass memoryless nonlinearity characterized by

$$f^{eq}(R) = g_1 f_1(R) + g_2 f_2(R) = (g_1 + g_2) f_c(R) + j(g_1 - g_2) f_e(R), \quad (2.135)$$

where $f_c(R)$ and $f_e(R)$ are given by (2.127) and (2.128), respectively. The "equivalent block" effectively transmitted in the frequency domain is $\{S_k^{Tx}; k = 0,1,\ldots,N'-1\}$, with

$$S_k^{Tx} = (g_1 + g_2)S_k^c G_k + j(g_1 - g_2)E_k G_k, \quad (2.136)$$

where $\{S_k^c; k = 0,1,\ldots,N'-1\}$ is the DFT of $\{s_n^c; n = 0,1,\ldots,N'-1\}$ and $\{E_k; k = 0,1,\ldots,N'-1\}$ is the DFT of $\{e_n; n = 0,1,\ldots,N'-1\}$. Clearly, the block $\{S_k^{Tx}; k = 0,1,\ldots,N'-1\}$ replaces the block $\{S_k; k = 0,1,\ldots,N-1\}$ of conventional OFDM signals.

Again, the effective frequency-domain samples can be broken down into two uncorrelated terms, a "useful" term and a "self-interference" term, as in (2.66), where $\alpha$ is given by (2.68) with $f^{eq}(R)$ replacing $g_C(R)$. A "signal-to-interference ratio" can still be defined by (2.90), now depending on the "equivalent nonlinearity" given by (2.135). When $g_1 = g_2 \triangleq g$ (perfect balance condition), of course, $f^{eq}(R) = g s_n'$ for $|s_n'| \leq s_M$, and $f^{eq}(R) = g s_M$ for $|s_n'| \geq s_M$; in this case, the impact of the proposed signal processing scheme within a LINC transmitter structure is the same as that of a polar clipping within a conventional, single-branch transmitter structure. This means that we can employ the analytical approach presented above for the statistical characterization of that block.

Next, we present a set of results for the signal processing scheme proposed here for LINC "signal separation." We consider $N = 256$, a QPSK constellation on each subcarrier and a relative clipping level $s_M/\sigma = 2.0$. As for the reduced-PMEPR schemes for conventional transmitters, we adopt $G_k = 1$ "in-band," and $G_k$ values decreasing linearly to 0 for the $(N_1 - N)/2$ subcarriers at both sides of the

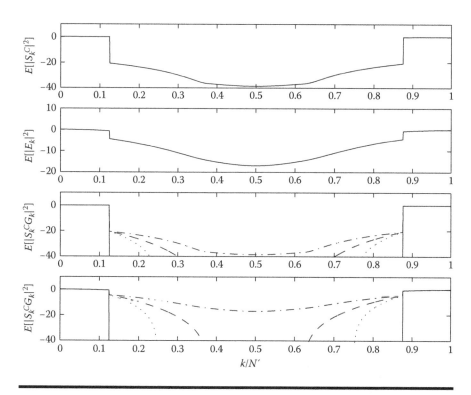

**Figure 2.34** Power distribution of $S_k^C$, $E_k$, $G_k$, and $E_kG_k$: solid line for $N_1 = N$, dotted line for $N_1 = 2N$, dashed line for $N_1 = 3N$ and dash-dotted line for $G_k = 1, k = 0,1,\ldots,N'-1$.

data band; an oversampling factor $M_{Tx} = 4$ is considered. The power amplifiers are assumed to be linear in the reduced dynamic range of the input envelopes.

Figure 2.34 shows the power distribution of $S_k^C$ and $E_k$, as well as the power distribution of $S_k^C G_k$ and $E_k G_k$ for $M_{Tx} = 4$ ($\{S_k^C; k = 0,1,\ldots,N'-1\} =$ DFT $\{s_n^C; n = 0,1,\ldots,N'-1\}$ and $\{E_k; k = 0,1,\ldots,N'-1\} =$ DFT $\{e_n; n = 0,1,\ldots,N'-1\}$). Clearly, the out-of-band radiation effects of the nonlinearity characterized by $f_c(\cdot)$ are very mild; those concerning $f_e(\cdot)$ are much more severe, but fortunately they only have a negative impact on performance when there are significant phase or gain imbalances: the degradation in the average SIR levels is below 3 dB as long as the phase imbalances are between –5° and 20° and the gain imbalances are lower than 20%. An interesting aspect is the fact that the best SIR performance is achieved for a nonzero phase imbalance ($\Delta\theta = 10°$). As with conventional transmitter structures, the frequency-domain filtering can be very effective in reducing the out-of-band radiation level. However, this filtering operation is associated with some envelope regrowth: the PMEPR of the samples $s_n^{1F}$ (and $s_n^{2F}$) is 4.0 dB for $N_1 = N$, 2.6 dB for $N_1 = 2N$, and 1.6 dB for $N_1 = 3N$, as opposed to the constant envelope when $G_k = 1, k = 0,1,\ldots,N'-1$.

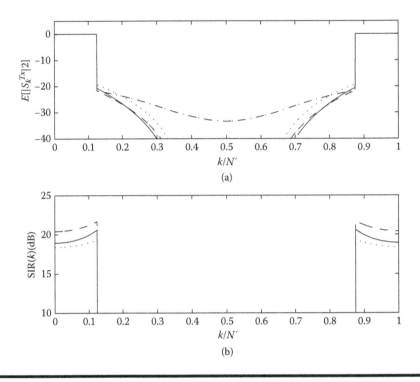

**Figure 2.35 Impact of the phase imbalances on the power distribution (a) and the SIR distribution (b), for $N_1 = 3N$: solid line for $\Delta\theta = 0°$, dashed line for $\Delta\theta = 10°$, and dotted line for $\Delta\theta = 20°$ (dash-dotted line for $\Delta\theta = 10°$ with $G_k = 1, k =, 1, ..., N' - 1$, for the sake of comparison).**

Figure 2.35 shows the impact of the phase imbalances on the distribution of the power among the subcarriers, as well as the evolution of the resulting SIR for $M_{Tx} = 4$ and $N_1 = 3N$, when a perfect gain balance is assumed. Once again, the best SIR is associated with a phase imbalance of about 10° (however, this imbalance leads to increased out-of-band radiation levels). If we employ the frequency-domain filtering with $N_1 = 3N$, the out-of-band radiation decreases significantly, as expected.

Again, we can derive analytically the BER performances from the SIR distribution, since the self-interference component is quasi-Gaussian at the subcarrier level. The corresponding BERs are presented in Figure 2.36, when an AWGN channel is assumed. Clearly, there is a close matching between the analytical and the simulated results. The case where $\Delta\theta = 0$ corresponds to an ideal polar (envelope) clipping.

### 2.2.9 Complementary Remarks

Several methods for reducing the envelope fluctuations of OFDM signals were described for both conventional and two-branch LINC transmitter structures.

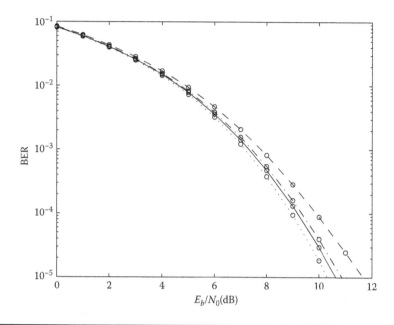

**Figure 2.36** **Impact of phase imbalances on the BER performance, for $N_1 = 3N$: solid line for $\Delta\theta = 0°$, dashed line for $\Delta\theta = -10°$, dotted line for $\Delta\theta = 10°$, and dash-dotted line for $\Delta\theta = 20°$ (The simulated points are denoted by (O).).**

Appropriate models for the statistical characterization of the frequency-domain blocks effectively transmitted were presented.

Within a conventional, single-branch transmitter structure, we considered the use of both cartesian and polar nonlinear operations of the "clipping" type. It was shown that the polar clipping has better performance, although the difference is not too high. By iteratively repeating the clipping and filtering operations, we can further reduce the envelope fluctuations of the transmitted signals, although at the expense of slightly higher nonlinear distortion levels (not to mention the additional signal processing complexity).

The two-branch LINC transmitter structures are an interesting alternative when high power efficiency is required, provided that we can ensure low phase and gain imbalances between the amplifiers. LINC decomposition is also the basis of the more exotic CEPB-OFDM schemes (constant-envelope paired-burst OFDM) that allow a single grossly nonlinear power amplifier at the transmitter [Dinis and Gusmão 1996b, 1998, 2003b]. Moreover, CEPB-OFDM signals have intrinsic characteristics that help the carrier synchronization procedure [Dinis and Gusmão 1997].

Although we have considered only OFDM schemes, our signal processing techniques and analytical approaches can also be employed in the analysis and optimization of other multicarrier schemes and different nonlinear devices [Dinis and Silva 2006; Araújo and Dinis 2007, 2011, 2012a].

## 2.2.9.1 Single Carrier with Frequency-Domain Equalization

The techniques described in the previous section can significantly reduce the envelope fluctuations of OFDM signals. However, this is achieved at the expense of increased signal processing requirements at the transmitting chain and, eventually, increased nonlinear distortion levels. Moreover, even for the most sophisticated techniques, the transmitted signals still exhibit higher envelope fluctuations than single-carrier (SC) signals based on similar constellations. This is the motivation behind SC-FDE schemes [Sari et al. 1994].

## 2.2.9.2 Linear FDE

To understand SC-FDE schemes as well as SC-FDMA schemes adopted for the uplink of 3GPP LTE [3GPP 2005], let us consider an SC-based block transmission with $N$ useful modulation symbols per block $\{s_n; n = 0,1,...,N-1\}$ resulting from a direct mapping of the original data into a selected signal constellation, plus a suitable CP that has essentially the same function as the OFDM schemes. The receiver can have the structure represented in Figure 2.37a, where it is assumed that after being down-converted and filtered, the signal is sampled and A/D converted. The resulting signal is S/P converted and the CP samples are removed, leading to the received time-domain samples $\{y_n; n = 0,1,...,N-1\}$. These samples are passed to the frequency-domain by an $N$-point DFT, leading to the corresponding

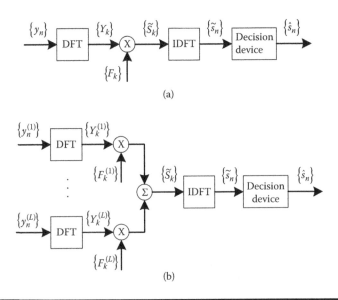

(a)

(b)

**Figure 2.37   SC-FDE receiver structure with no space diversity (a) and with an *L*-order space diversity receiver (b).**

frequency-domain samples $\{Y_k; k = 0,1,...,N-1\}$, where $Y_k = H_k S_k + N_k$ (as with OFDM schemes).

For an FDE optimized under the ZF criterion, the equalized frequency-domain samples, given by (2.41) can be obtained with the set of coefficients $F_k$ given by (2.32). However, for a typical frequency-selective channel, deep notches in the channel frequency response lead to significant noise enhancement effects when the ZF criterion is employed. To minimize the combined effect of ISI and channel noise, the equalized samples $\{\tilde{S}_k; k = 0,1,...,N-1\}$ are obtained with the coefficients $\{F_k; k = 0,1,...,N-1\}$ usually optimized under the minimum mean square error (MMSE) criterion, which leads to the set of optimized FDE coefficients

$$F_k = \frac{H_k^*}{\beta + |H_k|^2}, \qquad k = 0,1,...,N-1 \tag{2.137}$$

where $\beta$ is the inverse of the signal-to-noise ratio (SNR), given by

$$\beta = \frac{\sigma_N^2}{\sigma_S^2}, \tag{2.138}$$

with

$$\sigma_N^2 = \frac{E[|N_k|^2]}{2} \tag{2.139}$$

and

$$\sigma_S^2 = \frac{E[|S_k|^2]}{2} \tag{2.140}$$

denoting the variances of the real and imaginary parts of the channel noise components $\{N_k; k = 0,1,...,N-1\}$ and the data samples $\{S_k; k = 0,1,...,N-1\}$, respectively.

Since, for SC modulations the data contents of a given block are transmitted in the time domain, the equalized samples $\{\tilde{S}_k; k = 0,...,N-1\}$ are converted back to the time domain by an IDFT operation leading to the block of time-domain equalized samples $\{\tilde{s}_n; n = 0,...,N-1\}$. These equalized samples will then be used to make decisions about the transmitted symbols.

The SC-FDE receiver can be easily extended to an *L*-branch diversity scenario as depicted in Figure 2.37b. In this case, the frequency-domain samples at the FDE's output are given by (2.30), where the set $\{F_k^{(l)}; k = 0,1,...,N-1\}$ ($l = 1,..,L$) can be selected under the ZF or MMSE criterion. Under "equal noise level" conditions (i.e., $\sigma_n^{(1)} = ... = \sigma_n^{(L)} \triangleq \sigma_n$, with $(\sigma_n^{(l)})^2$ denoting the variance of the input

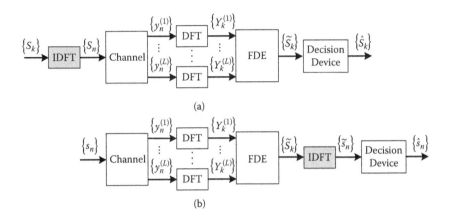

**Figure 2.38   Generic transmission chain for OFDM (a) and SC-FDE (b).**

noise samples at the *l*th branch) and an MMSE criterion, it can be easily shown that the optimized FDE coefficients are given by [Gusmão et al. 2003]

$$F_k^{(l)} = \frac{H_k^{(l)^*}}{\beta + \sum_{l'=1}^{L} |H_k^{(l')}|^2} \quad l = 1, 2, ..., L. \tag{2.141}$$

Contrary to the OFDM schemes where ZF and MMSE criterions yield the same performance [Gusmão et al. 2003], for the SC case the performance with ZF and MMSE criteria is only identical when the channel frequency response across the transmission bandwidth is practically constant (i.e., for $H_0^{(l)} = ... = H_{N-1}^{(l)} \triangleq H^{(l)}, l = 1, ..., L$). An FDE optimized under the MMSE criterion does not attempt to fully invert the channel when we have a deep fade, reducing noise enhancement effects and allowing better performance.

For comparison purposes, Figure 2.38 shows the block diagrams for the transmission chains of OFDM and SC-FDE options. From this figure, it is clear that both schemes are closely related. Their overall signal processing effort, measured in terms of DFT/IDFT blocks, are similar, the only difference being the shift of the IDFT block from the OFDM transmitter to the SC-FDE receiver. However, there are pros and cons to both of these transmission techniques. One of the most important issues that must be taken into consideration is the already mentioned strong envelope fluctuations and high PMEPR of OFDM signals with a large number of subcarriers, leading to power amplification difficulties. This requires the use of highly linear power amplifiers at the transmitter or more power backoff than with comparable SC signals. This is particularly important for the uplink transmission since low-cost and low-consumption power amplifiers are desirable at the MT/UE. In fact, even when suitable signal processing techniques are employed to reduce the envelope fluctuations of OFDM signals as the ones described in the previous section the resulting envelope fluctuations are still higher than with the corresponding SC schemes.

Nevertheless, when the wireless network includes a fixed terminal (e.g., within BSs or for broadcasting systems), OFDM schemes are good candidates. Bearing in mind the compatibility between SC-FDE and OFDM options, we can choose an SC-FDE scheme, exhibiting low envelope fluctuations, for the uplink and an OFDM scheme for the downlink. This implies an implementation advantage for the MTs/UEs, where simple SC transmissions and OFDM reception functions are then carried out. The implementation charge is concentrated in the BSs (where increased power consumption and cost are not so critical), concerning both the signal processing effort and the power amplification difficulties [Gusmão et al. 2000; Falconer et al. 2002].

### 2.2.9.3 IB-DFE Receivers

It is well known that decision feedback equalizers (DFE) [Proakis 2001] can significantly outperform linear equalizers (in fact, DFEs include as a special case linear equalizers). Time-domain DFE have good performance/complexity trade-offs provided that the channel impulse response is not too long. However, if the channel impulse responses expand over a large number of symbols (such as in the case of severely time-dispersive channels), conventional time-domain DFEs are too complex. For this reason, a hybrid time-frequency SC-DFE was proposed in [Benvenuto and Tomasin 2002], employing a frequency-domain feedforward filter and a time-domain feedback filter. This hybrid time-frequency-domain DFE performs better than a linear FDE. However, as with conventional time-domain DFEs, it can suffer from error propagation, especially when the feedback filters have a large number of taps. As an alternative, we can employ the iterative block-DFE (IB-DFE) approach for SC transmission [Benvenuto et al. 2010]. Within these IB-DFE schemes, both the feedforward and the feedback parts are implemented in the frequency domain, as depicted in Figure 2.39.

Let us consider an $L$-order space diversity IB-DFE. For a given $i$th iteration, the output samples are given by

$$\tilde{S}_k^{(i)} = \sum_{l=1}^{L} F_k^{(l,i)} Y_k^{(l)} - B_k^{(i)} \hat{S}_k^{(i-1)}, \qquad (2.142)$$

where $\{F_k^{(l,i)}; k = 0,1,...,N-1\}$ $l = 1,...,L)$ and $\{B_k^{(i)}; k = 0,1,...,N-1\}$ denote the feedforward and feedback equalizer coefficients, respectively, and $\{S_k^{(i-1)}; k = 0,1,...,N-1\}$ is the DFT of the hard-decision block $\{s_n^{(i-1)}; n = 0,1,...,N-1\}$, of the $(i - 1)$th iteration, associated to the transmitted time-domain block $\{s_n; n = 0,1,...,N-1\}$.

The forward and backward IB-DFE coefficients $\{F_k^{(l,i)}; k = 0,1,...,N-1\}$ $(l = 1,...,L)$ and $\{B_k^{(i)}; k = 0,1,...,N-1\}$, respectively, are chosen so as to maximize

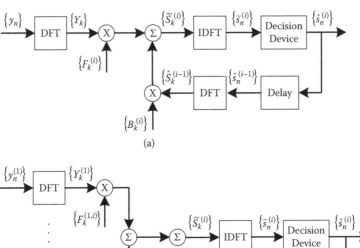

(a)

(b)

**Figure 2.39    IB-DFE receiver structure with no diversity (a) and with an *L*-branch space diversity (b).**

the signal-to-interference plus noise ratio (SINR). The optimum feedforward and feedback coefficients are given by

$$F_k^{(l,i)} = \frac{\kappa_F^{(i)} H_k^{(l)^*}}{\beta + \left[1 - (\rho^{(i-1)})^2\right] \sum_{l'=1}^{L} |H_k^{(l')}|^2}, \quad l = 1, 2, \ldots, L, \tag{2.143}$$

and

$$B_k^{(i)} = \rho^{(i-1)} \left( \sum_{l'=1}^{L} F_k^{(l',i)} H_k^{(l')} - 1 \right), \tag{2.144}$$

respectively, where $\kappa_F^{(i)}$ is selected to ensure that $\gamma^{(i)} = 1$, $\beta$ is given by (2.38), and

$$\rho^{(i)} = \frac{E[s_n^* \hat{s}_n^{(i)}]}{E[|s_n|^2]} \tag{2.145}$$

is a measure of the reliability of the decisions used in the feedback loop. Since the IB-DFE coefficients take into account the overall block reliability, the error propagation problem is significantly reduced. Consequently, the IB-DFE techniques offer much better performance than the non-iterative methods. In fact, the IB-DFE schemes can be regarded as low complexity turbo equalizers [Tüchler and Hagenauer 2000, 2001] since the feedback loop uses the equalizer outputs instead of the channel decoder outputs. For the first iteration, we do not have any information about $s_n$, which means that $\rho = 0$, $B_k^{(0)} = 0$, and $F_k^{(l,0)}$ are given by (2.141). Therefore, the IB-DFE reduces to a linear FDE.

Clearly, for the first iteration ($i = 0$), no information exists about $S_k$ and the correlation coefficient in (2.145) is zero. This means that $B_k^{(0)} = 0$ and

$$F_k^{(l,0)} = \frac{\kappa_F^{(0)} H_k^{(l)*}}{\beta + \sum_{l'=1}^{L} |H_k^{(l')}|^2}, \quad l = 1, 2, \ldots, L, \tag{2.146}$$

corresponding to the optimum frequency-domain equalizer coefficients under the MMSE criterion. After that first iteration, and if the residual BER is not too high, we can use the feedback coefficients to eliminate a significant part of the residual interference. When $\rho \approx 1$ (after several iterations or moderate-to-high SNR), we have an almost full cancellation of the residual ISI through these coefficients, while the feedforward coefficients perform an approximate matched filtering.

Clearly, (2.142) could be written as

$$\tilde{S}_k^{(i)} = \sum_{l=1}^{L} F_k^{(l,i)} Y_k^{(l)} - B_k^{(i)} \overline{S}_k^{(i-1)}, \tag{2.147}$$

with

$$\overline{S}_k^{(i-1)} = \rho^{(i-1)} \hat{S}_k^{(i-1)}. \tag{2.148}$$

Since $\rho^{(i-1)}$ can be regarded as the blockwise reliability of the estimates $\hat{S}_k^{(i-1)}$, $\overline{S}_k^{(i-1)}$ is the overall block average of $S_k^{(i-1)}$ at the FDE output. To improve the performance, we could replace the "blockwise averages" by "symbol averages," which can be done as described in the following.

If we assume that the transmitted symbols are selected from a QPSK constellation under a Gray mapping rule (the generalization to other cases is straightforward), that is, $s_n = \pm 1 \pm j = s_n^I + j s_n^Q$, with $s_n^I = \text{Re}\{s_n\}$ and $s_n^Q = \text{Im}\{s_n\}$ (and similar definitions for $\tilde{s}_n$, $\overline{s}_n$, and $\hat{s}_n$), then it can be shown that the LLRs (log likelihood ratios) of the "in-phase bit" and the "quadrature bit," associated with $s_n^I$ and $s_n^Q$, respectively, are given by

$$L_n^I = 2\tilde{s}_n^I / \sigma_p^2 \tag{2.149}$$

and

$$L_n^Q = 2\tilde{s}_n^Q/\sigma_p^2, \qquad (2.150)$$

respectively, where

$$\sigma_p^2 = \frac{1}{2} E\left[|s_n - \tilde{s}_n|^2\right] \approx \frac{1}{2N} \sum_{n=0}^{N-1} E\left[|\hat{s}_n - \tilde{s}_n|^2\right]. \qquad (2.151)$$

Under a Gaussian assumption, it can be shown that the mean value of $s_n$ conditioned to the FDE output $\tilde{s}_n$ is

$$\bar{s}_n = \tanh\left(\frac{L_n^I}{2}\right) + j\tanh\left(\frac{L_n^Q}{2}\right)$$

$$= \rho_n^I \hat{s}_n^I + j\rho_n^Q \hat{s}_n^Q, \qquad (2.152)$$

where the hard decisions $\hat{s}_n^I = \pm 1$ and $\hat{s}_n^Q = \pm 1$ are defined according to the signs of $L_n^I$ and $L_n^Q$, respectively, and $\rho_n^I$ and $\rho_n^Q$ can be regarded as the reliabilities associated with the "in-phase" and "quadrature" bits of the $n$th symbol, given by

$$\rho_n^I = \frac{E[s_n^{I*}\hat{s}_n^I]}{E[|s_n^I|^2]} = \tanh\left(\frac{|L_n^I|}{2}\right) \qquad (2.153)$$

and

$$\rho_n^Q = \frac{E[s_n^{Q*}\hat{s}_n^Q]}{E[|s_n^Q|^2]} = \tanh\left(\frac{|L_n^Q|}{2}\right) \qquad (2.154)$$

(for the first iteration, $\rho_n^I = \rho_n^Q = 0$ and $\bar{s}_n = 0$).

The feedforward coefficients are still obtained from (2.143), with the blockwise reliability given by

$$\rho^{(i)} = \frac{1}{2N} \sum_{n=0}^{N-1} \left(\rho_n^{I(i)} + \rho_n^{Q(i)}\right). \qquad (2.155)$$

Therefore, the receiver with "blockwise reliabilities," denoted in the following as IB-DFE with hard decisions, and the receiver with "symbol reliabilities," denoted in the following as IB-DFE with soft decisions, employ the same feedforward coefficients; however, in the first the feedback loop uses the "hard-decisions" on each data block, weighted by a common reliability factor, while in the second the reliability factor changes from symbol to symbol (in fact, the reliability factor is different in the real and imaginary components of each symbol).

It is also possible to define a Turbo FDE receiver based on IB-DFE receivers that, as conventional turbo equalizers, employ the channel decoder outputs instead of the uncoded "soft decisions" in the feedback loop [Gusmão et al. 2006, 2007]. The receiver structure is similar to the IB-DFE with soft decisions, but with a SISO channel decoder (soft-in, soft-out) employed in the feedback loop. The SISO block, which can be implemented as defined in Vucetic and Yuan [2002], provides the LLRs of both the "information bits" and the "coded bits." The inputs of the SISO block are the LLRs of the "coded bits" at the FDE output. Once again, the feedforward coefficients are obtained from (2.143), with the blockwise reliability given by (2.155).

### 2.2.9.4 Performance Comparisons between OFDM and SC-FDE

To provide some insight into the BER performance of OFDM and SC-FDE modulations, we now present a set of simulation results for an uncoded transmission over the strong frequency-selective channel. We consider perfect synchronization and channel estimation conditions, as well as linear power amplification. Blocks have length $N = 512$ useful modulation symbols that are selected from a QPSK constellation, under a Gray mapping rule. For both modulation choices a suitable CP is added to each time-domain block. We also consider an $L$-order space diversity receiver with uncorrelated receive antennas.

Figures 2.40 and 2.41 show the average uncoded BER performance (averaged over all blocks), for both OFDM and SC-FDE modulations (under the MMSE criterion) with $L = 1$, 2, and 4, respectively. The BERs are expressed as a function of $E_b/N_0$, where $E_b$ denotes the average bit energy and $N_0$ the one-sided power spectral density of the channel noise. For the sake of comparison, we also include the corresponding AWGN channel performance with $L$th order diversity

$$P_b = Q\left(\sqrt{\frac{2LE_b}{N_0}}\right),$$  (2.156)

the matched filter bound (MFB) performance, defined as

$$P_{b,MFB} = E\left[Q\left(\sqrt{\frac{2E_b}{N_0}\frac{1}{N}\sum_{k=0}^{N-1}\sum_{l=1}^{L}\left|H_k^{(l)}\right|^2}\right)\right],$$  (2.157)

where the expectation is over the set of channel realizations (it is assumed that $E[|H_k^{(l)}|^2] = 1$, for any $k$), and the performance for a Rayleigh fading channel with $L$-order space diversity, given by [Proakis 2001]

$$P_{b,Ray} = \left(\frac{1-\mu}{2}\right)^L \sum_{l=0}^{L-1}\binom{L-1+l}{l}\left(\frac{1+\mu}{2}\right)^l,$$  (2.158)

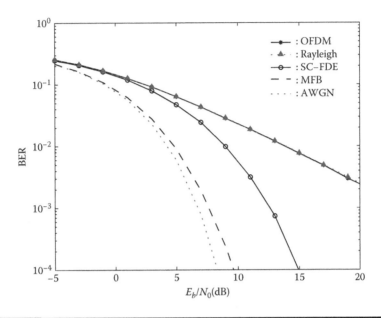

**Figure 2.40**  **Uncoded BER performance for OFDM and SC-FDE without receive diversity ($L = 1$) (for comparison purposes, the corresponding performance for an AWGN channel, for the MFB, and for a Rayleigh fading channel are also included).**

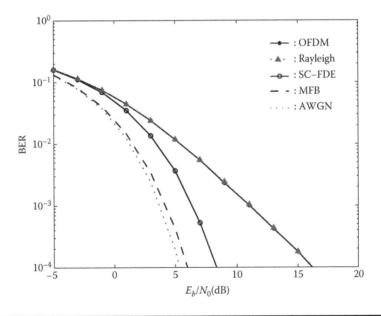

**Figure 2.41**  **As in Figure 2.40 but for two-branch receive diversity.**

with

$$\mu = \sqrt{\frac{\frac{E_b}{N_0}}{1 + \frac{E_b}{N_0}}}. \tag{2.159}$$

From these figures, it is clear that SC-FDE schemes outperform OFDM schemes in all cases. As expected, the use of diversity considerably improves the performance of both modulations schemes. We can also observe that the performance of OFDM schemes is very close to the performance under Rayleigh fading conditions. Naturally, the poor performance of OFDM schemes results from the fact that the overall performance is conditioned by the subchannels in deep fades. To overcome this problem, we can employ suitable channel coding schemes.

Let us consider now the impact of channel coding. We consider the rate-1/2 64-state convolutional code with generators $1 + D^2 + D^3 + D^5 + D^6$ and $1 + D + D^2 + D^3 + D^6$. For the OFDM scheme, an appropriate intra-block interleaving is employed. Figure 2.42 shows the average coded BER for both modulations, when $L = 1$, 2, and 4. From this figure, the benefits of channel coding become clear, specially for the OFDM choice, yielding similar performances for SC-FDE and OFDM. This means that channel coding can compensate the worse uncoded performance of OFDM modulations, as long as the code

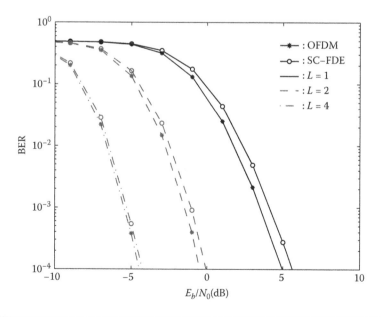

**Figure 2.42  Coded BER performance for OFDM and SC-FDE with a rate-1/2 convolutional code, with $L$ = 1, 2, and 4 branch receive diversity.**

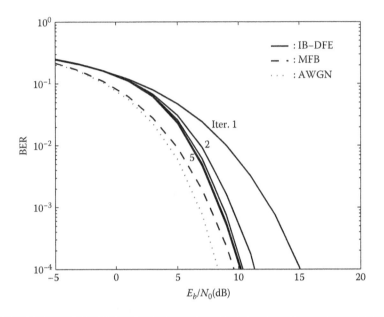

**Figure 2.43** Uncoded BER performance for an IB-DFE with one to five iterations without receive diversity ($L = 1$) (for comparison purposes, the performance for an AWGN channel and for the MFB are also included).

rate is low enough (as expected, for higher code rates SC-FDE can significantly outperform OFDM).

Finally, let us consider now an uncoded transmission with SC modulation where the linear FDE is replaced by an IB-DFE (once again, under the MMSE criterion). Figure 2.43 shows the average uncoded BER performance for iterations one to five with $L = 1$, as well as the corresponding performance of the MFB and AWGN channel. From this figure, we can observe that the iterative receiver allows a significant improvement in the BER performance: the required $E_b/N_0$ for BER $= 10^{-4}$ is about 15 dB for the first iteration (corresponding to the conventional linear SC-FDE), dropping to about 10.5 dB after just three iterations; iterations four and five provide less significant improvements. Moreover, the asymptotic BER performance becomes very close to the corresponding MFB after just a few iterations. These results show the potential of IB-DFE receivers.

# Chapter 3

# Enhancement Techniques for 4G Systems

## 3.1  Channel Coding

In the design of the first digital communication systems, the solution commonly employed to obtain low-bit-error probabilities consisted of transmitting at high powers and using bandwidths larger than what is strictly necessary. The obvious problem with these approaches was that the performance improvements were accomplished at the cost of using more of the most precious resources: spectral bandwidth and/or power. In Shannon [1948] showed that performance improvement can be obtained by calling into play a third resource: the system complexity. In his paper, Shannon showed that, as long as the rate at which information is transmitted is lower than the channel capacity, there are error correcting codes that can provide arbitrary levels of reliability for the received information bits. In this theorem, the concept of channel capacity is defined as

$$C = B_{wd} \log_2 \left( 1 + \frac{E_s}{N_0} \right) \ \left[ \text{bps} \right], \tag{3.1}$$

where $B_{wd}$ is the bandwidth of the channel (in hertz), $E_s$ is the signal energy, and $N_0$ is the single-sided noise power spectral density. Shannon proved the existence of these error-correcting codes but did not present a way to build them. Nevertheless, considerable research work has been done since then with the aim of developing error-correcting codes that could be used to increase the reliability of the received information. There are two main classes of codes: block codes and convolutional codes. Codes from both classes have found use in real-life applications, for example, the use of Reed–Solomon codes on CDs and DVDs and convolutional codes in deep space communications. It is also common to implement a concatenation of two different codes, for example, the concatenation of an inner convolutional code with an outer Reed–Solomon code.

Until 1993 the most powerful error-correcting codes achieved a performance of about 2 dB from the theoretical limit. The difficulty in approaching this limit resides in the fact that as the codes become more powerful, their complexity increases and the respective maximum likelihood decoder (MLD) becomes computationally intractable. A breakthrough occurred with the development of turbo codes, which made it possible to significantly reduce the gap to the theoretical limit. These codes, introduced in Berrou et al. [1993], were shown to achieve near-Shannon limit performance on additive white Gaussian noise (AWGN) and Rayleigh flat fading channels. Although the maximum likelihood decoder for these codes has a prohibitively high complexity, its structure allows the implementation of a suboptimal iterative decoder.

### 3.1.1 Convolutional Codes

### 3.1.1.1 Encoder

Convolutional codes owe their name to the fact that the encoding process corresponds to the convolution of the information sequence with the impulse responses of the encoder. Figure 3.1 presents the general block diagram of a binary convolutional encoder.

It is composed of $m$ shift registers that define its state at any given time. The encoder receives tuples of $k$ input bits, outputting $n$ bits. The output bits are computed through modulo-2 additions using the contents of the shift registers together with the result of the addition of the input bits with the feedback bits. The additions

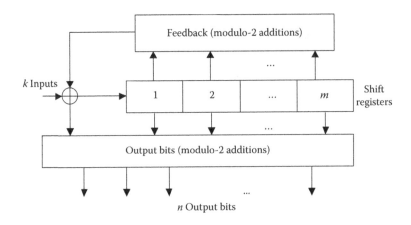

**Figure 3.1　Block diagram of a convolutional encoder.**

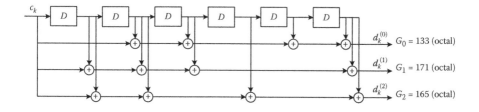

**Figure 3.2 LTE-Advanced rate 1/3 tail biting convolutional encoder. (From 3GPP, Evolved Universal Terrestrial Radio Access (E-UTRA); Multiplexing and channel coding TS 36.212 v10.4.0, December 2011.)**

are performed for each output bit according to the respective generator polynomial. If $k$ output bits are the same as the $k$ input bits, then the code is systematic. The feedback bits are obtained through modulo-2 additions using the contents of the shift registers according to the feedback polynomial. Feedback bits only exist if the code is a recursive convolutional code. The $k$ input bits are then stored on the first positions of the shift registers, and a new computation is performed using the next $k$ input bits. Since for each tuple of $k$ input bits the encoder generates $n$ output bits, the coding rate is $R = k/n$. However, the coding rate can be increased when puncturing is applied on the output sequence, that is, when some coded bits are intentionally deleted. When a sequence is encoded using a convolutional encoder, its starting and ending states can be predefined (for example, in the all-zero state), but this approach requires the insertion of tail bits, resulting in a small loss in the coding rate. As an alternative, one can demand only that the start and ending states are the same, independent of a specific state. This type of approach can be implemented without adding extra tail bits, thus avoiding any loss in the coding rate. The resulting codes are called tail-biting convolutional codes and are specified for LTE-Advanced [3GPP 2011a] and WirelessMan-Advanced [IEEE 2011a] (for control channels). As an example, Figure 3.2 shows the tail-biting convolutional encoder defined for LTE-Advanced where, to guarantee that the initial and ending states of the encoder are the same, the contents of the shift registers are initialized with the values of the last six bits of the input sequence.

### 3.1.1.2 Decoder

A convolutional encoder is a finite state machine and, therefore, the coding process along time can be represented using a trellis. An efficient decoding algorithm for solving the problem of estimating the information sequence encoded by a trellis-based code is the Viterbi algorithm (VA) [Viterbi 1967]. The VA is an MLD that finds the transmitted code sequence which is closest to the received one. The algorithm will be described next.

The likelihood probability of a received sequence $\mathbf{r}$ with length $N$, conditioned on a code sequence $\mathbf{d}$ transmitted through a memoryless noisy channel is given by

$$p(\mathbf{r}|\mathbf{d}) = \prod_{i=0}^{N-1} p(r_i|d_i). \tag{3.2}$$

An MLD selects the code sequence $\mathbf{d}$ that maximizes the likelihood probability:

$$\hat{\mathbf{d}} = \arg\max_{\mathbf{d}} p(\mathbf{r}|\mathbf{d}). \tag{3.3}$$

In the case of an AWGN channel, the likelihood probability (3.2) is expressed as

$$p(\mathbf{r}|\mathbf{d}) = \prod_{i=0}^{N-1} \frac{1}{\sqrt{2\pi\sigma^2}} \exp\left[-\frac{(r_i - d_i)^2}{2\sigma^2}\right], \tag{3.4}$$

which can be written in the log domain as

$$\log p(\mathbf{r}|\mathbf{d}) = \sum_{i=0}^{N-1}\left[-\frac{(r_i - d_i)^2}{2\sigma^2} - \frac{1}{2}\log(2\pi\sigma^2)\right] \tag{3.5}$$

$$= -\frac{1}{2\sigma^2} \sum_{i=0}^{N-1}(r_i - d_i)^2 - C,$$

where $C$ is a constant given as $C = \sum_{i=0}^{N-1} \frac{1}{2}\log(2\pi\sigma^2)$. Finding the sequence $\mathbf{d}$ that for a received sequence $\mathbf{r}$ maximizes $p(\mathbf{r}|\mathbf{d})$ is equivalent to maximizing $\log p(\mathbf{r}|\mathbf{d})$ which, according to (3.5), is equivalent to finding the sequence that minimizes the squared Euclidean distance

$$\sum_{i=0}^{N-1}(r_i - d_i)^2 = \|\mathbf{r} - \mathbf{d}\|^2. \tag{3.6}$$

Considering a code represented by a trellis, the VA has the goal of selecting the path along the trellis that minimizes the total Euclidean distance. If the code has memory $v$, then there will be $2^v$ possible states $\Lambda_j^f$ ($f$ is the state index with $f = 0,\ldots, 2^v-1$) in each trellis stage $j$.

To achieve the selection of the best-path candidate in an efficient manner, the VA performs the following steps:

1. Compute the metric for each branch transition.

   Considering that for each branch transition the encoder outputs $n$ bits, then for each possible trellis state transition $\Lambda_{j-1}^f \to \Lambda_j^{f'}$ at stage $j$ the decoder computes the partial branch metrics $\mu(r_j, d_{j,\Lambda_{j-1}^f \to \Lambda_j^{f'}})$ between the partial received sequence $r_j$ and the possible encoder output $d_{j,\Lambda_{j-1}^f \to \Lambda_j^{f'}}$. Since the aim of the VA is to find the sequence that maximizes the likelihood probability or equivalently its logarithm, in the AWGN case this corresponds to using the Euclidean distance for the metrics:

$$\mu(r_j, d_{j,\Lambda_{j-1}^f \to \Lambda_j^{f'}}) = \sum_{i=0}^{n-1} (r_{j+i} - d_{j+i,\Lambda_{j-1}^f \to \Lambda_j^{f'}})^2 = \left\| r_j - d_{j,\Lambda_{j-1}^f \to \Lambda_j^{f'}} \right\|^2. \quad (3.7)$$

2. Add, compare, and select.
   For each state $\Lambda_j^{f'}$ compute the accumulated metrics of all paths converging to that state, select the one that results in the lowest value, and store that value as the accumulated metrics for that state:

$$M(\Lambda_j^{f'}) = \min\left\{ M(\Lambda_{j-1}^f) + \mu(r_j, d_{j,\Lambda_{j-1}^f \to \Lambda_j^{f'}}), \forall_{f: \text{ exist } \Lambda_{j-1}^f \to \Lambda_j^{f'}} \right\}. \quad (3.8)$$

3. Update path memory.
   For each state $\Lambda_j^{f'}$ store the corresponding survivor path by adding the winning transition branch to the survivor path of the departing state, $\Lambda_{j-1}^f$, in the previous stage:

$$\rho(\Lambda_j^{f'}) = \left\{ \rho(\Lambda_j^f), x_{j,\Lambda_{j-1}^f \to \Lambda_j^{f'}} \right\}. \quad (3.9)$$

4. Select the estimated code sequence.
   In Viterbi [1967] it was explained that after $K$ branch transitions, with $K > 5v$ it is possible to make a decision on the section of the coded sequence composed of the last $K \cdot n$ bits. The selected sequence is the one with the lowest accumulated metric. Update the trellis transition to $j = j + 1$ and go back to decoding step 1.

In step 3, the stored paths can correspond to the information sequence.
For a terminated convolutional code, the algorithm is initialized with

$$\begin{cases} j = 0 \\ M(\Lambda_0^{f'}) = 0. \\ \rho(\Lambda_0^{f'}) = \{\} \end{cases} \quad (3.10)$$

In the case of a tail-biting convolutional code, this initialization cannot be used. A possible approach is to apply the algorithm on an expanded trellis where the last transition sections of the original trellis are repeated and appended to the beginning while the first sections are repeated and attached to the end. The algorithm can then be applied to the new trellis with the assumption of equal probability for the starting states. The number of repeated sections that were attached to the trellis is called wrap depth.

## 3.1.2 *Turbo Codes*

In modern digital communication, the area of channel coding has attained an important role in accomplishing reliable transmissions. This importance increased with the appearance of turbo codes. Turbo codes are a class of codes presented in 1993 by Berrou et al. [Berrou et al. 1993] that are capable of achieving near-Shannon limit performance. The basic turbo encoder presented in that paper was composed of two convolutional encoders in parallel with an interleaver between them and are usually referred to as convolutional turbo codes (CTC). Nevertheless, the constituent codes can also be block codes (block turbo codes [BTC]), and they can also be concatenated in series. Figures 3.3 and 3.4 show the two basic concatenated encoder schemes that can be used for obtaining turbo codes. In these figures, represents the permutation function associated with an interleaver, **b** is an input sequence of bits, and **d** is an encoded sequence. Although these schemes only comprise two component codes, it is also possible to build turbo codes through the concatenation of more constituent codes, resulting in the so-called multiple turbo codes [Divsalar and Pollara 1995]. In general, the component encoders need not be identical with regard to constraint length and rate. It is important to highlight that the idea of combining or concatenating several codes for improving the performance of the resulting code had already appeared several decades before [Elias 1954; Forney 1966] the original paper on turbo codes. However, the most interesting aspect of the codes proposed in [Berrou et al. 1993] is that with a careful selection of constituent codes and interleaver, astonishing performances can be achieved using a suboptimal decoder with an acceptable complexity. This decoder is based on an iterative processing method where

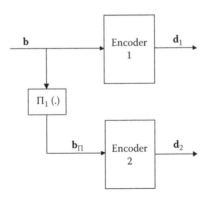

**Figure 3.3  Turbo encoder using two constituent codes concatenated in parallel.**

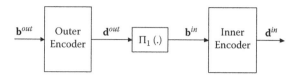

**Figure 3.4  Turbo encoder using two constituent codes concatenated in serial.**

two soft-input soft-output (SISO) component decoders are employed and exchange information between each other during the decoding process. The iterative decoding principle can be applied in other cases such as decoders for different channel codes or in the design of advanced receiver schemes.

Because of their astonishing performance, CTCs were introduced into 3G systems (UMTS [3GPP04a] and WiMax [IEEE 2006]) and are also specified for 4G systems, namely, LTE-Advanced [3GPP 2011A] and WirelessMan-Advanced [IEEE 2011a].

### 3.1.2.1 Convolutional Turbo Encoder

A CTC is formed by the parallel concatenation of Recursive Systematic Convolutional (RSC) codes. As an example, Figure 3.5 portrays the turbo encoder employed in LTE Advanced ([3GPP 2011A]), which is composed of two RSC encoders and an interleaver represented by the permutation function $\Pi_1(\cdot)$. The same information block is applied at the input of both constituent encoders, but one of them sees an

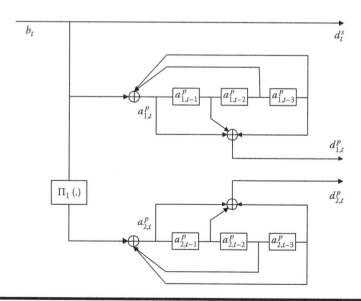

**Figure 3.5  LTE Advanced turbo encoder scheme.**

interleaved version of the block. This way, although both encoders are processing the same information bits, the output codewords are different. According to the encoder of Figure 3.5, if the input block has size $N_b$, the interleaver will have the same size, and the output codeword will be composed of $N_b$ systematic bits and $2N_b$ parity bits generated by the two constituent encoders. However, after encoding the input sequence, both constituent encoders are terminated into the all-zeros state (as detailed in [3GPP 2011A]), thus generating 12 additional tail bits. The overall coding rate is thus slightly lower than 1/3. Nevertheless, similarly to the convolutional codes, the output of the turbo encoder can be punctured, usually the parity bits, to increase the coding rate.

If the memory of each component encoder $e$ (in Figure 3.5 $e = \{1,2\}$) is $v$, the feed-forward and feed-back connections can be defined as $\mathbf{g}_e^f = [g_{e,0}^f ... g_{e,v}^f]$ and $\mathbf{g}_e^b = [g_{e,0}^b ... g_{e,v}^b]$, respectively, with $g_{e,i}^{f/b} = \{0,1\}$. For the specific case of Figure 3.5, and since both constituent encoders are identical, we have: $\mathbf{g}_e^f = [1 \quad 1 \quad 0 \quad 1]$ and $\mathbf{g}_e^b = [1 \quad 0 \quad 1 \quad 1]$. Considering that the input sequence of bits is $\mathbf{b} = [b_1 ... b_{N_b}]$, the systematic and parity sequences will be

$$
\begin{cases}
d_t^s = b_t \\[2mm]
d_{e,t}^p = \displaystyle\sum_{i=0}^{v} g_{e,i}^f a_{e,t-i}
\end{cases}
\tag{3.11}
$$

In (3.11) the coefficients $a_{e,t}$ are expressed as

$$
a_{e,t} = b_{\Pi_{e-1}^{-1}(t)} \oplus \sum_{i=1}^{v} g_{e,i}^b a_{e,t-i},
\tag{3.12}
$$

where $b_{\Pi_{e-1}^{-1}(t)}$ is the $t$th bit at the input of encoder $e$ (after the interleaver). The interleaver is denoted by the permutation function $\Pi_e(.)$. For the first encoder, since there is no interleaver, simply encodes the input sequence directly, which is equivalent to having a permutation function $\Pi_0(t) = t$. In all these expressions, the sums are modulo-2 additions.

Although the interleaver can be a block pseudorandom interleaver, other types are possible as described in Heegard and Wicker [1999]. The information sequence fed to the second component encoder is reordered in such a way that it becomes decorrelated from the original input sequence as much as possible. In fact, the use of this type of concatenation of two convolutional encoders separated by an interleaver has the objective of producing a set of codewords where very few of them have low weight. This does not necessarily mean that the resulting code will have

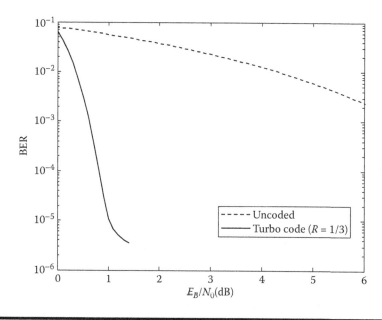

**Figure 3.6    BER performance of 3GPP rate 1/3 turbo code (from 3GPP, Evolved Universal Terrestrial Radio Access (E-UTRA); Multiplexing and channel coding TS 36.212 v10.4.0, December 2011) in AWGN.**

a particularly large free distance, but at least the number of nearest neighbors of each codeword will be substantially reduced. The substantially improved weight distribution of the resulting codewords is one of the main reasons for the good performances of turbo codes. This weight distribution is also responsible for the typical turbo code performance curve in AWGN, which can be seen in Figure 3.6. These curves are usually divided into two main zones. The first is located in the low-SNR region and is characterized by an abrupt decrease in the BER, which is mostly caused by the existence of a low number of nearest neighbors surrounding each codeword. The second zone is located in higher-SNR regions and is characterized by the slower descent of the performance curve. This is caused by the not particularly high free distance of the turbo code due to the few low weight codewords.

## 3.1.2.2 Turbo Decoder

Although the codeword weight distribution of the turbo code is important to achieve good performance, it is also necessary to have a decoder algorithm that is not excessively complex to implement. In fact, this is one the key ideas behind the success of these codes. Let us consider a turbo code composed of a two-component RSC (such as the one in Figure 3.5). In theory, the turbo code can be modeled using a single Markov process but due to the existence of the interleaver at the

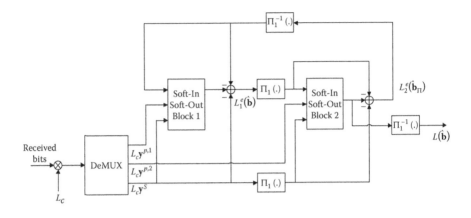

**Figure 3.7  Turbo decoder block diagram.**

input of one of the constituent codes, this representation is extremely complex and does not allow the direct implementation of computationally tractable algorithms. Consequently, a maximum likelihood decoder for turbo codes cannot be employed, and some suboptimal decoder must be used in its place.

Instead of modeling the whole code using a single Markov process, each of the constituent codes can be associated with an individual Markov process, and both can be linked by an interleaver. A trellis-based decoding algorithm can be used for estimating each of the Markov processes and, to improve the estimates, these two processes can exchange information iteratively. A basic turbo decoder accomplishing this is shown in Figure 3.7, where $\hat{b}$ denotes the information bits sequence estimate, $\hat{b}_\Pi$ the respective interleaved sequence, $y^s$ the systematic observations, and $y^{p,1}$ and $y^{p,2}$ the two parity observations sequences. The two soft-input soft-output blocks implement a trellis decoding algorithm for the respective component RSC encoders. Each of these blocks has three inputs and one output. Two of the inputs correspond to the systematic and parity observations, both weighted by the channel reliability factor $L_c$ ($L_c = 4E_c/N_0$). The third input is the a priori information obtained from the other decoder, which is represented in the form

$$L_{ap}(b_t) = \log \frac{\text{Prob}(b_t = +1)}{\text{Prob}(b_t = -1)}. \tag{3.13}$$

The output of the soft-input soft-output decoders is the sequence of log-likelihood ratios of the information bits, which can be written as the sum of three terms:

$$L(b_t) = \log \frac{\text{Prob}(b_t = +1 | y)}{\text{Prob}(b_t = -1 | y)} = L_c y_t^s + L_{ap}(b_t) + L_{1|2}^e(b_t). \tag{3.14}$$

In this expression $L^e_{1|2}(b_t)$ is the extrinsic information produced by either decoder 1 or 2. While the first two terms are inputs to the component decoders, this third term represents new information derived in the decoder, which is used as a priori information, $L_{ap}(b_t)$, by the other decoder. To obtain $L^e_{1|2}(b_t)$ subtract the weighted systematic observation and the a priori information from the output (3.14) of the decoder. After passing this information to the interleaver (or de-interleaver, depending on the target decoder), the other decoder uses this extrinsic information as new a priori information and computes new estimates for the log-likelihoods of the information bits. The exchange of extrinsic information between the two decoders can proceed until the information sequence is correct (checked using CRC) or until a certain number of iterations is achieved. The output of the second decoder is then interleaved, and some decision function can be applied for obtaining the final estimates of the information bits. This decoding process can be easily extended to multiple turbo codes as presented in Divsalar and Pollara [1995].

Each soft-input soft-output decoder can be implemented using a trellis decoding algorithm such as the optimal maximum a posteriori (MAP) algorithm [Bahl et al. 1974], its simplification max-log-MAP [Robertson and Worz 1995], or the soft output Viterbi algorithm (SOVA) [Hagenauer and Hoher 1989]. In [Robertson and Worz [1995], these three algorithms were compared, and it was concluded that the MAP algorithm clearly has the best performance but at the cost of substantial complexity. The SOVA exhibits a slight performance degradation compared with the max-log-MAP but has a lower decoding complexity. Nevertheless, it was shown in Fossorier et al. [1998] that the SOVA can become equivalent to the max-log-MAP algorithm after a simple modification.

### 3.1.2.3 General MAP Derivation

An optimal decoding algorithm for linear block and convolutional codes, in the sense that it minimizes the symbol error probability, was proposed in [Bahl et al. 1974]. This algorithm estimates the a posteriori probabilities (APPs) of a Markov source observed through a noisy discrete memoryless channel (DMC). It can be applied to any code that can be represented as a discrete-time finite-state Markov process and was employed in Berrou et al. [1993] as the component soft-input soft-output decoders for the turbo-decoder. In the following, the MAP algorithm will be derived in detail for a more general case of nonbinary trellises. A similar derivation was done in Robertson and Worz [1998], with only some minor differences since the objective of that derivation was to employ the algorithm in turbo trellis coded modulation schemes.

Let us consider a component encoder with memory $v$ modeled as a discrete-time finite-state Markov process. The Markov source starts in the initial state $\Lambda_0 = 0$ and ends in the terminal state $\Lambda_n = 0$, producing a sequence of $N$ output codeword symbols $D_1^N = d_1, d_2, d_3, ..., d_N$. This sequence of symbols is then input to a noisy DMC whose output is the sequence of words $Y_1^N = Y_1, Y_2, Y_3, ..., Y_N$. The objective of the decoder is to examine $Y_1^N$ and estimate the a posteriori probabilities of the transitions of the Markov source, that is,

$$\text{Prob}\{\Lambda_{t-1} = \delta'; \Lambda_t = \delta \mid Y_1^N\} = p(\Lambda_{t-1} = \delta'; \Lambda_t = \delta; Y_1^N) / p(Y_1^N), \quad (3.15)$$

where $\delta$ and $\delta'$ are the indexes of the Markov source's $2^v$ possible distinct states $(\delta = \delta' = 0... 2^v-1)$. Each of the transition probabilities is associated with the corresponding transition branch in the trellis of the code.

For a given $Y_1^N$, $p(Y_1^N)$ is just a normalizing constant, and thus it is not relevant for maximizing (3.15). Hence, it is only required to derive an algorithm for computing $p(\Lambda_{t-1} = \delta'; \Lambda_t = \delta, Y_1^N)$. This probability can be written as

$$p(\Lambda_{t-1} = \delta'; \Lambda_t = \delta; Y_1^N)$$

$$= p(\Lambda_{t-1} = \delta'; Y_1^{t-1}) \cdot p(\Lambda_t = \delta; Y_t^N \mid \Lambda_{t-1} = \delta'; Y_1^{t-1})$$

$$= p(\Lambda_{t-1} = \delta'; Y_1^{t-1}) \cdot p(\Lambda_t = \delta; Y_t^N \mid \Lambda_{t-1} = \delta')$$

$$= p(\Lambda_{t-1} = \delta'; Y_1^{t-1}) \cdot p(\Lambda_t = \delta; Y_t \mid \Lambda_{t-1} = \delta') \cdot p(Y_{t+1}^N \mid \Lambda_{t-1} = \delta'; \Lambda_t = \delta; Y_t)$$

$$= p(\Lambda_{t-1} = \delta'; Y_1^{t-1}) \cdot p(\Lambda_t = \delta; Y_t \mid \Lambda_{t-1} = \delta') \cdot p(Y_{t+1}^N \mid \Lambda_t = \delta)$$

$$= \psi_{t-1}(\delta') \cdot \gamma_t(\delta', \delta) \cdot \beta_t(\delta).$$

$$(3.16)$$

The second and fourth equalities follow from the Markov property that if at an instant $t$ the state $\Lambda t\vert$ is known, events after that time do not depend on nor $Y_1^t$ on $\Lambda t\vert_{-1}$. The three probability functions in (3.16) can be computed as

$$\psi_t(\delta) = \sum_{\delta'=0}^{2^v-1} p(\Lambda_{t-1} = \delta'; \Lambda_t = \delta; Y_1^t)$$

$$= \sum_{\delta'=0}^{2^v-1} p(\Lambda_{t-1} = \delta'; Y_1^{t-1}) \cdot p(\Lambda_t = \delta; Y_t \mid \Lambda_{t-1} = \delta') \qquad (3.17)$$

$$= \sum_{\delta'=0}^{2^v-1} \alpha_{t-1}(\delta') \cdot \gamma_t(\delta', \delta)$$

$$\beta_t(\delta) = \sum_{\delta'=0}^{2^v-1} p(\Lambda_{t+1} = \delta'; Y_{t+1}^N | \Lambda_t = \delta)$$

$$= \sum_{\delta'=0}^{2^v-1} p(\Lambda_{t+1} = \delta'; Y_{t+1} | \Lambda_t = \delta) \cdot p(Y_{t+2}^N | \Lambda_{t+1} = \delta'; Y_{t+1}; \Lambda_t = \delta)$$

$$\quad (3.18)$$

$$= \sum_{\delta'=0}^{2^v-1} p(\Lambda_{t+1} = \delta'; Y_{t+1} | \Lambda_t = \delta) \cdot p(Y_{t+2}^N | \Lambda_{t+1} = \delta')$$

$$= \sum_{\delta'=0}^{2^v-1} \gamma_{t+1}(\delta, \delta') \cdot \beta_{t+1}(\delta')$$

These recursive computations must be combined with the initial conditions

$$\psi_0(0) = 1 \text{ and } \psi_0(\delta \neq 0) = 0 \quad (3.19)$$

$$\beta_N(0) = 1 \text{ and } \beta_N(\delta \neq 0) = 0. \quad (3.20)$$

To avoid numeric problems when implementing the algorithm, some modifications are usually made to the basic algorithm. First, (3.15) is rewritten as

$$\text{Prob}\{\Lambda_{t-1} = \delta'; \Lambda_t = \delta | Y_1^N\} = \psi_{t-1}(\delta') \cdot \gamma_t(\delta', \delta) \cdot \beta_t(\delta) / p(Y_1^N)$$

$$= \frac{1}{p(Y_t | Y_1^{t-1})} \cdot \frac{\psi_{t-1}(\delta')}{p(Y_1^{t-1})} \cdot \frac{\beta_t(\delta)}{p(Y_{t+1}^N | Y_1^t)} \cdot \gamma_t(\delta', \delta) \quad (3.21)$$

$$= \frac{1}{p(Y_t | Y_1^{t-1})} \cdot \tilde{\psi}_{t-1}(\delta') \cdot \tilde{\beta}_t(\delta) \cdot \gamma_t(\delta', \delta),$$

where $\dfrac{1}{p(Y_t | Y_1^{t-1})}$ is a normalizing constant so that $\sum_{\delta' \to \delta} \text{Prob}\{\Lambda_{t-1} = \delta'; \Lambda_t = \delta | Y_1^N\} = 1$. The expression for $\tilde{\psi}_t(\delta)$ can be derived taking into account that

$$p(Y_1^{t-1}) = \sum_{\delta} p(\Lambda_{t-1} = \delta'; Y_1^{t-1}) = \sum_{\delta} \psi_{t-1}(\delta), \quad (3.22)$$

which results in

$$\tilde{\psi}_t(\delta) = \frac{\psi_t(\delta)}{\sum_\delta \psi_t(\delta)} = \frac{\psi_t(\delta)}{\sum_\delta \psi_t(\delta)} \cdot \frac{1/p(Y_1^t)}{1/p(Y_1^t)} = \frac{\tilde{\psi}_t(\delta)}{\sum_\delta \tilde{\psi}_t(\delta)}$$

$$= \frac{\displaystyle\sum_{\delta'=0}^{2^\nu-1} \tilde{\psi}_{t-1}(\delta') \cdot \gamma_t(\delta',\delta)}{\displaystyle\sum_{\delta=0}^{2^\nu-1}\sum_{\delta'=0}^{2^\nu-1} \tilde{\psi}_{t-1}(\delta') \cdot \gamma_t(\delta',\delta)}.$$

(3.23)

Before obtaining the expression for $\tilde{\beta}_t(\delta)$ the following equality is deduced:

$$p(Y_{t+1}^N|Y_1^t) = \frac{p(Y_1^N)}{p(Y_1^t)} = \frac{p(Y_1^{t+1})p(Y_{t+2}^N|Y_1^{t+1})}{p(Y_1^t)}$$

$$= \frac{\displaystyle\sum_\delta \psi_{t+1}(\delta)}{p(Y_1^t)} \cdot p(Y_{t+2}^N|Y_1^{t+1})$$

$$= \frac{\displaystyle\sum_\delta\sum_{\delta'} \psi_t(\delta')\gamma_{t+1}(\delta',\delta)}{p(Y_1^t)} \cdot p(Y_{t+2}^N|Y_1^{t+1})$$

$$= \sum_\delta\sum_{\delta'} \tilde{\psi}_t(\delta')\gamma_{t+1}(\delta',\delta) \cdot p(Y_{t+2}^N|Y_1^{t+1}).$$

(3.24)

The expression for $\tilde{\beta}_t(\delta)$ is then obtained as

$$\tilde{\beta}_t(\delta) = \frac{\beta_t(\delta)}{p(Y_{t+1}^N|Y_1^t)} = \frac{\beta_t(\delta)}{\displaystyle\sum_\delta\sum_{\delta'} \tilde{\psi}_t(\delta') \cdot \gamma_{t+1}(\delta',\delta) \cdot p(Y_{t+2}^N|Y_1^{t+1})}$$

(3.25)

$$= \frac{\displaystyle\sum_{\delta'} \beta_{t+1}(\delta') \cdot \gamma_{t+1}(\delta',\delta)}{\displaystyle\sum_\delta\sum_{\delta'} \tilde{\psi}_t(\delta') \cdot \gamma_{t+1}(\delta',\delta) \cdot p(Y_{t+2}^N|Y_1^{t+1})}$$

$$= \frac{\displaystyle\sum_{\delta'=0}^{2^{\nu}-1} \tilde{\beta}_{t+1}(\delta') \cdot \gamma_{t+1}(\delta',\delta)}{\displaystyle\sum_{\delta=0}^{2^{\nu}-1}\sum_{\delta'=0}^{2^{\nu}-1} \tilde{\psi}_t(\delta') \cdot \gamma_{t+1}(\delta',\delta)}.$$

The initializations are the same:

$$\tilde{\psi}_0(0) = 1 \text{ and } \tilde{\psi}_0(\delta \neq 0) = 0 \tag{3.26}$$

$$\tilde{\beta}_N(0) = 1 \text{ and } \tilde{\beta}_N(\delta \neq 0) = 0. \tag{3.27}$$

$\gamma_t$, can be written as

$$\gamma_t(\delta',\delta) = p\left(\Lambda_t = \delta; Y_t | \Lambda_{t-1} = \delta'\right)$$

$$= \text{Prob}\{\Lambda_t = \delta | \Lambda_{t-1} = \delta'\} \cdot p\left(Y_t | \Lambda_{t-1} = \delta'; \Lambda_t = \delta\right)$$

$$= \text{Prob}\{\Lambda_t = \delta | \Lambda_{t-1} = \delta'\} \cdot \sum_D \text{Prob}\{d_t = D | \Lambda_{t-1} = \delta'; \Lambda_t = \delta\} \cdot p\left(Y_t | D\right),$$

$$\tag{3.28}$$

where **D** represents possible transition output codewords. Although this expression allows parallel transitions between two states, in most of the applications there can be only one transition and (3.28) simplifies to

$$\gamma_t(\delta',\delta) = \begin{cases} 0, & \text{if } \sim \exists_{\delta'\to\delta} \\ \text{Prob}\{b_t^{\delta'\to\delta}\} \cdot p\left(Y_t | d_t^{\delta'\to\delta}\right), & \text{if } \exists_{\delta'\to\delta,} \end{cases} \tag{3.29}$$

where $\delta' \to \delta$ denotes a transition from state $\delta'$ to $\delta$, $b_t^{\delta'\to\delta}$ is the input word at instant $t$, and $d_t^{\delta'\to\delta}$ is the encoded word associated with state transition $\delta' \to \delta$ (it only depends on $t$ if the trellis is time variant, which is the usual case for block codes). $\text{Prob}\{b_t^{\delta'\to\delta}\}$ is the a priori probability associated with input word $b_t^{\delta'\to\delta}$ which, in the case of a turbo decoder, corresponds to the extrinsic information produced by the other component decoder. Usually, this extrinsic information refers to the individual symbols composing the input word and not the whole word. Nevertheless, since in most of the applications the input symbols can be considered independent, $\text{Prob}\{b_t^{\delta' \to \delta}\}$ can be computed from the individual symbols' a priori probabilities as

$$\text{Prob}\{b_t^{\delta' \to \delta}\} = \prod_{j=0}^{k-1} P_{ap}\{b_t^{\delta' \to \delta}(j)\}, \tag{3.30}$$

where $k$ is the size of each input word $b_t$.

The APP for each possible input symbol $B$ in position $j$ ($j = 0, .., k-1$) in input word $b_t$ can be computed by summing all the state transition probabilities associated with input words that have the $j$th symbol equal to $B$, that is, using

$$\text{Prob}\{\mathbf{b}_t(j) = B \mid Y_1^N\}$$

$$= \sum_{\substack{\delta' \to \delta \\ b_t(j)=B}} \text{Prob}\{\Lambda_{t-1} = \delta'; \Lambda_t = \delta | Y_1^N\}$$

$$= \frac{1}{p(Y_t | Y_1^{t-1})} \cdot \sum_{\substack{\delta' \to \delta \\ b_t(j)=B}} \tilde{\psi}_{t-1}(\delta') \cdot \beta_t(\delta) \cdot \gamma_t(\delta', \delta)$$

$$= \frac{1}{p(Y_t | Y_1^{t-1})} \cdot \sum_{\substack{\delta' \to \delta \\ b_t(j)=B}} \tilde{\psi}_{t-1}(\delta') \cdot \beta_t(\delta) \cdot \text{Prob}\{b_t^{\delta' \to \delta}\} \cdot p(Y_t | d_t^{\delta' \to \delta}),$$

$$= \frac{1}{p(Y_t | Y_1^{t-1})} \cdot P_{ap}\{\mathbf{b}_t(j) = B\} \cdot \sum_{\substack{\delta' \to \delta \\ b_t(j)=B}} \tilde{\psi}_{t-1}(\delta') \cdot \beta_t(\delta) \cdot \prod_{\substack{j'=0 \\ j' \neq j}}^{k-1} P_{ap}\{b_t^{\delta' \to \delta}(j')\}$$

$$\cdot p(Y_t | d_t^{\delta' \to \delta})$$

$$= \frac{1}{p(Y_t | Y_1^{t-1})} \cdot P_{ap}\{\mathbf{b}_t(j) = B\} \cdot P_{ext}\{\mathbf{b}_t(j) = B\}, \tag{3.31}$$

where $P_{ext}\{\mathbf{b}_t(j) = B\}$ is the extrinsic information produced in the decoder that can be used as a priori information by another component decoder, as in the case of a turbo decoder.

Let us now consider the use of the MAP algorithm just derived in each of the two soft-input soft-output decoders of a typical rate 1/3 turbo code, such as the one used in LTE-Advanced [3GPP 2011a]. The trellis processed by the MAP algorithm corresponds to the one defined by each of the component convolutional encoders where each input symbol is composed of one binary element with polar mapping ($b_t \in \{-1,1\}$). Each output word has two elements, one systematic, $d^s$, and one parity, $d^p$, that is, $\mathbf{d}_t = (d_t^s \; d_t^p) = (b_t \; d_t^p)$. The corresponding received word can be written as $\mathbf{Y}_t = (y_t^s \; y_t^p)$. Since the input symbol can only take two different values, it is more convenient to work with logarithms of probability ratios. Therefore, the objective of the algorithm is the computation of

$$
L(b_t) = \log\left(\frac{\text{Prob}\{b_t = +1|\mathbf{Y}_1^N\}}{\text{Prob}\{b_t = -1|\mathbf{Y}_1^N\}}\right) = \log\left(\frac{\displaystyle\sum_{\substack{\delta' \to \delta \\ b_t = +1}} p\left(\Lambda_{t-1} = \delta'; \Lambda_t = \delta, \mathbf{Y}_1^N\right)/p\left(\mathbf{Y}_1^N\right)}{\displaystyle\sum_{\substack{\delta' \to \delta \\ b_t = -1}} p\left(\Lambda_{t-1} = \delta'; \Lambda_t = \delta, \mathbf{Y}_1^N\right)/p\left(\mathbf{Y}_1^N\right)}\right)
$$

$$
= \log\left(\frac{\displaystyle\sum_{\substack{\delta' \to \delta \\ b_t = +1}} \tilde{\psi}_{t-1}(\delta').\gamma_t(\delta',\delta).\tilde{\beta}_t(\delta)}{\displaystyle\sum_{\substack{\delta' \to \delta \\ b_t = -1}} \tilde{\psi}_{t-1}(\delta').\gamma_t(\delta',\delta).\tilde{\beta}_t(\delta)}\right).
$$

(3.32)

Considering an AWGN channel, the likelihood probability $p(\mathbf{Y}_t \mid \mathbf{d}_t^{\delta' \to \delta})$ can be written as

$$
p\left(\mathbf{Y}_t \mid \mathbf{d}_t^{\delta' \to \delta}\right) = \frac{1}{2\pi\sigma^2}\exp\left[-\frac{(y_t^s - b_t)^2}{2\sigma^2} - \frac{(y_t^p - d_t^{\delta' \to \delta, p})^2}{2\sigma^2}\right]
$$

$$
= \frac{1}{2\pi\sigma^2}\exp\left(-\frac{(y_t^s)^2 + b_t^2 + (y_t^p)^2 + (d_t^{\delta' \to \delta, p})^2}{2\sigma^2}\right)
$$

(3.33)

$$
.\exp\left(\frac{b_t.y_t^s + d_t^{\delta' \to \delta, p}.y_t^p}{\sigma^2}\right),
$$

$$
= A_t.\exp\left(\frac{b_t.y_t^s + d_t^{\delta' \to \delta, p}.y_t^p}{\sigma^2}\right),
$$

where $\sigma^2$ is the noise variance and

$$
A_t = \frac{1}{2\pi\sigma^2}\exp\left(-\frac{(y_t^s)^2 + b_t^2 + (y_t^p)^2 + (d_t^{\delta' \to \delta, p})^2}{2\sigma^2}\right)
$$

(3.34)

is just a normalizing constant that does not depend on the state transition (since polar mapping is employed). Considering that the a priori information is in the form of a log ratio, $L_{ap}(b_t) = \log\left(\dfrac{\text{Prob}\{b_t = +1\}}{\text{Prob}\{b_t = -1\}}\right)$, then the a priori probability can be expressed as

$$
P_{ap}\{b_t\} = \left(\frac{\exp\left[-\dfrac{L_{ap}(b_t)}{2}\right]}{1+\exp\left[-L_{ap}(b_t)\right]}\right).\exp\left[b_t.\frac{L_{ap}(b_t)}{2}\right]
$$

$$
= C_t.\exp\left[b_t.\frac{L_{ap}(b_t)}{2}\right].
$$

(3.35)

From (3.29), the branch transition function $\gamma_t(\delta',\delta)$ can then be written as

$$
\gamma_t(\delta',\delta) = A_t.C_t.\exp\left(b_t.\frac{L_{ap}(b_t)}{2}\right).\exp\left(\frac{b_t.y_t^s + d_t^{\delta'\to\delta,P}.y_t^P}{\sigma^2}\right)
$$

$$
= A_t.C_t.\exp\left(\frac{1}{2}.b_t.\left(L_{ap}(b_t)+L_c.y_t^s\right)+\frac{1}{2}.L_c.d_t^{\delta'\to\delta,P}.y_t^P\right)
$$

$$
= A_t.C_t.\exp\left(\frac{1}{2}.b_t.\left(L_{ap}(b_t)+L_c.y_t^s\right)\right).\exp\left(\frac{1}{2}.L_c.d_t^{\delta'\to\delta,P}.y_t^P\right)
$$

$$
= A_t.C_t.\exp\left(\frac{1}{2}.b_t.\left(L_{ap}(b_t)+L_c.y_t^s\right)\right).\gamma_t^e(\delta',\delta)
$$

(3.36)

with the reliability factor, $L_c$, computed as

$$
L_c = \frac{2}{\sigma^2} = \frac{2}{\dfrac{N_0}{2.E_c}} = \frac{4.E_c}{N_0}
$$

(3.37)

and

$$
E_c = R.E_b
$$

(3.38)

($R$ is the code rate and $E_b$ is the bit energy). According to (3.36), constants $A_t$ and $C_t$ do not depend on the value of $b_t$ and they both cancel inside the log a posteriori ratio (3.32). Hence, it is not necessary to compute these two constants. Inserting (3.37) in (3.32) gives

$$
L(b_t) = \log \left( \frac{\exp\left(\frac{1}{2}.(+1).\left(L_{ap}(b_t) + L_c.y_t^s\right)\right).\sum_{\substack{\delta' \to \delta \\ b_t = +1}} \tilde{\psi}_{t-1}(\delta').\gamma_t^e(\delta',\delta).\tilde{\beta}_t(\delta)}{\exp\left(\frac{1}{2}.(-1).\left(L_{ap}(b_t) + L_c.y_t^s\right)\right).\sum_{\substack{\delta' \to \delta \\ b_t = -1}} \tilde{\psi}_{t-1}(\delta').\gamma_t^e(\delta',\delta).\tilde{\beta}_t(\delta)} \right)
$$

$$
= L_{ap}(b_t) + L_c.y_t^s + \log\left( \frac{\sum_{\substack{\delta' \to \delta \\ b_t=+1}} \tilde{\psi}_{t-1}(\delta').\gamma_t^e(\delta',\delta).\tilde{\beta}_t(\delta)}{\sum_{\substack{\delta' \to \delta \\ b_t=-1}} \tilde{\psi}_{t-1}(\delta').\gamma_t^e(\delta',\delta).\tilde{\beta}_t(\delta)} \right).
$$

(3.39)

In this expression, the third term is the extrinsic information that can be fed to the other component decoder as a priori information

$$
L^e(b_t) = \log\left( \frac{\sum_{\substack{\delta' \to \delta \\ b_t=+1}} \tilde{\psi}_{t-1}(\delta').\gamma_t^e(\delta',\delta).\tilde{\beta}_t(\delta)}{\sum_{\substack{\delta' \to \delta \\ b_t=-1}} \tilde{\psi}_{t-1}(\delta').\gamma_t^e(\delta',\delta).\tilde{\beta}_t(\delta)} \right)
$$

(3.40)

and

$$
L(b_t) = L_{ap}(b_t) + L_c.y_t^s + L^e(b_t).
$$

(3.41)

The MAP algorithm can be completely implemented in the logarithm domain. This type of implementation is usually referred to as the log-MAP algorithm. In this version of the algorithm, the APP probability is computed as

$$
\log\left(\text{Prob}\{\mathbf{b}_t(j) = B | Y_1^N\}\right)
$$

$$
= \log\left( \frac{1}{p\left(\mathbf{Y}_t | \mathbf{Y}_1^{t-1}\right)} \cdot \sum_{\substack{\delta' \to \delta \\ b_t(j)=B}} \tilde{\psi}_{t-1}(\delta') \cdot \tilde{\beta}_t(\delta) \cdot \gamma_t(\delta',\delta) \right)
$$

(3.42)

$$
= \log\left( \sum_{\substack{\delta' \to \delta \\ b_t(j)=B}} \exp\left(\log\left(\tilde{\psi}_{t-1}(\delta')\right) + \log\left(\tilde{\beta}_t(\delta)\right) + \log\left(\gamma_t(\delta',\delta)\right)\right) \right)
$$

$$
- \log\left( p\left(\mathbf{Y}_t | \mathbf{Y}_1^{t-1}\right) \right).
$$

The logarithms of $\tilde{\psi}_t(\delta)$, $\tilde{\beta}_t(\delta)$, and $\gamma_t(\delta',\delta)$ are computed as

$$\log(\tilde{\psi}_t(\delta)) = \log\left(\frac{\displaystyle\sum_{\delta'=0}^{2^v-1}\tilde{\psi}_{t-1}(\delta').\gamma_t(\delta',\delta)}{\displaystyle\sum_{\delta=0}^{2^v-1}\sum_{\delta'=0}^{2^v-1}\tilde{\psi}_{t-1}(\delta').\gamma_t(\delta',\delta)}\right)$$

$$= \log\left(\sum_{\delta'=0}^{2^v-1}\exp\left(\log(\tilde{\psi}_{t-1}(\delta'))+\log(\gamma_t(\delta',\delta))\right)\right)$$

$$- \log\left(\sum_{\delta=0}^{2^v-1}\sum_{\delta'=0}^{2^v-1}\exp\left(\log(\tilde{\psi}_{t-1}(\delta'))+\log(\gamma_t(\delta',\delta))\right)\right); \quad (3.43)$$

$$\log(\tilde{\beta}_t(\delta)) = \log\left(\frac{\displaystyle\sum_{\delta'=0}^{2^v-1}\tilde{\beta}_{t+1}(\delta').\gamma_{t+1}(\delta',\delta)}{\displaystyle\sum_{\delta=0}^{2^v-1}\sum_{\delta'=0}^{2^v-1}\tilde{\psi}_t(\delta').\gamma_{t+1}(\delta',\delta)}\right)$$

$$= \log\left(\sum_{\delta'=0}^{2^v-1}\exp\left(\log(\tilde{\beta}_{t+1}(\delta'))+\log(\gamma_{t+1}(\delta',\delta))\right)\right)$$

$$- \log\left(\sum_{\delta=0}^{2^v-1}\sum_{\delta'=0}^{2^v-1}\exp\left(\log(\tilde{\psi}_t(\delta'))+\log(\gamma_{t+1}(\delta',\delta))\right)\right), \quad (3.44)$$

$$\log(\gamma_t(\delta',\delta)) = \begin{cases} 0, & \text{if } \sim\exists_{\delta'\to\delta} \\ \log\left(\text{Prob}\{b_t^{\delta'\to\delta}\}\right)+\log\left(p\left(Y_t\,|\,d_t^{\delta'\to\delta}\right)\right) & \text{if } \exists_{\delta'\to\delta}. \end{cases} \quad (3.45)$$

A significant degree of the complexity of the log-MAP algorithm is due to the calculations of $\log(e^{a_1}+\cdots+e^{a_n})$. The complexity necessary for this computation can be reduced by employing the Jacobian logarithm [Robertson and Worz 1995]:

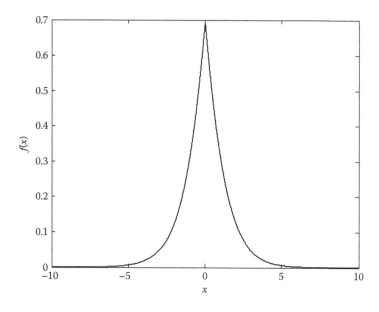

**Figure 3.8** Plot of the function $f(x) = \log(1 + e^{-|x|})$.

$$\log(e^{a_1} + e^{a_2}) = \max(a_1, a_2) + \log(1 + e^{-|a_2 - a_1|}). \qquad (3.46)$$

Using this relation, it is possible to compute $\log(e^{a_1} + \cdots + e^{a_n})$ recursively as

$$\log(e^{a_1} + \cdots + e^{a_n}) = \log\left( e^{\log\left(e^{a_1} + \cdots + e^{a_{n-1}}\right)} + e^{a_n} \right)$$

$$= \log(e^{\Delta_1^{n-1}} + e^{a_n}) \text{ with } \Delta_1^{n-1} = \log(e^{a_1} + \cdots + e^{a_{n-1}}) \qquad (3.47)$$

$$= \max(\Delta_1^{n-1}, a_2) + \log\left( 1 + e^{-\left|\Delta_1^{n-1} - a_1\right|} \right).$$

The function $f(x) = \log(1 + e^{-|x|})$ in the last expression can be approximated by a pre-computed table with only very few values stored (usually eight is enough) bearing in mind that the function is symmetric, as can be seen in Figure 3.8.

As another alternative, the Max-Log-MAP algorithm can be used. It corresponds to a small modification of the Log-MAP algorithm that results in a more computationally efficient decoder at the cost of small performance degradation. This algorithm is obtained by simplifying the computation of the logarithm of the sum of exponentials in the Log-MAP algorithm using the following approximation [Robertson and Worz 1995]:

$$\log(e^{a_1} + e^{a_2}) = \max(a_1, a_2) + \log\left(1 + e^{-|a_2 - a_1|}\right) \approx \max(a_1, a_2). \qquad (3.48)$$

The MAP algorithm can also be used to compute the APP of the encoder output symbols, Prob$\{d_t(i) = B|Y_1^N\}$ ($d_t(i)$ is the $i$th symbol of the encoded word at trellis transition $t$) as was shown in Benedetto and Montorsi [1997]. The procedure is very similar to the one used for computing the probabilities of the encoder input bits. The APP probability can be computed by summing all the state transition probabilities associated with encoded words that have the $i$th symbol equal to $B$, that is., using

$$\text{Prob}\{d_t(i) = B \mid Y_1^N\} = \sum_{\substack{\delta' \to \delta \\ d_t(i)=B}} \text{Prob}\{\Lambda_{t-1} = \delta'; \Lambda_t = \delta \mid Y_1^N\}$$

(3.49)

$$= \frac{1}{p\left(Y_t \mid Y_1^{t-1}\right)} \cdot \sum_{\substack{\delta' \to \delta \\ d_t(i)=B}} \tilde{\psi}_{t-1}(\delta') \cdot \beta_t(\delta) \cdot \gamma_t(\delta',\delta).$$

The summation is performed over all possible state transitions associated with $d_t(i) = B$. If working on the log domain, this probability is computed as

$$\log\left(\text{Prob}\{d_t(i) = B|Y_1^N\}\right)$$

$$= \log\left[\sum_{\substack{\delta' \to \delta \\ d_t(i)=B}} \exp\left(\log(\tilde{\psi}_{t-1}(\delta')) + \log(\tilde{\beta}_t(\delta)) + \log(\gamma_t(\delta',\delta))\right)\right] - \log\left(\text{Prob}\{Y_t \mid Y_1^{t-1}\}\right).$$

(3.50)

The possibility of computation of the probabilities of the encoded symbols allows the implementation of the MAP algorithm as a soft-input soft-output module with four ports, as shown in Figure 3.9. The two input ports correspond to the a priori probabilities of the information symbols, $P_{in}\{b_t\}$, and the probabilities of the coded symbols, $P_{in}\{d_t\}$. The other two ports output the probabilities associated with the input symbols, $P_{out}\{b_t\}$ and with the coded symbols, $P_{out}\{d_t\}$. This soft-input soft-output module capable of outputting encoded symbols probabilities is useful in several

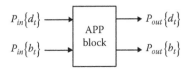

**Figure 3.9   Soft-input soft-output APP module.**

applications where feedback from the turbo-decoder to other blocks of the receiver is required, for example, when implementing iterative joint estimation schemes.

### 3.1.2.4 Block Turbo Encoder

As previously explained, instead of concatenating RSC it is possible to build turbo codes using block codes as constituents. In this case, the resulting codes are called block turbo codes (BTCs).

A block code consists of a set of fixed-length vectors called codewords whose elements are selected from an alphabet of $\varsigma$ symbols. If the alphabet consists of two symbols, 0 and 1, then it is called a binary code, and the elements of the codewords are called bits. If the symbols of the alphabet have more than two different elements ($\varsigma > 2$), then the code is nonbinary. For a binary block code of length $n$, there are $2^n$ possible words, from which $2^k$ ($k < n$) codewords are selected as the set of words to form an $(n,k)$ code with coding rate $R = k/n$. A block code can be characterized as being either linear or nonlinear. If $\mathbf{d}^i$ and $\mathbf{d}^j$ are two codewords belonging to an $(n,k)$ block code, then the code is considered to be linear if and only if $\alpha_1 \mathbf{d}^i + \alpha_2 \mathbf{d}^j$ ($\alpha_1$ and $\alpha_2$ are two of the possible $\varsigma$ symbols from the alphabet) is also a codeword. According to this definition, a linear code must contain the all-zero codeword. The encoding process of linear codes can be accomplished through

$$\mathbf{d} = \mathbf{b} \cdot \mathbf{G}, \tag{3.51}$$

where $\mathbf{G}$ is a $k \times n$ generator matrix, $\mathbf{b} = [b_0 \quad b_1 \quad ... \quad b_{k-1}]$ is the information word, and $\mathbf{d} = [d_0 \quad d_1 \quad ... \quad d_{k-1}]$ is the output codeword. According to this, matrix $\mathbf{G}$ can be used to describe the code, although it can also be described using parity check matrix $\mathbf{H}$, which is an $(n - k) \times n$ matrix that satisfies

$$\mathbf{H} \cdot \mathbf{d}^T = 0 \tag{3.52}$$

for any codeword $\mathbf{d}$.

Through the combination of two block codes, it is possible to improve the performance of the resulting codes, as was initially proposed in Elias [1954], which introduced the concept of product codes. The construction of these codes is shown in Figure 3.10. The process is as follows. The information bits are grouped into $k_V$ information words with length $k_H$ and inserted along the rows of an $n_V \times n_H$ matrix from left to right, thus filling the first $k_H$ columns. An $(n_H, k_H)$ systematic encoder is applied along each of the first $k_V$ rows generating $n_H - k_H$ check bits, which are used for filling in the remaining matrix columns. Then an $(n_V, k_V)$ block code is applied along each column, including the columns containing the check bits of the first encoder, thus generating $n_V - k_V$ check bits for filling completely the remaining rows. The resulting code is a product code or BTC with a coding rate given by

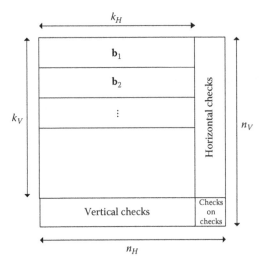

**Figure 3.10  BTC construction.**

$$k = \frac{k_H \cdot k_V}{n_H \cdot n_V}.$$ (3.53)

It is easy to verify that the construction of these codes follows the general scheme of serial concatenated turbo codes as was shown in Figure 3.4, with the interleaver corresponding to a block interleaver. For the decoding process, an iterative decoder similar to the one employed for CTC can be used with simple modifications. The resulting decoder structure (assuming the use of log likelihood ratios) is presented in Figure 3.11.

BTCs based on two component codes, either parity check codes or extended Hamming codes, are used as optional coding in mobile WiMax [IEEE 2006].

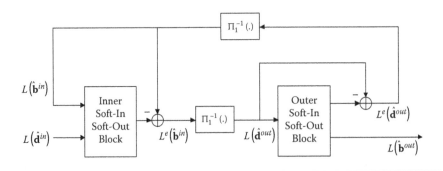

**Figure 3.11  Iterative decoder for a serial concatenated turbo code (for example, BTC).**

Since the information packet size may not match the BTC array, [IEEE 2006] allows some rows and columns to be removed, as well as some bits of the first row.

### 3.1.3 LDPC Codes

Low density parity check (LDPC) codes were initially proposed by Robert Gallager [1962]; he showed that with a careful design of the parity check matrix, they could achieve near-Shannon limit performance using iterative probabilistic-based decoders. However, the high complexity associated with the encoding and decoding processes made them computationally impractical at the time, and for 30 years they did not attract the attention of the research community. After the advent of the turbo codes, interest in these codes was rekindled with the rediscovery of Gallager codes in MacKay and Neal [1995] and MacKay and Neal [1996], who showed that they could achieve similar performance. For this reason, LDPCs have been incorporated into several standards: WiMax [IEEE 2006], DVB-T2 [ETSI 2009b], DVB-S2 [ETSI 2009a], IEEE Std 802.11n-2009 [IEEE 2011b], etc.

LDPC codes are $(n,k)$ linear block codes (although it is also possible to build LDPC convolutional codes) defined through very sparse $(n-k) \times n$ parity check matrices **H**. According to (3.52), each line of matrix **H** represents an equation that must be verified by a sequence **d** in order to be a valid codeword. For example, regarding a (7,4) Hamming code whose parity check matrix is displayed in Figure 3.12, its parity check equations can be written as

$$\begin{cases} c_1 = d_1 \oplus d_4 \oplus d_6 \oplus d_7 \\ c_2 = d_2 \oplus d_4 \oplus d_5 \oplus d_6, \\ c_3 = d_3 \oplus d_5 \oplus d_6 \oplus d_7 \end{cases} \tag{3.54}$$

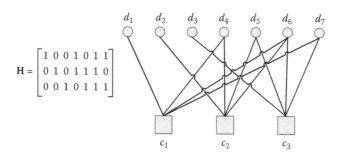

**Figure 3.12** Hamming code (7,4) parity check matrix (left) and respective Tanner graph (right).

where $\oplus$ represents modulo-2 addition. In this case, $\mathbf{d}$ is a codeword as long as $c_1 = c_2 = c_3 = 0$.

Since LDPC codes are defined through a parity check matrix $\mathbf{H}$, it is necessary to obtain the generator matrix $\mathbf{G}$ from this one in order to perform the encoding process using (3.51). The traditional method used in block codes consists of transforming matrix $\mathbf{H}$ into a systematic form using the Gaussian elimination method and column permutations resulting in

$$\mathbf{H}' = \begin{bmatrix} \mathbf{P} & \vdots & \mathbf{I}_{n-k} \end{bmatrix}, \tag{3.55}$$

where $\mathbf{P}$ is a $(n - k) \times k$ parity submatrix and $\mathbf{I}_{n-k}$ is the $(n - k) \times (n - k)$ identity matrix. From (3.55), it is simple to obtain the generator matrix as

$$\mathbf{G} = \begin{bmatrix} \mathbf{I}_k & \vdots & \mathbf{P}^T \end{bmatrix}. \tag{3.56}$$

Although this method seems straightforward, it does not take into account the sparseness of the original matrix $\mathbf{H}$ and, as such, the resulting generator matrix $\mathbf{G}$ is likely to be dense. As a consequence, the encoding complexity can become significant, especially for large matrices (which are usually associated with better performances). Other lower-complexity approaches have been proposed that take advantage of the sparseness of $\mathbf{H}$. For example, as proposed in Richardson and urban [2001b], it is possible to transform $\mathbf{H}$ into an almost lower triangular form using only row and column permutations, thus preserving the sparseness of the matrix. Then, Gaussian elimination is applied to the rows that do not match the triangular form. These rows are called the *gap* since they will be associated with the high-density part (higher complexity) of the encoding process. The presence of the other low-density rows allows the use of back substitution for the corresponding parity bits.

The LDPC code specified for WiMax [IEEE 2006] is defined using a parity check matrix $\mathbf{H}$ built from circulant submatrices and allows the encoding process to be efficiently implemented using back substitution with a gap of $n/24$.

In Tanner [1981], the use of bipartite graphs, named Tanner graphs, was proposed as a simple approach to characterizing an $(n,k)$ linear code. As an example, Figure 3.12 shows the Tanner graph associated with a (7,4) Hamming code. The graph is composed of two types of nodes: parity check nodes $(c_j)$ and variable nodes $(d_i)$. Each parity check node is associated with a parity check equation while each variable node is associated with a code bit. The construction of the graphs is simple. Each parity check node $(c_j)$ is connected to all the variable nodes $(d_i)$ that are involved in the respective parity check equation (according to (3.54)).

From the Tanner graph it is possible to know the degree of each node by simply counting the number of connections to that node. Since LDPC codes are defined through very sparse matrices, the degrees of the nodes are much smaller than $n$. If the degrees of all variable nodes are equal and the same happens to the degrees of all parity check nodes, then the LDPC code is called regular, otherwise it is an irregular LDPC code [Richardson and Urbanke 2001a]. Tanner graphs are useful to find the length of closed cycles. For example, in Figure 3.12, the existence of a length four closed cycle associated with path $c_1 \rightarrow d_4 \rightarrow c_2 \rightarrow d_6 \rightarrow c_1$ is visible. When constructing LDPC codes, it is important to avoid short length cycles, which are associated with worse performances of the decoder algorithm.

### 3.1.3.1 Sum-Product Algorithm

Tanner graphs are also useful for implementing iterative probabilistic decoding algorithms where the different types of nodes exchange messages (probabilities) according to the graph connections. In the following, we will briefly describe the iterative sum-product algorithm (also known as belief propagation algorithm) assuming binary codes and working with log likelihood ratios (LLRs). The sum-product algorithm is a message-passing-based algorithm with two types of messages:

■ Message from variable node $d_j$ to parity check node $c_i$, which takes into account all the messages coming from the other check nodes connected with $d_j$, with the exception of the message coming from the target check node $c_i$:

$$L(q_{ij}) = L(p_j) + \sum_{\substack{i'=1 \\ i' \neq i}}^{n-k} h_{i'j} L(r_{i'j}) \tag{3.57}$$

■ Message from parity check node $c_i$ to variable node $d_j$, which takes into account all the messages coming from the variable nodes connected to $c_i$, with the exception of the message coming from the target variable node $d_j$:

$$L(r_{ij}) = 2 \ \text{atanh}\left[ -\prod_{\substack{j'=1 \\ j' \neq j \wedge h_{ij'} \neq 0}}^{n} \tanh(-L(q_{ij'})/2) \right] \tag{3.58}$$

In these expressions $h_{i'j}$ represents an element of parity check matrix **H**. Furthermore, the following logarithms of probability ratios are used:

$$L(p_j) = \log\left( \frac{p_j^1}{1 - p_j^1} \right) \tag{3.59}$$

$$L(q_{ij}) = \log\left(\frac{q_{ij}^1}{1-q_{ij}^1}\right) \tag{3.60}$$

$$L(r_{ij}) = \log\left(\frac{r_{ij}^1}{1-r_{ij}^1}\right), \tag{3.61}$$

where $P_j^1$ is the likelihood probability associated with bit $d_j$ being 1, that is,

$$p_j^1 = p(y_j \mid d_j = 1). \tag{3.62}$$

In AWGN, and similarly to what was explained previously regarding the turbo decoder, (3.59) can be written as

$$L(p_j) = \frac{2}{\sigma^2} y_j = \frac{2}{\dfrac{N_0}{2.E_c}} y_j = \frac{4.E_c}{N_0} y_j = L_c y_j, \tag{3.63}$$

with $L_c$ corresponding to the reliability factor defined in (3.37).

In (3.58), the following hyperbolic functions are used:

$$\tanh(x) = \frac{e^{2x}-1}{e^{2x}+1} \tag{3.64}$$

and

$$\text{atanh}(x) = \frac{1}{2}\log\frac{1+x}{1-x}, |x| < 1. \tag{3.65}$$

The algorithm starts with $L(r_{ij}) = 0$ for all $i$ and $j$, followed by the computation of $L(q_{ij})$ using (3.57) and then $L(r_{ij})$ using (3.58). The a posteriori log probability ratios for each variable node $d_j$ can then be computed as

$$L(d_j) = \log\frac{\text{Prob}(d_i = 1 \mid observations)}{\text{Prob}(d_i = 0 \mid observations)} = L(p_j) + \sum_{i=1}^{n-k} h_{ij} L(r_{ij}). \tag{3.66}$$

If the Tanner graph did not have closed cycles, these a posteriori log probabilities would correspond to the exact ones. With the existence of cycles, it is necessary to repeat the computation of $L(q_{ij})$ and $L(r_{ij})$ for several iterations and the estimates computed using (3.66) will be only approximations. These approximations are accurate as long as the cycles have long lengths.

In the computation of the message from parity check node $c_i$ to variable node $d_j$, it is possible to avoid the use of hyperbolic functions as well as of the products of terms present in (3.58) by using the following approximation:

$$L(r_{ij}) = (-1)^{\gamma_i} \min_{\substack{j'=1,..,n \\ j' \neq j \wedge h_{ij'} \neq 0}} (|L(q_{ij'})|) \prod_{\substack{j'=1 \\ j' \neq j \wedge h_{ij'} \neq 0}}^{n} \text{sign}(L(q_{ij'})), \qquad (3.67)$$

where $\gamma_i$ is the degree of check node $i$ and the sign(.) represents the sign function defined as

$$\text{sign}(x) = \begin{cases} 1, & x > 0 \\ 0, & x = 0 \\ -1, & x < 0 \end{cases} \qquad (3.68)$$

When using this approximation, the algorithm is named the min-sum algorithm.

Figure 3.13 presents some simulation results obtained using DVB-S2 [ETSI 2009a] LDPC codes and employing the sum product algorithm with a maximum of 50 iterations. Different coding rates are shown.

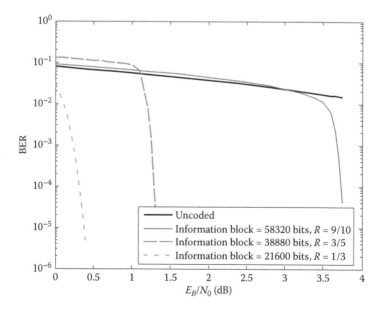

**Figure 3.13** **BER performance of several DVB-S2 [ETSI 2009a] LDPC codes.**

## 3.2 High-Order Modulations

In the design of wireless communication networks, the limited spectrum resources available are one of the main restrictions to achieving high-bit-rate transmissions. Multilevel quadrature amplitude modulation (M-QAM) is considered an attractive technique to achieve this objective due to its high spectral efficiency, and it has been employed in several wireless systems. 16-QAM and 64-QAM modulations are already used in HSDPA mode of UMTS [3GPP 2004], in WiMax [IEEE 2006], and DVB-T2 [ETSI 2009b], and they have been incorporated into 4G standards LTE-Advanced [3GPP 2011a] and WirelessMan-Advanced [IEEE 2011a].

Besides their typical applications in point-to-point transmissions for high spectral efficiency, QAM constellations can also be used as hierarchical constellations, which constitute a simple and flexible technique for achieving multi-resolution in a wireless system. Multi-resolution is important for some types of transmissions, for example, for supporting multimedia broadcast and multicast services, and as such 16-QAM and 64-QAM hierarchical constellations are used in some standards such as DVB-T2 [ETSI 2009b]. Although several modulations can be used for obtaining multi-resolution and higher spectral efficiency, such as M-PSK (M-ary Phase Shift Keying) and M-DAPSK (Multilevel Differential Amplitude and Phase Shift Keying), in this section we focus especially on M-QAM, which we will refer to as M-HQAM (Multilevel Hierarchical Quadrature Amplitude Modulation). Although in the following descriptions we will be referring mainly to hierarchical constellations, the explanations are valid for conventional constellations since they can be treated as a special case of the former ones.

### 3.2.1 Constellation Design

M-HQAM constellations can be mapped by two or more classes of bits with different error protection levels onto which different streams of information can be mapped. By using nonuniformly spaced signal points (where the distances along the I or Q axis between adjacent symbols are different), it is possible to modify the different error protection levels. Figure 3.14 shows an example of a 16-HQAM constellation.

In this constellation, four bits are required for selecting a symbol. Two bits select one of the four inner QPSK constellations, while the other two bits select the symbol inside that inner QPSK constellation. Since we are assuming a square constellation, there are two classes of bits with different error protection. The basic idea is that the constellation can be viewed as a 16-QAM constellation if the channel conditions are good enough or as a QPSK constellation otherwise. In the latter case, the received bit rate is reduced to half. Uniform or nonuniformly spaced signal points can be used depending on the desired differentiation between the error protection levels. Working with inner distances $D_1$ and $D_2$ defined in Figure 3.14, each symbol $s$ of the constellation can be written as

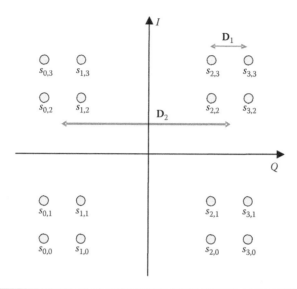

**Figure 3.14  Signal constellation for 16-QAM hierarchical modulation.**

$$s = \left( \pm \frac{D_2}{2} \pm \frac{D_1}{2} \right) + i \left( \pm \frac{D_2}{2} \pm \frac{D_1}{2} \right). \tag{3.69}$$

Instead of defining a hierarchical constellation through the inner constellation distances, it is usually more convenient to characterize it using ratios between these distances. For example, the 16-HQAM constellation of Figure 3.14 can be defined using $k_1 = D_1/D_2$ $(0 < k_1)$. Changing this parameter is equivalent to changing distances $D_1$ and $D_2$ modifying the degree of protection of the different bits. If $k_1 = 0.5$, the resulting constellation corresponds to a uniform 16-QAM constellation. This approach can easily be extended to any square M-QAM constellation, resulting in the following general expression for a symbol:

$$s = \sum_{l=1}^{1/2 \cdot \log_2 M} \left( \pm \frac{D_l}{2} \right) + i \sum_{l=1}^{1/2 \cdot \log_2 M} \left( \pm \frac{D_l}{2} \right), \tag{3.70}$$

where the number of possible classes of bits with different error protection is $1/2 \cdot \log_2 M$. These constellations can be characterized by the ratios between the distances of the inner nested constellations as

$$k_i = \frac{D_i}{D_{i+1}}, \quad i = 1,...,1/2 \cdot \log_2 M - 1. \tag{3.71}$$

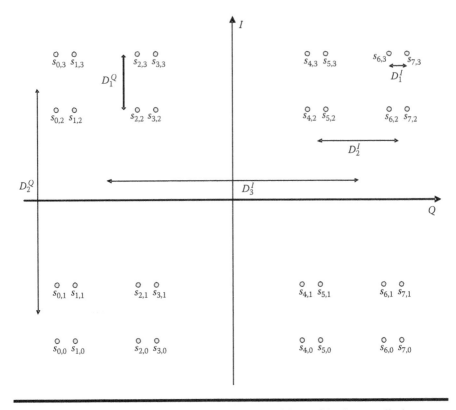

**Figure 3.15    Signal constellation for a 32-HQAM hierarchical constellation.**

The previous description is adequate for square constellations. However, we can also employ rectangular constellations. A rectangular constellation is obtained when the number of bits mapped to the I and Q branches is not the same or when the inner distances along these axis are different. In any case, the number of different error protection levels becomes $\log_2 M$. As an example, Figure 3.15 shows a rectangular 32-HQAM constellation. It can be constructed as a hierarchical 8-PAM constellation with three different protection levels along the I axis combined with a hierarchical 4-PAM constellation with two different protection levels along the Q axis. The constellation can be characterized using different nonuniformity ratios for the I and Q branches, namely, $k_1^I = D_1^I/D_2^I$, $k_2^I = D_2^I/D_3^I$ and $k_1^Q = D_1^Q/D_2^Q$ ($0 < k_1^I, k_2^I, k_1^Q$) with the case $k_1^I = k_2^I = k_1^Q = 0.5$ corresponding to a uniform constellation. Each symbol can be written as

$$s = \left( \pm \frac{D_3^I}{2} \pm \frac{D_2^I}{2} \pm \frac{D_1^I}{2} \right) + i \left( \pm \frac{D_2^Q}{2} \pm \frac{D_1^Q}{2} \right). \tag{3.72}$$

Extending this construction approach to any *M*-QAM rectangular constellation, the required nonuniformity ratios become

$$
\begin{cases}
k_i^I = \dfrac{D_i^I}{D_{i+1}^I}, & i = 1,\ldots,\log_2 M_{PAM}^I - 1 \\[4mm]
k_j^Q = \dfrac{D_j^Q}{D_{j+1}^Q}, & j = 1,\ldots,\log_2 M_{PAM}^Q - 1
\end{cases}
\tag{3.73}
$$

with the symbols expressed as

$$
s = \sum_{l=1}^{\log_2 M_{PAM}^I} \left( \pm \frac{D_i^I}{2} \right) + i \sum_{l=1}^{\log_2 M_{PAM}^Q} \left( \pm \frac{D_i^Q}{2} \right),
\tag{3.74}
$$

where $M_{PAM}^I$ and $M_{PAM}^Q$ are the number of projected symbols onto the I and Q axes, respectively of the constellation, that is, the sizes of the equivalent PAM constellations. For a square constellation, we have $M_{PAM}^I = M_{PAM}^Q = \sqrt{M}$. After defining a square/rectangular constellation through these ratios, we can build the constellation with a specified average symbol energy $E_s$ by computing the corresponding inner distances $D_i^I$ and $D_i^Q$. To show how these are obtained, we must start by writing the expression for the average symbol energy, which, due to symmetry, only requires taking into account the symbols inside one quadrant, resulting in

$$
E_s = E\left[ |s|^2 \right] = \frac{4}{M} \sum_{l=1}^{M/4} |s_l|^2 = \frac{2}{M_{PAM}^I} \sum_{l=1}^{M_{PAM}^I/2} \left( s_l^I \right)^2 + \frac{2}{M_{PAM}^Q} \sum_{l=1}^{M_{PAM}^Q/2} \left( s_l^Q \right)^2
$$

$$
= \frac{2}{M_{PAM}^I} \sum_{l=1}^{M_{PAM}^I/2} \left( \sum_{i=1}^{\log_2 M_{PAM}^I} \left( \pm \frac{D_i^I}{2} \right) \right)^2 + \frac{2}{M_{PAM}^Q} \sum_{l=1}^{M_{PAM}^Q/2} \left( \sum_{i=1}^{\log_2 M_{PAM}^Q} \left( \pm \frac{D_i^Q}{2} \right) \right)^2
$$

$$
= \frac{1}{2M_{PAM}^I} \left[ \sum_{l=1}^{M_{PAM}^I/2} \sum_{i=1}^{\log_2 M_{PAM}^I} \left( D_i^I \right)^2 + \sum_{l=1}^{M_{PAM}^I/2} \sum_{k=1}^{\log_2 M_{PAM}^I} \sum_{i=1}^{\log_2 M_{PAM}^I} \left( \pm D_k^I \right)\left( \pm D_i^I \right) \right]
$$

$$
+ \frac{1}{2M_{PAM}^Q} \left[ \sum_{l=1}^{M_{PAM}^Q/2} \sum_{i=1}^{\log_2 M_{PAM}^Q} \left( D_i^Q \right)^2 + \sum_{l=1}^{M_{PAM}^Q/2} \sum_{k=1}^{\log_2 M_{PAM}^Q} \sum_{i=1}^{\log_2 M_{PAM}^Q} \left( \pm D_k^Q \right)\left( \pm D_i^Q \right) \right]
$$

$$
= \frac{1}{4} \left[ \sum_{i=1}^{\log_2 M_{PAM}^I} \left( D_i^I \right)^2 + \sum_{i=1}^{\log_2 M_{PAM}^Q} \left( D_i^Q \right)^2 \right].
$$

$$
\tag{3.75}
$$

We can combine (3.75) with the definition of the nonuniformity ratios (3.73) and obtain an expression that only depends on one of the inner distances on each axis:

$$
E_s = \frac{1}{4} \left[ \left( D^I_{\log_2 M^I_{PAM}} \right)^2 + \sum_{i=1}^{\log_2 M^I_{PAM}-1} \left( D^I_{\log_2 M^I_{PAM}} \prod_{j=i}^{\log_2 M^I_{PAM}-1} k^I_j \right)^2 \right.
$$

$$
\left. + \left( D^Q_{1/2 \cdot \log_2 M^Q_{PAM}} \right)^2 + \sum_{i=1}^{\log_2 M^Q_{PAM}-1} \left( D^Q_{\log_2 M^Q_{PAM}} \prod_{j=i}^{\log_2 M^Q_{PAM}-1} k^Q_j \right)^2 \right].
$$

(3.76)

The first two terms in (3.76) correspond to the average symbol energy in the I branch, while the other two refer to the average symbol energy in the Q branch. Therefore, we can also write

$$
E_s = E_s^I + E_s^Q ,
$$

(3.77)

with

$$
E_s^{I/Q} = \frac{\left( D^{I/Q}_{\log_2 M^{I/Q}_{PAM}} \right)^2}{4} \left[ 1 + \sum_{i=1}^{\log_2 M^{I/Q}_{PAM}-1} \prod_{j=i}^{\log_2 M^{I/Q}_{PAM}-1} \left( k^{I/Q}_j \right)^2 \right].
$$

(3.78)

The desired inner distances can then be easily computed using

$$
\begin{cases}
D^{I/Q}_{\log_2 M^{I/Q}_{PAM}} = \sqrt{\dfrac{4 E_s^{I/Q}}{1 + \displaystyle\sum_{l=1}^{\log_2 M^{I/Q}_{PAM}-1} \prod_{j=l}^{\log_2 M^{I/Q}_{PAM}-1} \left( k^{I/Q}_j \right)^2}} \\[4ex]
D^{I/Q}_i = D^{I/Q}_{\log_2 M^{I/Q}_{PAM}} \displaystyle\prod_{j=i}^{\log_2 M^{I/Q}_{PAM}-1} k^{I/Q}_j , \quad i < \log_2 M^{I/Q}_{PAM}
\end{cases}
$$

(3.79)

Although in this chapter we focus on QAM constellation, we will briefly show how we can apply a similar construction approach and obtain other hierarchical constellations such as hierarchical PSK (M-HPSK). PSK symbols lie on the

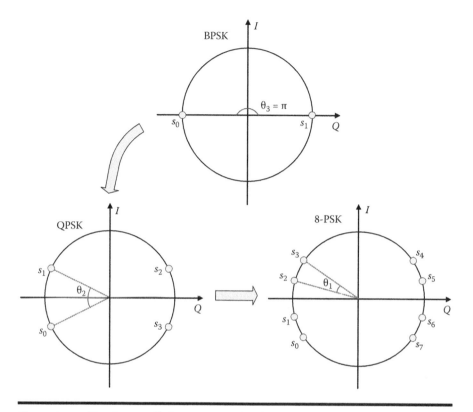

**Figure 3.16    Signal constellation construction for an 8-PSK hierarchical modulation.**

same circumference on the complex plan, only the phase differing between them. Following the procedure described in Pursley [1999], we can start with a BPSK constellation with two symbols spaced by $\pi$ radians, and split each symbol into two new ones spaced by some angle $\theta_2$ as shown in Figure 3.16. This results in a hierarchical QPSK constellation where the closest symbols are spaced by an angle $\theta_2$ and the others are spaced by at least $\pi - \theta_2$. We can then split each symbol again into two new ones and space them by an angle $\theta_1$, in order to obtain an 8-HPSK constellation with three differently protected bits.

A complex 8-HPSK constellation symbol can be written as

$$s = \sqrt{E_s}\, \exp\left[ i\left( \frac{\pi}{2} \pm \frac{\theta_3}{2} \pm \frac{\theta_2}{2} \pm \frac{\theta_1}{2} \right) \right], \tag{3.80}$$

with the nonuniformity ratios defined following a similar approach to HQAM as $k_1 = \theta_1/\theta_2$ and $k_2 = \theta_2/\theta_3$. Typically, a constellation is defined with $\theta_3 = \pi$ to minimize the error probability for the most protected bit. Generalizing to higher-order constellations, we can write an M-HPSK symbol as

$$
s = \sqrt{E_s}\, \exp\left[ i\left( \frac{\pi}{2} + \sum_{l=1}^{\log_2 M} \left( \pm\frac{\theta_l}{2} \right) \right) \right]
$$

$$
= \sqrt{E_s}\, \exp\left[ i\left( \frac{\pi}{2} \pm \frac{\pi}{2} + \sum_{l=1}^{\log_2 M-1} \left( \pm\frac{\theta_l}{2} \right) \right) \right],
$$

(3.81)

with

$$
k_i = \frac{\theta_i}{\theta_{i+1}}, \quad i = 1,\ldots,1/2 \cdot \log_2 M - 1.
$$

(3.82)

## 3.2.2 BER Analysis of M-HQAM Constellations in AWGN

In this section, we will derive expressions for computing the bit error probability of M-HQAM constellations in several channels. We will assume independent mapping of the bits to the I and Q branches of the constellation so that half of each stream goes for the in-phase branch and the other for the quadrature branch of the modulator. This means that we can analyze an M-QAM constellation as two independent $\sqrt{M}$-PAM constellations for obtaining the bit error probability. Due to the separation between the I and Q branches, the following expressions are also directly applicable to rectangular constellations, only requiring the substitution of $\sqrt{M}$ by either $M_{PAM}^I$ or $M_{PAM}^Q$ as well as using the corresponding inner constellation distances $D_i^I$ and $D_i^Q$. We will also admit that the resulting bit sequence in each branch is Gray encoded and mapped to the respective $\sqrt{M}$-PAM constellation symbols. This encoding is performed according to the procedure described in Vitthaladevuni [2003]. First, the constellation symbols are represented in an horizontal axis and are labeled from left to right with integers starting from 0 to $\sqrt{M}-1$. These labels are subsequently converted to their binary representation so that each symbol $s_j$ can be represented by a $1/2 \cdot \log_2 M$-digit binary sequence: $b_j^1, b_j^2, \ldots, b_j^{1/2 \cdot \log_2 M}$. The corresponding Gray code is then computed using

$$
g_j^1 = b_j^1
$$

$$
g_j^i = b_j^i \oplus b_j^{i-1}, \qquad i=2,3,\ldots,\ 1/2 \cdot \log_2 M.
$$

(3.83)

Considering the transmission of a PAM symbol, $s$, the received signal in the presence of AWGN can be modeled as

$$r = s + n, \tag{3.84}$$

where $n$ is a Gaussian variable with zero mean and variance $\sigma^2 = N_0/2$ ($N_0/2$ is the two-sided noise power spectrum density). In Vitthaladevuni and Alouini [2003], it was shown that for an $M$-HQAM constellation in AWGN, the individual BER of the different bits can always be computed as a weighted sum of $Q(x)$ functions. From the constellation construction pattern, an explicit closed-form expression for the individual BERs was derived in Vitthaladevuni and Alouini [2004]. Using the characterization of M-HQAM constellations given in Section 3.2.1 and assuming equiprobable symbols, we can rewrite the general expression from [Vitthaladevuni 2004] as

$$P_b(b_m) = \frac{2}{\sqrt{M}} \sum_{j=0}^{\sqrt{M}/2-1} \left[ g_j^m + (-1)^{g_j^m} \sum_{l=1}^{2^{(m-1)}} \left[ (-1)^{l+1} \times Q\left( \sqrt{\frac{2}{N_0}} \left[ \mathbf{B}_m(l) - \mathbf{D}_s(j) \right] \right) \right] \right]$$

$$m = 1,\ldots,\log_2 M/2,$$

$$\tag{3.85}$$

where

$$\mathbf{B}_m(l) = \frac{\mathbf{D}_s\left( (2l-1)2^{1/2 \cdot \log_2 M - m} - 1 \right) + \mathbf{D}_s\left( (2l-1)2^{1/2 \cdot \log_2 M - m} \right)}{2} \tag{3.86}$$

represents the decision borders and

$$\mathbf{D}_s(j) = \sum_{i=1}^{1/2 \cdot \log_2 M} \left( 2b_j^i - 1 \right) \frac{D_{1/2 \cdot \log_2 M - i + 1}}{2}. \tag{3.87}$$

are coordinates of the symbols.

Based on these expressions, Figure 3.17 shows the BER performance of the different bit types for a 64-HQAM constellation with $k_1 = 0.3$ and $k_2 = 0.4$. As a reference, performance curves of conventional QPSK, 16-QAM, and 64-QAM are also shown. In the graph legends, MPB designates most protected bits, IPB denotes intermediate protected bits, while LPB corresponds to least protected bits. The use of a nonuniform constellation with a hierarchical configuration allows us to obtain two thirds of the bits with better performance than the average 64-QAM BER at the cost of a degradation of the remainder one third (the least protected ones).

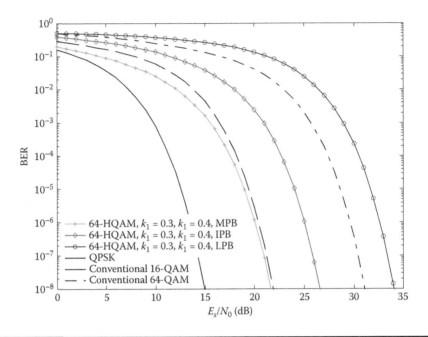

**Figure 3.17**  **BER performance of 64-HQAM, $k_1 = 0.3$ and $k_2 = 0.4$, in AWGN.**

### 3.2.3 BER Analysis of M-HQAM Constellations in Flat Fading Channels

The presence of a fading channel will affect the BER performance of the constellations. We will start by obtaining BER expressions for a flat Rayleigh fading channel and then extend it to other fading statistics. Let us consider that the received signal is distorted in amplitude and phase by the channel according to

$$r = \alpha \cdot s + n, \tag{3.88}$$

where $s$ is the $M$-HPAM transmitted symbol and $n$ is a Gaussian variable with mean 0 and variance $\sigma^2 = N_0/2$ . Regarding the channel coefficient, $\alpha$, we will start by modeling it as a zero mean complex Gaussian variable, which means $|\alpha|$ has a Rayleigh distribution, its phase follows a uniform distribution, and $|\alpha|^2$ has an exponential distribution. Since the BER expressions obtained previously for the AWGN channel are written as a weighted sum of $Q(.)$ functions, it is simple to extend them for flat Rayleigh fading channels. We only need to note that, apart from a phase rotation that can be compensated for assuming perfect channel knowledge and coherent detection at the receiver, the channel will result in a multiplication of the constellation symbols by $|\alpha|$, which is equivalent to a multiplication of the instantaneous symbol energy by $|\alpha|^2$. Therefore, the $Q(.)$ function in (3.85) can be rewritten as and $Q\left(\sqrt{\dfrac{2}{N_0}|\alpha|^2}\left[\mathbf{B}_m(l)-\mathbf{D}_s(j)\right]\right)$ the resulting overall expression will correspond to the conditional probability $P_b(b_m|\gamma)$ with $\gamma = |\alpha|^2$. To obtain the average BER for each bit, it is only necessary to average that expression over the PDF of $|\alpha|^2$, i.e.,

$$P_b(b_m) = \int\limits_0^{+\infty} P_b(b_m|\gamma)p(\gamma)d\gamma. \tag{3.89}$$

Since $\gamma = |\alpha|^2$ follows an exponential distribution, we can write

$$p(\gamma) = \frac{1}{\bar{\gamma}}e^{\frac{\gamma}{\bar{\gamma}}}, \quad \gamma \geq 0, \tag{3.90},$$

where $\bar{\gamma} = E[\gamma]$. It can be shown that

$$\int\limits_0^{+\infty} Q\left(z \cdot \sqrt{\gamma}\right)p(\gamma)d\gamma = \begin{cases} \dfrac{1}{2} - \dfrac{1}{2}\sqrt{\dfrac{z^2\,\bar{\gamma}/2}{1+z^2\,\bar{\gamma}/2}}, & z > 0 \\[4mm] \dfrac{1}{2} + \dfrac{1}{2}\sqrt{\dfrac{z^2\,\bar{\gamma}/2}{1+z^2\,\bar{\gamma}/2}}, & z < 0 \end{cases}, \tag{3.91}$$

which, combined with (3.89) and (3.85), results in the following BER expression for the individual bits:

$$P_b(b_m|\kappa) = \frac{2}{\sqrt{M}}\sum_{j=0}^{\sqrt{M}/2-1}\left[g_j^m + (-1)^{g_j^m}\sum_{l=1}^{2^{(m-1)}}\left[(-1)^{l+1}\left(\frac{1}{2} + (-1)^{\frac{1}{2}+\frac{1}{2}\text{sign}(\mathbf{B}_m(l)-\mathbf{D}_s(j))}\xi\right)\right]\right], \tag{3.92}$$

with

$$\xi = \frac{1}{2}\sqrt{\dfrac{\dfrac{1}{N_0}\left[\mathbf{B}_m(l)-\mathbf{D}_s(j)\right]^2\bar{\gamma}}{1+\dfrac{1}{N_0}\left[\mathbf{B}_m(l)-\mathbf{D}_s(j)\right]^2\bar{\gamma}}} \tag{3.93}$$

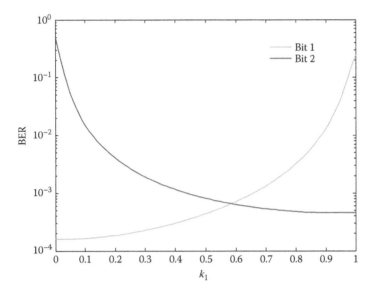

**Figure 3.18 BER performances of hierarchical 16-QAM constellations as a function of the nonuniformity ratios $k_1$ for $E_s/N_0 = 35$ dB.**

and $\text{sign}(x)$ being the sign function defined as

$$\text{sign}(x) = \begin{cases} 1, & x > 0 \\ 0, & x = 0 \,. \\ -1, & x < 0 \end{cases} \tag{3.94}$$

As examples, Figures 3.18 and 3.19 show the BER performances of 16-HQAM and 64-HQAM constellations in flat Rayleigh fading channels as a function of the non-uniformity parameters $k_1$ (16-HQAM) and $k_1$ and $k_2$ (64-HQAM). It is clear that by changing the values of the nonuniformity parameters it is possible to improve the error protection of some of the bit streams at the cost of some performance degradation of the others. When $k_1 = 0$ (16-QAM) or $k_1 = k_2 = 0$ (64-QAM), the constellations reduce to QPSK, and only the most protected bit stream can be reliably extracted (the other bits have BERs of 0.5). It is important to note also that $k_1 = 0.5$ (16-QAM) or $k_1 = k_2 = 0.5$ (64-QAM) does not correspond to equal protection of all the bits; it only means that the constellations are uniform. For example, for the conditions of Figure 3.18, equal protection for the different bit streams is obtained when $k_1$ is close to 0.6.

We can also extend the BER expressions for other fading environments. In a Nakagami fading environment, the PDF of $a = |\alpha|$ can be written as

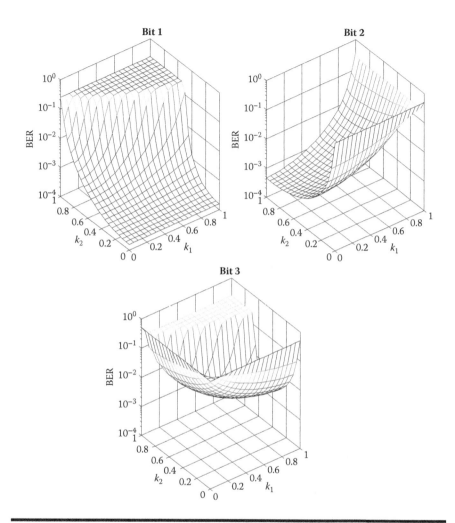

**Figure 3.19  BER performances of hierarchical 64-QAM constellations as a function of the nonuniformity ratios $k_1$ and $k_2$ for $E_s/N_0 = 35$ dB.**

$$p(a) = \frac{2}{\Gamma(m)}\left(\frac{m}{\bar{\gamma}}\right)a^{2m-1}e^{-\frac{m \cdot a^2}{\bar{\gamma}}}, \quad a \geq 0, \tag{3.95}$$

where $\bar{\gamma} = E\left[|\alpha|^2\right]$, $m$ is the Nakagami severity of the fading parameter

$$m = \frac{\bar{\gamma}^2}{E\left[\left(|\alpha|^2 - \bar{\gamma}\right)^2\right]}, \quad m \geq \frac{1}{2}, \tag{3.96}$$

and $\Gamma(\cdot)$ is the gamma function defined as

$$\Gamma(x) = \int_0^{+\infty} t^{x-1} e^{-t} \, dt, \quad x > 0. \tag{3.97}$$

It is easy to verify that in this case $\gamma = |\alpha|^2$ will follow a gamma distribution described as

$$p(\gamma) = \frac{m^m}{\Gamma(m)\bar{\gamma}^m} \gamma^{m-1} e^{-\frac{m\cdot\gamma}{\bar{\gamma}}}, \quad \gamma \geq 0. \tag{3.98}$$

Using this distribution, it is possible to show that

$$\int_0^{+\infty} Q\left(z \cdot \sqrt{\gamma}\right) p(\gamma) d\gamma$$

$$= \begin{cases} \dfrac{1}{2}\sqrt{\dfrac{z^2\,\bar{\gamma}/2}{\pi}} \dfrac{m^m}{\left(m+z^2\,\bar{\gamma}/2\right)^{m+1/2}} \dfrac{\Gamma\left(m+1/2\right)}{\Gamma(m+1)}\, {}_2F_1 \\ \left(1, m+1/2; m+1; \dfrac{m}{m+z^2\,\bar{\gamma}/2}\right), \quad z > 0 \\[2em] 1 - \dfrac{1}{2}\sqrt{\dfrac{z^2\,\bar{\gamma}/2}{\pi}} \dfrac{m^m}{\left(m+z^2\,\bar{\gamma}/2\right)^{m+1/2}} \dfrac{\Gamma\left(m+1/2\right)}{\Gamma(m+1)}\, {}_2F_1 \\ \left(1, m+1/2; m+1; \dfrac{m}{m+z^2\,\bar{\gamma}/2}\right), \quad z < 0 \end{cases} \tag{3.99}$$

where ${}_2F_1(\cdot,\cdot;\cdot;\cdot)$ denotes the Gauss hypergeometric function. As a result, the BER expression for the individual bits becomes

$$P_b\left(b_m | \kappa\right) = \frac{2}{\sqrt{M}} \sum_{j=0}^{\sqrt{M}/2-1} \left[ g_j^m + (-1)^{g_j^m} \sum_{l=1}^{2^{(m-1)}} \frac{1}{2} (-1)^{l+1} \left( 1 - \text{sign}\left(B_m(l) - D_s(j)\right) \right. \right.$$

$$+ (-1)^{\frac{1}{2}-\frac{1}{2}\operatorname{sign}(B_m(l)-D_s(j))} \frac{\sqrt{\dfrac{1}{N_0}\left[B_m(l)-D_s(j)\right]^2 \bar{\gamma}}}{\pi}$$

$$\frac{m^m}{\left(m + \dfrac{1}{N_0}\left[B_m(l)-D_s(j)\right]^2 \bar{\gamma}\right)^{m+1/2}}$$

$$\times \frac{\Gamma(m+1/2)}{\Gamma(m+1)}\, {}_2F_1\left(1, m+1/2; m+1; \frac{m}{m + \dfrac{1}{N_0}\left[B_m(l)-D_s(j)\right]^2 \bar{\gamma}}\right)\Bigg)\Bigg]\Bigg]$$

$$(3.100)$$

Another important distribution frequently used for characterizing fading channels is the Rice distribution. In this case, $a = |\alpha|$ is composed of two components: one fixed (representing a line-of-sight component), $a_0$, and one with Rayleigh distribution, $a - a_0$. The PDF of a Rice random variable is expressed as

$$p(a) = \frac{2a}{\sigma_a^2} e^{-\frac{(a^2+a_0^2)}{\sigma_a^2}} I_0\left(\frac{a \cdot a_0}{\sigma_a^2/2}\right), \quad a \geq 0, \tag{3.101}$$

with $I_0(\cdot)$ being the zeroth-order modified Bessel function of the first kind and $\sigma_a^2 = E\left[(a-a_0)^2\right]$. As a consequence $\gamma = |\alpha|^2$ follows a noncentral chi-square distribution given by

$$p(\gamma) = \frac{1}{\sigma_a^2} e^{-\frac{(\gamma+a_0^2)}{\sigma_a^2}} I_0\left(\frac{\sqrt{\gamma} \cdot a_0}{\sigma_a^2/2}\right), \quad \gamma \geq 0. \tag{3.102}$$

Averaging function $Q(\cdot)$ over this distribution results in

$$\int_0^{+\infty} Q\left(z \cdot \sqrt{\gamma}\right) p(\gamma)\, d\gamma = \begin{cases} \dfrac{1}{2}Q(u,v) - \dfrac{1}{4}\left[1+\sqrt{\dfrac{w}{1+w}}\right]e^{-\frac{(u^2+v^2)}{2}} I_0(u \cdot v), & z > 0 \\[4mm] 1 - \dfrac{1}{2}Q(u,v) + \dfrac{1}{4}\left[1+\sqrt{\dfrac{w}{1+w}}\right]e^{-\frac{(u^2+v^2)}{2}} I_0(u \cdot v), & z < 0 \end{cases},$$

$$(3.103)$$

with

$$w = \frac{z^2 \bar{\gamma}/2}{K+1},$$

(3.104)

$$u = \sqrt{K \left( \frac{1+2w}{2+2w} - \left( \frac{w}{1+w} \right)^{\frac{1}{2}} \right)}$$

(3.105)

and

$$v = \sqrt{K \left( \frac{1+2w}{2+2w} + \left( \frac{w}{1+w} \right)^{\frac{1}{2}} \right)}.$$

(3.106)

$Q(u,v)$ is the Marcum Q-function and $K$ is the Ricean factor defined as $K = a_0^2 / \sigma_a^2$. Therefore, for a Rice fading channel, the BER expression for the individual bits becomes

$$P_b(b_m | \kappa) = \frac{2}{\sqrt{M}} \sum_{j=0}^{\sqrt{M}/2-1} \left[ g_j^m + (-1)^{g_j^m} \sum_{l=1}^{2^{(m-1)}} \left[ \frac{1}{2}(-1)^{l+1} \left( 1 - \text{sign}\left(B_m(l) - D_s(j)\right) \right) \right. \right.$$

$$\left. \left. + (-1)^{\frac{1}{2} - \frac{1}{2}\text{sign}(B_m(l) - D_s(j))} \left( Q(u,v) - \frac{1}{2}\left[ 1 + \sqrt{\frac{w}{1+w}} \right] e^{-\frac{(u^2+v^2)}{2}} I_0(u \cdot v) \right) \right] \right]$$

(3.107)

with

$$w = \frac{\frac{1}{N_0} \left[ B_m(l) - D_s(j) \right]^2 \bar{\gamma}}{K+1}.$$

(3.108)

## 3.2.4 MFB in Multipath Rayleigh Fading Channels

Although it is possible to obtain exact expressions for the performance of hierarchical constellations in flat fading environments, as we have shown in the previous section, extending them to multipath channels becomes more difficult due to the problem of dealing with ISI. Nevertheless, instead of trying to obtain exact

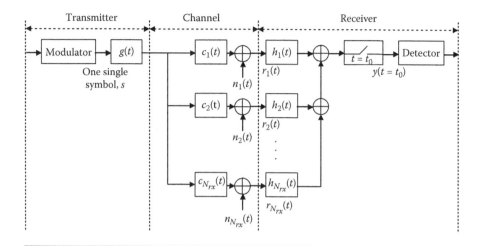

**Figure 3.20 Baseband transmit/receive scheme.**

expressions for this type of environment, it can be satisfactory just to have some analytical expressions representing an ideal performance that may not be possible to achieve in practice but can work as a very important benchmark for the evaluation of a specific receiver. The matched filter bound (MFB) can be regarded as a lower limit on the BER for a particular communication channel and is derived assuming perfect channel knowledge and the transmission of only one pulse, that is, no ISI occurs. In this section, we will derive the MFB for the BERs of the different types of bits (in terms of error protection level) of $M$-HQAM constellations for diversity reception in time-discrete multipath Rayleigh fading channels with correlated paths.

Let us consider the case of a transmission over an $N_{rx}$-th order diversity branch multipath Rayleigh fading channel where all branches can have different fading powers. The corresponding transmit/receive block diagram is shown in Figure 3.20.

Assuming a discrete multipath channel for each diversity branch $n$ composed of $L_n$ discrete taps, the respective response at time $t$ to an impulse applied at $t$-$\tau$ can be modeled as

$$c_n(\tau,t) = \sum_{i=1}^{L_n} \alpha_{i,n}(t)\delta(\tau - \tau_{i,n}), \quad n = 1...N_{rx}, \tag{3.109}$$

where $\alpha_{i,n}(t)$ is a zero-mean complex Gaussian random process, $\delta(t)$ is the Dirac function and $\tau_{i,n}$ is the delay (assumed constant) of the $i^{th}$ tap. Each channel tap autocovariance function can be expressed as [Stuber 2001]

$$R_{\alpha_{i,n}}(\tau) = E\left\{\alpha_{i,n}(t)\alpha_{i,n}(t+\tau)\right\} = \Omega_{i,n}^2 J_0(2\pi f_D \tau), \tag{3.110}$$

where $J_0(\cdot)$ is the zeroth-order Bessel function of the first kind, $f_D$ is the Doppler frequency, and $\Omega_{i,n}$ is the root mean square value of the magnitude of tap $i$ in diversity branch $n$. Regarding the cross-correlation between different taps belonging to the same or to different diversity branches, no restriction is imposed; that is, all the taps can be correlated.

For the derivation of the MFB, we assume the transmission of only one pulse $sg(t)$, where $s$ is an $M$-HQAM symbol and $g(t)$ is the impulse response of the transmit filter. Therefore, the signal received at each diversity branch can be written as

$$r_n(t) = s \sum_{i=1}^{L_n} \alpha_{i,n}(t) g(t - \tau_{i,n}) + n_n(t),$$  (3.111)

where $n_n(t)$ is a zero-mean complex Gaussian random process with power spectral density $N_0$, representing AWGN. As previously stated, the MFB is obtained considering perfect channel knowledge. The receiver filters employed in each diversity branch are expressed as

$$h_n(t) = \sum_{i=1}^{L_n} \alpha_{i,n}^*(t_0 - t) g^*(t_0 - t - \tau_{i,n}),$$  (3.112)

with $t_0$ representing the sampling instant. After the filter, the sampled signals of the different diversity branches are added, leading to

$$y(t = t_0) = s \cdot \sum_{n=1}^{N_{rx}} \sum_{i=1}^{L_n} \sum_{i'=1}^{L_n} \int_{-\infty}^{+\infty} \alpha_{i,n}(\tau) \alpha_{i',n}^*(\tau) g(\tau - \tau_{i,n}) g^*(\tau - \tau_{i',n}) d\tau + \sum_{n=1}^{N_{rx}} w_n,$$

(3.113)

where $w_n$ represents noise samples, that is,

$$w_n = \sum_{i=1}^{L_n} \int_{-\infty}^{+\infty} n_n(\tau) \alpha_{i,n}^*(\tau) g^*(\tau - \tau_{i,n}) d\tau.$$  (3.114)

Although the transmitter pulse $g(t)$ can have unlimited duration, typically it becomes almost zero outside a narrow interval, usually related to the symbol period. If the channel is slowly time-varying, it can be considered approximately constant inside this interval, and (3.113) becomes

$$y(t = t_0) = s \cdot \sum_{n=1}^{N_{rx}} \sum_{i=1}^{L_n} \sum_{i'=1}^{L_n} \alpha_{i,n} \alpha_{i',n}^* R(\tau_{i,n} - \tau_{i',n}) + \sum_{n=1}^{N_{rx}} w_n$$  (3.115)

$$= y_{signal} + y_{noise},$$

with $R(\tau)$ representing the autocorrelation function of the transmit filter,

$$R(\tau) = \int_{-\infty}^{+\infty} g(\tau') g^*(\tau' + \tau) d\tau'. \tag{3.116}$$

The instantaneous received signal-to-noise power ratio is given by the ratio between the expected value of the signal component over the expected value of the noise component of (3.115) conditioned to the channel and can be written as

$$SNR = \frac{E\left[\left|y_{signal}\right|^2 \Big| \alpha_{\cdot,\cdot}\right]}{E\left[\left|y_{noise}\right|^2 \Big| \alpha_{\cdot,\cdot}\right]} = \frac{E_s}{N_0} \kappa, \tag{3.117}$$

with

$$\kappa = \sum_{n=1}^{N_{rx}} \sum_{i=1}^{L_n} \sum_{i'=1}^{L_n} \alpha_{i,n} \alpha_{i',n}^* R\left(\tau_{i,n} - \tau_{i',n}\right) \tag{3.118}$$

$$= z^H \Sigma z$$

and

$$E_s = E\left[|s|^2\right]. \tag{3.119}$$

The last equality in (3.118) corresponds to the matrix representation of $\kappa$. In this matrix format, $z$ is an $L_{total} \times 1$ vector (with $L_{total} = \sum_{n=1}^{N_{rx}} L_n$) containing random variables $\alpha_{i,n}$, $^H$ denotes conjugate transpose, and $\Sigma$ is an $L_{total} \times L_{total}$ matrix constructed according to

$$\Sigma = \begin{bmatrix} R(0) & \cdots & R(\tau_{L_1,1} - \tau_{1,1}) & & & & \\ \vdots & \ddots & \vdots & & & \mathbf{0} & \\ R(\tau_{1,1} - \tau_{L_1,1}) & \cdots & R(0) & & & & \\ & & & \ddots & & & \\ & & & & R(0) & \cdots & R(\tau_{L_{Nrx},N_{rx}} - \tau_{1,N_{rx}}) \\ & \mathbf{0} & & & \vdots & \ddots & \vdots \\ & & & & R(\tau_{1,N_{rx}} - \tau_{L_{Nrx},N_{rx}}) & \cdots & R(0) \end{bmatrix}. \tag{3.120}$$

Since $R(\tau_{i,n} - \tau_{i',n}) = R^*(\tau_{i',n} - \tau_{i,n})$, we can easily recognize that $\Sigma$ is Hermitian. For obtaining the MFB, it is necessary to find the probability density function (PDF) of $\kappa$. To accomplish this, we will apply a characteristic function-based method that requires writing $\kappa$ in the form of a sum of uncorrelated random variables with known PDFs. Since different taps can be correlated, we need to transform z into a vector of uncorrelated components. Denoting $\Psi$ as the covariance matrix of z ($\psi = \text{Cov}[z]$), and knowing that this matrix is Hermitian and positive-semidefinite, it is possible to find at least one matrix Q so that $\Psi$ can be broken down into $\Sigma = QQ^H$. In particular, if we apply the Cholesky decomposition, Q will be a lower triangular matrix. After finding Q, we can define a new vector $z' = Q^{-1}z$ whose covariance matrix corresponds to the identity matrix, as is easy to verify. As a consequence, all components of $z'$ will be uncorrelated unit-variance complex Gaussian variables. Introducing this transformation into (3.118) results in

$$\kappa = z'^H Q^H \Sigma Q z'$$
$$= z'^H \Sigma' z',$$

(3.121)

with

$$\Sigma' = Q^H \Sigma Q.$$

(3.122)

Matrix $\Sigma'$ is still Hermitian, which means it can be as

$$\Sigma' = \Phi \Lambda \Phi^H,$$

(3.123)

where $\Lambda$ is a diagonal matrix whose elements are the eigenvalues $\lambda_i$ ($i = 1,..,L_{total}$) of $\Sigma'$ and $\Phi$ is a matrix whose columns are the eigenvectors of $\Sigma'$. These eigenvectors are orthogonal ($\Phi\Phi^H = I$). We can then rewrite (3.121) as

$$\kappa = z'^H \Phi \Lambda \Phi^H z''$$
$$= z'^H \Lambda z''$$
$$= \sum_{i=1}^{L_{total}} \lambda_i \left| z_i'' \right|^2,$$

(3.124)

where we have defined another vector, $z'' = \Phi^H z'$, whose components are still uncorrelated unit-variance complex Gaussian variables (as is simple to verify). According to (3.124) $\kappa$ can be expressed as a sum of independent random variables with exponential distributions ($z_i''$ is Gaussian and, therefore, $\lambda_i \left| z_i'' \right|^2$ follows an exponential distribution). The characteristic function of $\kappa$ is then simply the product of the respective individual characteristic functions

$$E\left\{e^{-j\upsilon\kappa}\right\} = \prod_{i=1}^{L_{total}} \frac{1}{1+j\lambda_i\upsilon}. \tag{3.125}$$

If there are $L'$ distinct eigenvalues, each with a multiplicity of $\theta_i$, $i = 1...L'$, then by partial fractions decomposition and using the change of variables $s = j\upsilon$ we can rewrite (3.125) as

$$E\left\{e^{-s\kappa}\right\} = \prod_{i=1}^{L'} \frac{1}{\left(1+s\lambda_i\right)^{\theta_i}}$$

$$= \sum_{i=1}^{L'} \sum_{k=1}^{\theta_i} \frac{A_{i,k}}{\lambda_i^{\theta_i-k}\left(\theta_i-k\right)!\left(1+s\lambda_i\right)^{k}}, \tag{3.126}$$

with

$$A_{i,k} = \left[\frac{\partial^{\theta_i-k}}{\partial s^{\theta_i-k}}\left(\prod_{\substack{j=1\\j\neq i}}^{L'} \frac{1}{\left(1+s\lambda_j\right)^{\theta_j}}\right)\right]_{s=-\frac{1}{\lambda_i}}. \tag{3.127}$$

Reversing the change of variables and computing the inverse Fourier transform, we obtain the PDF of $\kappa$ as

$$p(\kappa) = \sum_{i=1}^{L'} \sum_{k=1}^{\theta_i} \frac{A_{i,k}}{\lambda_i^{\theta_i}\left(\theta_i-k\right)!(k-1)!}\kappa^{k-1}e^{-\frac{\kappa}{\lambda_i}}. \tag{3.128}$$

Based on the BER expressions for AWGN presented in Section 3.2.2 and observing that $\kappa$ acts as a simple scaling factor on the constellation symbols, we can write the individual BERs conditioned to $\kappa$ as

$$P_b\left(b_m|\kappa\right) = \frac{2}{\sqrt{M}} \sum_{j=0}^{\sqrt{M}/2-1} \left[g_j^m + (-1)^{g_j^m} \sum_{l=1}^{2^{(m-1)}} \left[(-1)^{l+1} \times Q\left(\sqrt{\frac{2}{N_0}}\kappa\left[B_m(l) - D_s(j)\right]\right)\right]\right],$$

$$m = 1,...,\log_2 M/2.$$

$$\tag{3.129}$$

The average BER can be computed averaging (3.129) over the PDF of $\kappa$:

$$P_b(b_m) = \int_{-\infty}^{+\infty} P_b(b_m|\kappa)\, p(\kappa)\, d\kappa. \tag{3.130}$$

To obtain an explicit expression for (3.130), we need to make use of the following integral expression:

$$\int_0^{+\infty} Q(\sqrt{x}) \frac{x^{L-1}}{(L-1)!\,\bar{x}_i^L}\, e^{-\frac{x}{\bar{x}_i}}\, dx = \left[\frac{1-\mu}{2}\right]^L \sum_{r=0}^{L-1}\binom{L-1+r}{r}\left[\frac{1+\mu}{2}\right]^r, \tag{3.131}$$

with

$$\mu = \sqrt{\frac{\bar{x}_i/2}{1+\bar{x}_i/2}}. \tag{3.132}$$

The average BER can subsequently be written as

$$P_b(b_m) = \frac{2}{\sqrt{M}} \sum_{j=0}^{\sqrt{M}/2-1}\left[ g_j^m + (-1)^{g_j^m}\sum_{l=1}^{2^{(m-1)}}\left[\frac{1}{2}(-1)^{l+1}\left\{1-\mathrm{sign}\big(\mathrm{B}_m(l)-\mathrm{D}_s(j)\big)\right\}\right. \right.$$

$$\left.\left. 2\cdot(-1)^{1/2-\mathrm{sign}(\mathrm{B}_m(l)-\mathrm{D}_s(j))/2}\sum_{i=1}^{L'}\sum_{k=1}^{\theta_i}\frac{A_{i,k}}{\lambda_i^{\theta_i-k}(\theta_i-k)!}\left[\frac{1-\mu_i}{2}\right]^k\sum_{r=0}^{k-1}\binom{k-1+r}{r}\left[\frac{1+\mu_i}{2}\right]^r\right\}\right]\right]$$

$$\tag{3.133}$$

where

$$\mu_i = \sqrt{\frac{\dfrac{E_s}{N_0}\big[\mathrm{B}_m(l)-\mathrm{D}_s(j)\big]^2 \lambda_i}{1+\dfrac{E_s}{N_0}\big[\mathrm{B}_m(l)-\mathrm{D}_s(j)\big]^2 \lambda_i}}. \tag{3.134}$$

Using the MFB expressions, we can evaluate the effect of the correlation coefficient between taps, $\rho = E[\alpha_i\alpha_{i'}^*]/\sqrt{E[|\alpha_{i'}|^2]E[|\alpha_{i'}|^2]}$ (assumed the same for all the taps), on the performance of M-HQAM constellations.

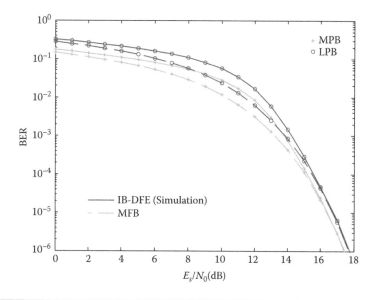

**Figure 3.21** **IB-DFE performance versus MFB for a 16-HQAM constellation with** $k_1 = 0.5$ **in a 64-tap environment with two receiver antennas.**

Figure 3.21 compares the MFB with the Monte Carlo simulated performance of an IB-DFE receiver (described in Chapter 2). We consider a 16-HQAM constellation with $k_1 = 0.5$, a root-square raised cosine (RRC) filter with a roll-off bandwidth factor $\beta = 0.22$, a channel with 64 equal-power taps and 2 receiver antennas. Curves for the most protected bits (MPB) and for the least protected bits (LPB) are presented. The results show that the IB-DFE performance can become asymptotically close to the MFB for severely time-dispersive channels with rich multipath propagation combined with diversity techniques.

## 3.3 Signal Space Diversity

### 3.3.1 Complex Rotation Matrices

When powerful channel coding schemes are employed, OFDM schemes can have excellent performance. However, the required code rate must be low, reducing the system's spectral efficiency. On the other hand, for uncoded systems or when high rate codes are employed, the performance of OFDM systems can be very poor. In this case, we can associate a given symbol with different subcarriers so as to take advantage of the diversity effects inherent in a severely frequency selective channel, which is typical in mobile communication environments. This alternative diversity technique is often named signal space diversity as proposed in Boutros and Viterbo [1998] and is accomplished without additional power or bandwidth. A simple way of doing this is to employ real rotation matrices (RRMs) [Rainish 1996], which allows significant gains. Unfortunately, RRMs were only designed to spread a symbol over two subcarriers, which is accomplished using

$$X = A_{RRM} \cdot S \tag{3.135}$$

with

$$A_{RRM} = \begin{bmatrix} \cos(\varphi) & \sin(\varphi) \\ -\sin(\varphi) & \cos(\varphi) \end{bmatrix} \tag{3.136}$$

and **S** being a $2 \times 1$ vector containing two modulated symbols.

Spreading a symbol over a larger number of subcarriers can be accomplished using the Hadamard matrix (HM) adopted in fully loaded multicarrier code division multiplexing schemes. The HM for spreading over two symbols can be obtained from RRM using $\varphi = \pi/4$. A more general alternative lies in the use of complex rotation matrices (CRMs). CRM is a technique for achieving signal space diversity (SSD) in SISO and MIMO OFDM/OFDMA systems and can be easily combined with turbo or LDPC codes in order to improve system performance without a substantial reduction of the spectral efficiency.

The process of applying CRM is similar to RRM, where a rotated super-symbol is obtained using

$$X = A_{M_{CRM}} \cdot S \tag{3.137}$$

with S being a $M_{CRM} \times 1$ vector with a set of modulated symbols constituting a super-symbol. Matrix $A_{M_{CRM}}$ belongs to the family of the orthonormal (OCRM) or non-orthonormal (NCRM) complex matrices, which are defined as follows:

$$(\text{OCRM}) \; A_{M_{CRM}} = \begin{cases} \begin{bmatrix} e^{j\varphi} & je^{-j\varphi} \\ -je^{j\varphi} & e^{-j\varphi} \end{bmatrix} / |A_2|^{1/2}, M_{CRM} = 2 \\ |A_2| = \det(A_2) = 2 \\ \begin{bmatrix} \Sigma_{M_{CRM}}/2 & \Sigma_{M_{CRM}}/2 \\ \Sigma_{M_{CRM}}/2 & -\Sigma_{M_{CRM}}/2 \end{bmatrix} / |\Sigma_{M_{CRM}}|^{1/M_{CRM}}, M_{CRM} > 2 \end{cases}$$

$$\tag{3.138}$$

and

$$
\text{(NCRM)} \quad \Sigma_{M_{CRM}} = \begin{cases} \begin{bmatrix} e^{j\varphi} & e^{-j\varphi} \\ -e^{-j\varphi} & e^{j\varphi} \end{bmatrix} / |\mathbf{A}_2|^{1/2}, M_{CRM} = 2 \\ |A_2| = \det(A_2) = 2\cos(\varphi) \\ \begin{bmatrix} \Sigma_{M_{CRM}}/2 & \Sigma_{M_{CRM}}/2 \\ \Sigma_{M_{CRM}}/2 & -\Sigma_{M_{CRM}}/2 \end{bmatrix} / |\Sigma_{M_{CRM}}|^{1/M_{CRM}}, M_{CRM} > 2 \end{cases} \tag{3.139}
$$

with $M_{CRM} = 2^n$, $(n \geq 1)$, $|\mathbf{A}_{M_{CRM}}| = \det(\mathbf{A}_{M_{CRM}})$ and $\phi$ being the rotation angle [Correia 2002].

### 3.3.2 Transmitter for OFDM Schemes with CRM

CRM can be easily incorporated into OFDM systems. Figure 3.22 shows the block diagram of an OFDM transmitter with CRM and multiple transmitting antennas.

An information block is encoded, interleaved, and mapped according to the constellation symbols. A rotation matrix (RM) is then applied by grouping the symbols into $M_{CRM}$-tuples and multiplying them by rotation matrix $\Sigma_{M_{CRM}}$. The resulting sequence is then interleaved in the symbol interleaver. The objective of the symbol interleaver is to explore the characteristics of OFDM transmissions in severe time-dispersive environments whose channel frequency response can change significantly between different subcarriers. The interleaver ensures that samples of a super-symbol are mapped to distant subcarriers, thus taking advantage of the diversity in the frequency domain. Finally, pilot symbols are inserted into the resulting sequence before it is converted to the time domain using a size-$N$ Inverse discrete Fourier transform (IDFT) and transmitted as a conventional OFDM transmission.

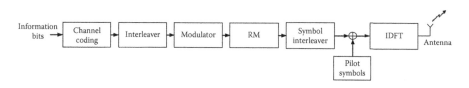

**Figure 3.22 Transmitter block diagram for MIMO-OFDM transmissions using CRM.**

### 3.3.3 Receiver for OFDM Schemes with CRM

Figure 3.23 represents the receiver block diagram for OFDM transmissions with CRM.

At the receiver, the signal is sampled, the cyclic prefix removed, and the resulting signal is converted to the frequency domain with an appropriate size-$N$ DFT operation. The sequence of symbols is then de-interleaved. If the cyclic prefix is longer than the overall channel impulse response, each received $M_{CRM}$-sized super-symbol can be expressed using matrix notation as

$$R = H \cdot X + N, \tag{3.140}$$

where $H$ is the frequency response channel matrix. Matrix $H$ can be defined as a diagonal matrix

$$H = \begin{bmatrix} H_1 & & 0 \\ & \ddots & \\ 0 & & H_{M_{CRM}} \end{bmatrix}. \tag{3.141}$$

Index $k$ represents a subcarrier position. It is important to note that due to the presence of the symbol interleaver, the different subcarriers denoted by index $k$ may not be necessarily adjacent. $N$ is an $M_{CRM} \times 1$ vector containing additive white Gaussian noise (AWGN) samples.

The super-symbol's samples enter the CRM inverter block, which performs channel equalization and inverts the rotation applied at the transmitter. Using a maximum likelihood based soft output (MLSO) criterion, each symbol estimate is computed as

$$S_l = E\left[S_l \middle| R\right]$$

$$= \sum_{s_i \in \Lambda} s_i \cdot P\left(S_l = s_i \middle| R\right) \tag{3.142}$$

$$= \sum_{s_i \in \Lambda} s_i \cdot \frac{P\left(S_l = s_i\right)}{p(R)} p\left(R \middle| S_l = s_i\right),$$

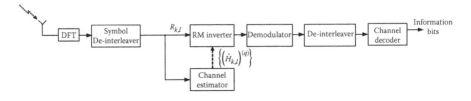

**Figure 3.23   Receiver structure for OFDM transmissions with CRM.**

with $s_i$ representing a constellation symbol from the modulation alphabet $\Lambda$, $E[\cdot]$ denoting the expected value, $P(\cdot)$ a discrete probability, $p(\cdot)$ and a probability density function (PDF). Considering equiprobable symbols, we have $P(S_l = s_i) = 1/M$, where $M$ is the constellation size. The PDF values required in (3.142) can be computed as

$$p\left(R|S_l = s_i\right) = \frac{1}{M^{M_{CRM}-1}} \sum_{S_l^{compl} \in \Lambda^{M_{CRM}-1}} p\left(R|S_l = s_i, S_l^{compl}\right), \quad (3.143)$$

with

$$p\left(R|S_l = s_i, S_l^{compl}\right) = \frac{1}{\left(2\pi\sigma^2\right)^{M_{CRM}}} \exp\left[\sum_{n=1}^{M_{CRM}} -\frac{\left|R_n - H(n,:)\cdot A_{M_{CRM}} \cdot s\right|^2}{2\sigma^2}\right], \quad (3.144)$$

where $S_l^{compl}$ is an $(M_{CRM} - 1) \times 1$ vector representing a possible combination of symbols transmitted together with $S_l$ in the same super-symbol, $s$ is an $M_{CRM} \times 1$ vector comprising $S_l^{compl}$ and $s_i$, $R_n$ is the $n$th received sample in (3.140), and $H(n,:)$ is the $n$th line of channel matrix H.

The resulting symbol estimates are serialized, demodulated, and de-interleaved before entering the channel decoder block that produces the final estimate of the information sequence.

As an example of the possible use of CRM, we present some simulation results considering an UTRA LTE-based system with a transmission bandwidth of 10 MHz. 200 subcarriers are occupied with QPSK-modulated data in each FFT block. The channel impulse response is based on a Typical Urban (TypU) environment [3GPP 2011c] with Rayleigh fading assumed for the different paths. A velocity of 50 km/h was considered. The channel encoder was a rate-1/3 turbo code based on two parallel recursive convolutional codes characterized by G(D) = [1 (1 + D2 + D3)/ (1 + D + D3)]. Puncturing is applied to the parity bits for achieving higher coding rates. At the receiver, a maximum of 12 turbo decoding iterations are applied. The results presented next will be shown as a function of $E_b/N_0$, where $E_b$ is the average information bit energy and $N_0$ is the single-sided noise power spectral density.

Figure 3.24 shows the BER performance of SISO uncoded OFDM transmission with different rotation matrices, namely, CRM, real rotation matrix (RRM), and Hadamard matrices (HMs), versus the angle for $E_b/N_0 = 20$ dB. It is clear that the performance depends on the chosen rotation angle for CRM and RRM (HM matrices have no angle dependence in their construction). To evaluate the performance gain due to the use of SSD, the results can be compared to the reference performance, which corresponds to angle 0° of RRM. According to the results of Figure 3.24, CRM matrices with dimensions 2, 4, and 8 provide increasing SSD gain with increasing matrix dimension. In fact, by increasing the dimension of CRM, it is possible to achieve a BER reduction from $2.5 \times 10^{-3}$ (RRM,0°) to 2.0

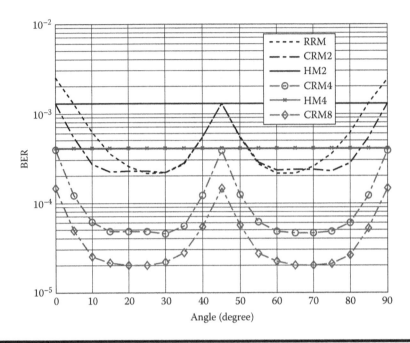

**Figure 3.24  BER performance of uncoded transmission for SISO with different rotation matrices, $E_b/N_0$ = 20 dB.**

× $10^{-5}$ (CRM8,22.5°). For CRM the preferred angle interval lies between 15° and 30°. For RRM the preferred interval is shorter and located around 30° ± 3°.

Figure 3.25 shows the BER performance of SISO OFDM transmissions employing turbo codes with coding rate ¾ and $E_b/N_0$ = 12 dB. As before, the SSD gain provided by CRM increases with the matrix size, but due to the presence of channel coding the dependence on the angle decreases and is lower compared to the uncoded case. According to the results, it is possible to reduce the BER value by one order of magnitude with the use of CRM8.

Figure 3.26 shows the BLER performance of a SISO OFDM transmission with LDPC codes of different coding rates and two different dimensions of the CRM matrix, $\varphi$ = 30°, $M_{CRM}$ = 2, and $M_{CRM}$ = 16. The SSD gain provided by CRM increases with the size of the CRM matrix but is strongly dependent on the coding rate of the LDPC code. The SSD gain is higher for higher coding rates because of the lower coding gain of these codes (less channel bit redundancy). When the redundancy of the channel code increases, the SSD gain is not so noticeable due to the higher diversity gains that channel coding can offer compared to SSD.

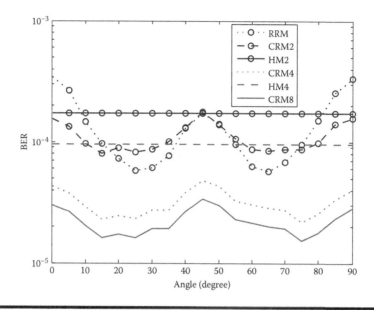

**Figure 3.25    BER performance of SISO transmission turbo coded with rate ¾ for different rotation matrices, $E_b/N_0 = 12$ dB.**

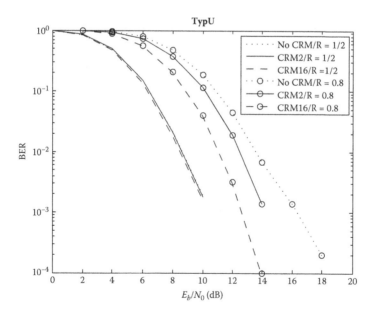

**Figure 3.26    BLER performance of a SISO LDPC transmission with CRM2, TypU channel.**

## 3.4 Channel Estimation Strategies

### 3.4.1 Issues of Imperfect Channel Estimation

In the previous sections, the performance was always obtained assuming perfect channel state information (CSI) at the receiver. However, in real systems, working with coherent detection channel estimation is always imperfect and can have a significant impact on their performances.

The simple existence of a phase error due to imperfect channel estimation causes a rotation of the constellation after channel compensation, which reduces the contribution of the correct components, adding interference from the quadrature components to the in-phase components and vice versa. Figure 3.27 shows how this cross-quadrature interference is produced ($\varphi$ is the phase error).

Although in QPSK it is only important to compensate for phase fluctuations caused by the channel, in QAM it is also required to know the amplitude fluctuations. Thus, $M$-QAM is more sensitive to the accuracy of channel estimation than QPSK.

In this section, we will analyze the impact of the channel estimation error on the performance of QPSK and $M$-QAM constellations.

In a flat Rayleigh channel, each received signal sample can be expressed as

$$r = \alpha \cdot s + n, \tag{3.145}$$

where $\alpha$ is the channel coefficient, $s$ is the transmitted QPSK or $M$-HQAM symbol, and $n$ represents additive white thermal noise. Both $\alpha$ and $n$ are modeled as complex Gaussian random variables with $E[\alpha] = 0$, $E[|\alpha|^2] = 2\sigma_\alpha^2$ (average fading power), $E[n] = 0$ and $E[|n|^2] = N_0$ ($N_0/2$ is the two-sided noise power spectrum density).

In Souto et al. [2007a] analytical BER expressions were derived for the different classes of bits $b_m$ ($m = 1,\ldots, 1/2 \cdot \log_2 M$) of $M$-QAM constellations in the presence of imperfect channel estimation. Considering a frame comprising $N_F$ symbols, these expressions can be written as

$$P_b(b_m) = \frac{1}{N_F} \frac{2}{\sqrt{M}} \sum_{t=1}^{N_F} \sum_{j=0}^{\sqrt{M}/2-1} \left[ 1 - g_j^m - (-1)^{g_j^m} \sum_{l=1}^{2^{(m-1)}} \left[ (-1)^{l+1} \frac{2}{\sqrt{M}} \right. \right.$$

$$\left. \left. \times \sum_{f=0}^{\sqrt{M}/2-1} \mathrm{Prob}\left( \mathrm{Re}\{z\} < |\alpha|^2 \, \mathbf{B}_m(l) \,|\, s_{j,f}, t \right) \right] \right], \tag{3.146}$$

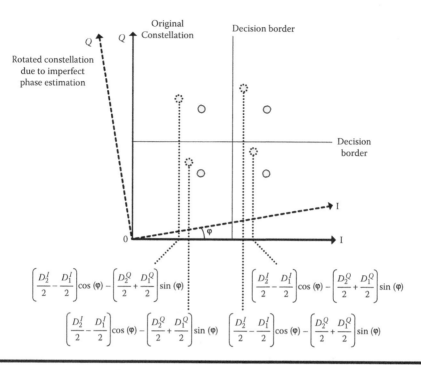

**Figure 3.27** **Impact of cross-quadrature interference originated by imperfect channel phase estimation in a 16-QAM constellation (upper-right quadrant shown).**

with

$$
\text{Prob}\left(\text{Re}\{z\} < |\hat{\alpha}|^2 \, B_m(l) \,|\, s_{j,f}, t\right) = \frac{1 - |\mu'|^2}{2\left(1 - |\mu'|^2 (\sin\varepsilon')^2 + |\mu'|\cos\varepsilon'\sqrt{1 - |\mu'|^2 (\sin\varepsilon')^2}\right)} \cdot
$$

$$(3.147)$$

In (3.147), $\mu'$ and $\varepsilon'$ are defined according to

$$
\mu' = \frac{E\left[\left(r - \alpha \cdot B_m(l)\right)\alpha^* \,|\, s\right]}{\sqrt{E\left[\left|r - \alpha \cdot B_m(l)\right|^2 \,|\, s\right] E\left[|\alpha|^2 \,|\, s\right]}} = |\mu'|e^{-\varepsilon'j},
$$

$$(3.148)$$

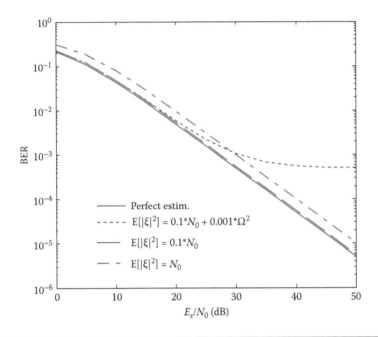

**Figure 3.28** QPSK BER performance for several channel estimation error variances.

where

$$E\left[\left|r - \hat{\alpha} \cdot B_m(l)\right|^2 | s\right] = 2|s|^2 \sigma_\alpha^2 + N_0 - 2B_m(l)\text{Re}\left\{E\left[r \cdot \hat{\alpha}^* | s\right]\right\} + B_m(l)^2 E\left[\left|\hat{\alpha}\right|^2\right]$$

(3.149)

and

$$E\left[\left(r - \hat{\alpha} \cdot B_m(l)\right)\hat{\alpha}^* | s\right] = s \cdot E\left[\alpha \cdot \hat{\alpha}^* | s\right] + E\left[n \cdot \hat{\alpha}^* | s\right] - B_m(l)E\left[\left|\hat{\alpha}\right|^2\right].$$ (3.150)

Using (3.146), we present in Figure 3.28 several BER curves for QPSK with imperfect channel estimation. Channel estimates are modeled as $\hat{\alpha} = a \cdot \alpha + \xi$, where $a$ is a complex value denoting the bias of the estimate (which will be assumed as $a = 1$) and $\xi$ is a zero-mean complex Gaussian variable representing the channel estimation error. We will consider that the channel estimation error variance is the same for all bit positions in the frame. Curves for perfect channel estimation as well as for imperfect channel estimation with $E[\Delta_{k,p} S_{k',p}] \approx 0$, $E\left[\left|\xi\right|^2\right] = N_0$, and $E\left[\left|\xi\right|^2\right] = 0.1 \cdot N_0 + 0.001 \cdot \Omega^2$ are presented. The latter case corresponds to channel estimates with irreducible errors and leads to performances with irreducible BER floors, as can be seen in the graph for high $E_s/N_0$ values. The other two cases correspond to channel estimates corrupted by thermal noise only and basically cause a shift in the curves relative to the perfect channel estimation one. Note that for low $E_S/N_0$ values it is the presence of thermal noise in the channel estimates that is the main source of performance degradation relative to the perfect channel estimation curve as can be seen by the nearly overlapping curves with $E\left[|\xi|^2\right]=0.1\cdot N_0$ and $E\left[|\xi|^2\right]=0.1\cdot N_0+0.001\cdot\Omega^2$.

Figure 3.29 shows several performance curves for a 16-HQAM constellation with $k_1=0.4$ and imperfect channel estimation. As observed for QPSK, channel estimates with irreducible errors lead to performances with irreducible BER floors which, for higher-order constellations such as 16-HQAM, is more critical due to the higher sensitivity of the least protected bits. The BER floors cease to exist if the channel estimates are only corrupted by a noise with variance proportional to $N_0$.

Figure 3.30 shows the effect of the nonuniformity parameter $k_1$ of a 16-HQAM on the $E_S/N_0$ degradation due to imperfect channel estimation, that is, the extra $E_S/N_0$ required for achieving a BER target of $10^{-3}$. The channel estimation error variance considered is $E\left[|\xi|^2\right]=0.1\cdot N_0+0.0001\cdot\Omega^2$. Note that in this legend we do not employ the terminology MPB and LPB for bit 1 and bit 2 since it might

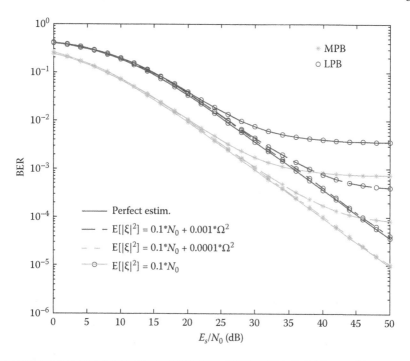

**Figure 3.29 BER performance of 16-HQAM constellation, $k_1 = 0.4$, in flat Rayleigh channel for several channel estimation error variances.**

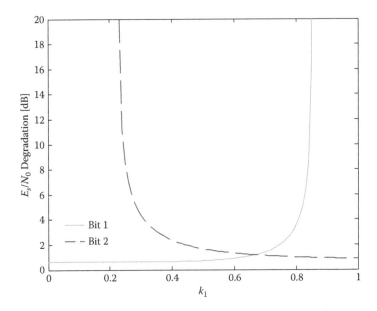

**Figure 3.30** $E_S/N_0$ degradation due to imperfect channel estimation versus nonuniformity parameter $k_1$ for a 16-HQAM. (Target BER = $10^{-3}$, $E\left[|\xi|^2\right] = 0.1 \cdot N_0 + 0.0001 \cdot \Omega^2$ ).

be misleading, because for $k_1 > 0.65$ the two bit types swap roles in terms of error protection. Nevertheless, independent of which particular bit is the most protected one, it is clear that it will have a lower sensitivity to channel estimation errors, with degradations often below 1 dB. As for the least protected bit, a highly nonuniform constellation (with $k_1 < 0.3$ or $k_1 > 0.8$) results in a very high sensitivity to channel estimation errors that can make a BER of $10^{-3}$ unattainable.

In the following two sections, we will study efficient channel estimation strategies for QPSK and M-QAM modulations using two different pilot symbols transmission methods: data-multiplexed pilots and superimposed pilots. Although we will often be referring to M-HQAM constellations, the following discussions are generic since QPSK and nonhierarchical M-QAM constellations can be treated as special cases of the former ones.

### 3.4.2 Channel Estimation Using Data-Multiplexed Pilots

In Figure 3.31 we show a possible configuration for an OFDM transmitter chain that incorporates M-HQAM constellations. In this scheme, there are $1/2 \cdot \log_2 M$ possible parallel chains for the different input bit streams. Each stream is encoded, interleaved, and mapped into the constellation symbols according to the importance attributed to it. Pilot symbols are inserted into the modulated symbols sequence,

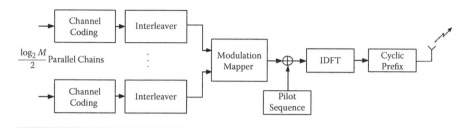

**Figure 3.31 OFDM transmitter chain incorporating hierarchical constellations.**

which is then converted to the time domain using an inverse discrete Fourier transform (IDFT). The cyclic prefix is added before transmission. Note that each individual chain is similar to a unicast transmission scheme based on QPSK. The main difference is the addition of the modulation mapper that joins the parallel streams.

One possible pilot transmission method consists of periodically inserting known pilot symbols into the data stream, both in time and in frequency. The respective frame is shown in Figure 3.32. According to this structure, in an OFDM system with $N$ carriers, pilot symbols are multiplexed with data symbols using a spacing of $\Delta N_T$ OFDM blocks in the time domain and $\Delta N_F$ subcarriers in the frequency domain. This method of pilot transmission is used in LTE-Advanced [3GPP 2011b].

Before being transmitted, the sequence of symbols is converted to the time domain through $\{x_{i,l}, i = 0,1,...,N-1\} = \text{IDFT}\{S_{k,l}, k = 0,1,...,N-1\}$, where $S_{k,l}^m$ is the symbol transmitted on the $k$th subcarrier of the $l$th OFDM block. The OFDM signal is then expressed as

$$x\,(t) = \sum_{l} \sum_{i=-N_G}^{N-1} x_{i,l} \cdot h_T\left(t - i \cdot T_s\right), \qquad (3.151)$$

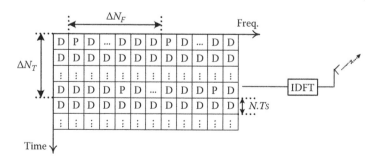

**Figure 3.32 Frame structure for an OFDM transmission with data-multiplexed pilots (P – pilot symbol, D – data symbol, $T_s$ – symbol duration).**

with $T_s$ denoting the symbol duration, $N_G$ the number of samples at the cyclic prefix (with $x_{-i,l} = x_{N-i,l}$, $i = 1,\ldots, N_G$) and $h_T(t)$ the adopted pulse shaping filter.

Due to a higher sensitivity to channel estimation errors of HQAM constellations, we can employ a receiver capable of jointly performing channel estimation and data detection through iterative processing. Figure 3.33 presents the structure of the iterative receiver for the case of data-multiplexed pilot transmission.

According to the figure, the signal, which is considered to be sampled and with the cyclic prefix removed, is converted to the frequency domain after an appropriate size-$N$ DFT operation. If the cyclic prefix is longer than the overall channel impulse response, the resulting received sequence can be expressed as

$$R_{k,l} = S_{k,l} H_{k,l} + N_{k,l}, \tag{3.152}$$

with $H_{k,l}$ denoting the overall channel frequency response for the $k$th frequency of the $l$th time block and $N_{k,l}$ denoting the corresponding channel noise sample. The sequence of received samples enter the equalizer, which computes symbol estimates as

$$\left(\tilde{S}_{k,l}\right)^{(q)} = \frac{\left(H_{k,l}\right)^{(q)*}\left(R_{k,l}\right)^{(q)}}{\left|\left(H_{k,l}\right)^{(q)}\right|^2}, \tag{3.153}$$

where $\left(\hat{H}_{k,l}\right)^{(q)}$ represents channel frequency response estimates and $q$ is the current iteration. After equalization the following processing steps are similar to the ones performed in the CDMA receiver presented previously. The estimated symbol sequence passes through the demodulator, which computes LLRs for the different bit stream, which are then de-interleaved and decoded in the channel decoder blocks. Each channel decoder outputs the respective estimated information sequence

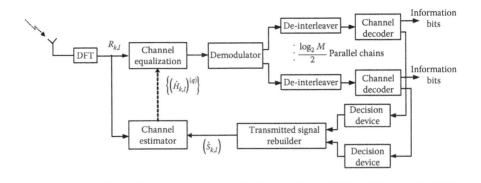

**Figure 3.33** Iterative receiver structure with enhanced channel estimation for OFDM systems employing M-HQAM modulations and data-multiplexed pilots.

along with the sequence of LLR estimates for the code symbols. These LLRs enter the decision device for producing either soft-decision or hard-decision estimates of the code symbols, which are then used by the transmitted signal rebuilder for reconstruction of the transmitted signal (in the frequency domain). The reconstructed symbol sequence can then be employed for a refinement of the channel estimates for the succeeding iteration.

Regarding the channel estimation process, the sequence of steps can be performed as follows:

1. Channel estimates for each pilot symbol position are simply computed as

$$\tilde{H}_{k,l} = \frac{\left(S_{k,l}^{Pilot}\right)^*}{\left|S_{k,l}^{Pilot}\right|^2} R_{k,l}, \tag{3.154}$$

where $S_{k,l}^{Pilot}$ corresponds to a pilot symbol transmitted in the $k$th subcarrier of the $l$th OFDM block. Obviously, not all indexes $k$ and $l$ will correspond to a pilot symbol because we have the constraint that $\Delta N_T > 1$ or $\Delta N_F > 1$.

2. Channel estimates for the same subcarrier $k$ but in time-domain positions (index $l$) that do not carry a pilot symbol can be obtained through interpolation using a finite impulse response (FIR) filter with length $W$ as follows:

$$\tilde{H}_{k,l+t} = \sum_{j=-\lfloor (W-1)/2 \rfloor}^{\lfloor W/2 \rfloor} b_t^j \tilde{H}_{k,l+j \cdot \Delta N_T}, \tag{3.155}$$

where $t$ is the OFDM block index relative to the last one carrying a pilot (whose index is $l$) and $b_t^j$ are the interpolation coefficients of the estimation filter, which depend on the channel estimation algorithm employed. There are several possible algorithms proposed in the literature such as the optimal Wiener filter interpolator [Cavers 1991] or the low pass sinc interpolator [Kim et al. 1997].

3. Interpolation in the frequency domain can be accomplished by ensuring that the corresponding impulse response in the time domain has a duration of $N_G$ samples. This is performed through the computation of the time-domain impulse response of (3.155) using $\{(\tilde{h}_{i,l})^{(q)} ; i = 0, 1, ..., N-1\} = \text{DFT}\{(\tilde{H}_{k,l})^{(q)}; k = 0, 1, ..., N-1\}$, followed by the truncation of this sequence

according to $\{(\hat{h}_{i,l})^{(q)} = w_i (\tilde{h}_{i,l})^{(q)}; i = 0, 1, ..., N-1\}$ with $w_i = 1$ if the $i$th time domain sample is inside the cyclic prefix duration and $w_i = 0$ otherwise. The final frequency response estimates are then simply computed using $\{(\hat{H}_{k,l})^{(q)}; k = 0, 1, ..., N-1\} = \text{IDFT}\{(\hat{h}_{i,l})^{(q)}; i = 0, 1, ..., N-1\}$.

4. After the first iteration, data estimates can also be used as additional pilots for channel estimation refinement. The respective channel estimates are computed as

$$\left(\hat{H}_{k,l}\right)^{(q)} = \frac{R_{k,l}^n \left(\hat{S}_{k,l}\right)^{(q-1)^*}}{\left|\left(\hat{S}_{k,l}\right)^{(q-1)}\right|^2}. \tag{3.156}$$

### 3.4.3 Channel Estimation Using Superimposed Multiplexed Pilots

The overhead due to training symbols for channel estimation leads to a decrease of system capacity. Therefore, it would be desirable to reduce the overhead for channel estimation purposes. A promising technique to overcome this problem is to use implicit training or implicit pilots, also called superimposed pilots, where the training block is added to the data block instead of being multiplexed with it [Lugo et al. 2004; Ho et al. 2001; Zhu et al. 2003; Lam et al. 2006; Meng et al. 2007; Lam et al. 2008]. This means that we can increase significantly the density of pilots (even to the extent of one pilot per data symbol), with zero pilot overhead.

Ohno and Giannakis [2002] provide a general framework for several approaches to low or zero pilot overhead in the context of OFDM. In one approach, periodic pilot sequences are added to data symbols in the time domain for single carrier systems Lugo et al. 2004; Meng et al. 2007; Tugnait and Meng 2006; Josiam and Rajan 2007], or in the frequency domain for OFDM systems [Ho et al. 2001; Zhu et al. 2003]. The power level of the added pilots is chosen so as to minimize error rate degradation due to channel estimation errors and to loss of data power. The interference from data to pilots (and therefore to channel estimates) can be mitigated by time-averaging over many pilot sequence repetitions [Lugo et al. 2004; Zhu et al. 2003; Meng et al. 2007; Tugnait and Meng 2006]. Once channel estimates are obtained in this way, pilots are subtracted from the received signal prior to equalization and data detection. Improved channel estimation and data detection performance can be obtained with iterative joint maximum likelihood or quasi-maximum likelihood data detection and channel estimation procedures [Meng et al. 2007; Josiam and Rajan 2007].

Another approach with zero pilot overhead is to add data-dependent pilot sequences to data such that interference on data is zero. Such data-dependent superimposed training (DDST) carried out in the time domain [Ghogho et al. 2005; Tugnait and He 2006] or frequency domain superimposed pilot techniques (FDSPT) carried out in the frequency domain [Lam et al. 2006, 2008] essentially replace a subset of data-carrying frequencies with pilots at those frequencies. Nulls are thus created in the apparent channel frequency response seen by the data, leading to performance degradation that must be dealt with by advanced equalization methods such as iterative equalization [Lam et al. 2006, 2008]. Moreover, those schemes lead to increased envelope fluctuations on the transmitted signals.

Assuming an OFDM transmission using the transmitter chain of Figure 3.31, it is simple to incorporate superimposed pilots. It is only necessary that the pilot insertion block adds the pilot symbols directly to the data symbols following the structure defined in Figure 3.34. According to this structure, superimposed pilots are generated using a grid with a spacing $\Delta N_T$ of symbols in the time domain and $\Delta N_F$ symbols in the frequency domain.

The transmitted sequence is given as

$$X_{k,l} = S_{k,l} + S_{k,l}^{Pilot}, \tag{3.157}$$

where $S_{k,l}^{Pilot}$ is the implicit pilot transmitted over the $k$th subcarrier, in the $l$th OFDM block, and $S_{k,l}$ is the respective data symbol. The resulting sequence is converted to the time domain through the usual process, $\{x_{i,l}, i = 0,1,...,N-1\} =$ IDFT$\{X_{k,l}, k = 0,1,...,N-1\}$, before being transmitted. OFDM signals can then also be expressed as (3.151).

The transmission of pilot symbols superimposed on data will clearly result in interference between them. To reduce the mutual interference and achieve reliable channel estimation and data detection, we can employ a receiver capable of jointly performing these tasks through iterative processing, which can be a modified version of the receiver studied for conventional pilots (Figure 3.33). The structure of the modified receiver is shown in Figure 3.35. The main differences lie in the addition of the "Remove Pilots" and "Remove Data" processing blocks.

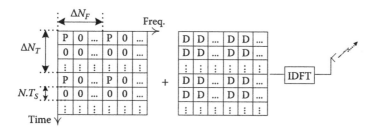

**Figure 3.34  Frame structure for an OFDM transmission with superimposed pilots (P – pilot symbol, D – data symbol, T_s – symbol duration).**

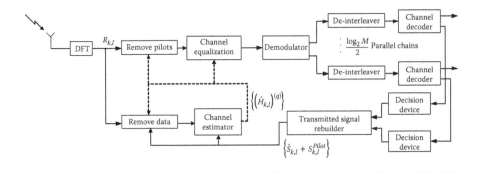

**Figure 3.35  Iterative receiver structure with enhanced channel estimation for OFDM systems employing M-HQAM modulations and superimposed pilots.**

In this case, after conversion to the frequency domain, the received sampled sequence can be expressed as

$$R_{k,l} = \left(S_{k,l} + S_{k,l}^{Pilot}\right)H_{k,l} + N_{k,l}. \tag{3.158}$$

The pilot symbols are removed from the sequence, resulting in

$$\left(Y_{k,l}\right)^{(q)} = R_{k,l} - S_{k,l}^{Pilot}\left(\hat{H}_{k,l}\right)^{(q)}, \tag{3.159}$$

where $(\hat{H}_{k,l}^{m,n})^{(q)}$ is the channel frequency response estimate and $q$ is the current iteration. The sequences of samples then follow the same processing sequence already described for the conventional pilots method.

For accomplishing reliable channel estimation with superimposed pilots, in each iteration the receiver applies the following steps:

1. Data symbols estimates are removed from the pilots. The resulting sequence becomes

$$\left(\tilde{R}_{k,l}\right)^{(q)} = R_{k,l} - \left(\hat{S}_{k,l}\right)^{(q-1)}\left(\hat{H}_{k,l}\right)^{(q-1)}, \tag{3.160}$$

where $(\hat{S}_{k,l})^{(q-1)}$ and $(\hat{H}_{k,l})^{(q-1)}$ are the data and channel response estimates of the previous iteration, respectively. This step can only be applied after the first iteration. In the first iteration, we set $(\tilde{R}_{k,l})^{(1)} = R_{k,l}$.

2. The channel frequency response estimates are computed using a moving average with size $W$ as follows:

$$\left(\tilde{H}_{k,l}\right)^{(q)} = \frac{1}{W} \sum_{l'=l-\lfloor W/2 \rfloor}^{l+\lceil W/2 \rceil - 1} \frac{\left(\tilde{R}_{k,l'}\right)^{(q-1)}}{S_{k,l'}^{Pilot}}. \tag{3.161}$$

3. After the first iteration, data estimates can also be used as additional pilots for channel estimation refinement. This is especially useful if a fully dense pilot distribution is not employed (i.e., $\Delta N_F \neq 1 \vee \Delta N_T \neq 1$). The respective channel estimates are computed as

$$\left(\tilde{H}_{k,l}\right)^{(q)} = \frac{\left(Y_{k,l}\right)^{(q-1)} \left(\hat{S}_{k,l}\right)^{(q-1)*}}{\left|\left(\hat{S}_{k,l}\right)^{(q-1)}\right|^2}. \tag{3.162}$$

These channel estimates can also be enhanced by ensuring that the corresponding impulse response has a duration $N_G$. This is accomplished by computing the time-domain impulse response of (3.161) and (3.162) through $\{(\tilde{h}_{i,l})^{(q)}; i = 0, 1, ..., N-1\} = \mathrm{DFT}\{(\tilde{H}_{k,l})^{(q)}; k = 0, 1, ..., N-1\}$, followed by the truncation of this sequence according to $\{(\hat{h}_{i,l})^{(q)} = w_i (\tilde{h}_{i,l})^{(q)}; i = 0, 1, ..., N-1\}$ with $w_i = 1$ if the $i$th time domain sample is inside the cyclic prefix duration and $w_i = 0$ otherwise. The final frequency response estimates are then simply computed using $\{(\hat{H}_{k,l})^{(q)}; k = 0, 1, ..., N-1\} = \mathrm{IDFT}\{(\hat{h}_{i,l})^{(q)}; i = 0, 1, ..., N-1\}$.

One of the advantages of using superimposed pilots is that it allows us to increase significantly the density of pilots without sacrificing system capacity. In fact, we can have a pilot for each data symbol, which can be important for fast fading channels. As an example, and using an UTRA LTE-based Monte Carlo simulator, Figures 3.36 and 3.37 show the performances of a 16-HQAM ($k_l = 0.4$) transmission for different velocities employing data-multiplexed pilots ($\Delta N_F = 4$, $\Delta N_T = 7$) and superimposed pilots ($\Delta N_F = 0$, $\Delta N_T = 0$, $E[|S_{k,l}^{m,Pilot}|^2]/E[|S_{k,l}^{m}|^2] = -16$ dB). A 10 MHz bandwidth configuration was assumed with the corresponding parameters shown in Table 3.1. The channel impulse response is based on the Vehicular A environment (from ETSI [1998a]) with Rayleigh fading adopted for the different paths.

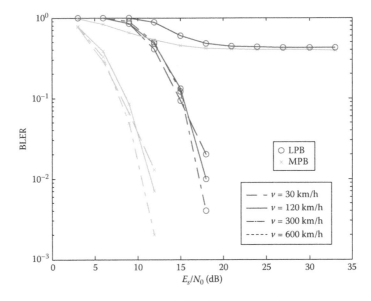

**Figure 3.36** **Performance of a 16-HQAM ($k_1 = 0.4$) transmission with data-multiplexed pilots ($\Delta N_F = 4$, $\Delta N_T = 7$) for different velocities.**

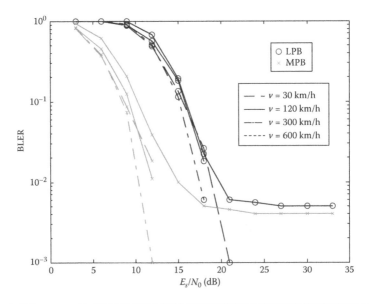

**Figure 3.37** **Performance of a 16-HQAM ($k_1 = 0.4$) transmission with superimposed pilots ($\Delta N_F = 0$, $\Delta N_T = 0$, $E[|S_{k,l}^{m,Pilot}|^2]/E[|S_{k,l}^{m}|^2] = -16$ dB) for different velocities.**

**Table 3.1  Simulation Parameters for 10 MHz Bandwidth**

| | |
|---|---|
| Transmission BW | 10 MHz |
| CP length (samples) | 72 |
| FFT size | 1024 |
| Number of occupied subcarriers | 600 |
| Subframe duration (ms) | 0.5 |
| Subcarrier spacing (kHz) | 15 |
| OFDM symbols per subframe | 7 |

All the results correspond to equivalent conditions in terms of overall percentage of transmitted energy spent on the pilots. The receivers are the iterative ones described previously (Figures 3.33 and 3.35) employing four main iterations with three inner turbo decoder iterations for each pass. In the graph legends, MPB designates most protected bits and LPB corresponds to least protected bits. Comparing both figures, we note that although both transmission techniques can attain low BLERs for velocities as high as 300 km/h, the implicit pilot transmission method shows a more robust behavior since it degrades less for very high velocities, such as 600 km/h. This is due to the fact that we are transmitting a pilot symbol in every position and, consequently, it is easier to track the channel in the time domain, whereas the data-multiplexed pilots method requires time-domain interpolation to try to estimate the channel for some positions. Note, however, that for lower velocities, superimposed pilots have a slightly worse performance than data-multiplexed pilots due to the mutual interference between data and pilots.

# Chapter 4

# Frequency-Domain Processing for Wireless Multimedia Services

This chapter focuses on frequency domain equalization (FDE) for Long Term Evolution (LTE), namely, SC-FDE (single carrier-FDE) and OFDM (orthogonal frequency division multiplexing).

Enhancement schemes such as turbo decoding, multi-resolution modulations, IB-DFE (iterative block-DFE), space time block coding, and CRMs (complex rotation matrices) will be analyzed, among others. Both SISO (single-input single-output) and MIMO (multiple input, multiple output) configurations will be used, and different channel estimation procedures will be discussed.

## 4.1 Turbo Decoding Schemes for SC-FDE with IB-DFE

The design of broadband wireless broadcast and multicast systems is a considerable challenge. In fact, broadband systems boast high bit rates, whose data is subject to severe time-dispersion effects due to to multipath propagation effects. Additionally, we need to transmit the same information to users that can have substantially different propagation conditions (e.g., users close to the transmitter and users at the edge of the cell).

Since OFDM (orthogonal frequency division multiplexing) schemes [Cimini 1985] are suitable for severely time-dispersive channels, they were the choice modulation for DAB (Digital Audio Broadcasting) [ETSI 2006], DVB (Digital Video

Broadcasting) [ETSI 2004a], and more recently for LTE and LTE Advanced [3GPP 2011b]. Since we need to send the same information to users with substantially different propagation conditions (and, inherently, different signal-to-noise ratio SNR levels), OFDM schemes are usually combined with hierarchical constellations [Cover 1972; Ramchandran et al. 1993; Souto et al. 2010]. By allowing multi-resolution schemes, different users will have different perceptions of the modulation according to their SNR conditions. Multi-resolution can be accomplished by sending a hierarchical constellation whose symbols are mapped in such a way that we have different error protection levels, which means that the distance between symbols associated to each bit may vary from bit to bit.

However, OFDM schemes have some drawbacks. One of the most important of these is the high envelope fluctuations and peak-to-average power ratio (PAPR) of the transmitted signals, which leads to amplification difficulties since we need amplifiers that are linear in a high range, which usually means high backoff and reduced amplification efficiency [Van Eetvelt et al. 1996]. To reduce the required backoff, we can employ clipping and filtering techniques [Di Zenobio et al. 1995; Li and Cimini 1998; Dinis and Gusmão 2004; Wu et al. 2008]. However, this leads not just to performance degradation, which is more serious for the least protected bits of hierarchical constellations, but also means the resulting envelope fluctuations are still larger than with conventional single-carrier modulations based on the same constellations. Another drawback of OFDM schemes is the sensitivity to frequency offset errors, which demand accurate carrier synchronization requirements (especially for larger constellations) [Li and Cimini 1998; Dinis et al. 2004].

A good alternative to OFDM is to use single-carrier frequency domain equalization, (SC-FDE) schemes as already described in Chapter 2 where a conventional, single-carrier signal is transmitted and a frequency equalization is performed at the receiver [Sari et al. 1994]. Due to the frequency-domain implementation of the equalizer, SC-FDE is as suitable for severely time-dispersive channels such as OFDM [Sari et al. 1994]. However, the transmitted signals have much lower envelope fluctuations than OFDM signals with similar constellations [Gusmão et al. 2000; Falconer et al. 2002]. Additionally, although the sensitivity to frequency errors is similar for SC-FDE and OFDM, the frequency offset in SC-FDE leads to a constellation rotation that increases along the block, which allows very simple and efficient methods to estimate and compensate residual carrier frequency errors [Dinis et al. 2010].

This section considers the transmission of broadcast and multicast services using SC-FDE with hierarchical constellations. As explained in Chapter 3, an important drawback of large constellations in general and nonuniform constellations in particular is that they are very sensitive to interference, namely, the residual ISI (intersymbol interference) at the output of a practical equalizer that does not invert completely the channel effects (e.g., a linear equalizer optimized under the minimum squared mean error (MMSE) [Proakis 1995]). To improve the performance of SC-FDE, we can replace the linear FDE by a nonlinear FDE. The most popular nonlinear equalizer is the DFE (decision feedback equalizer), whose complexity is not significantly higher than linear equalizers and allows significant performance improvements [Proakis 1995]. For this reason, several approaches have been proposed to design a DFE for SC-FDE [Benvenuto and Tomasin 2002a, 2002b]. The most promising is the IB-DFE (iterative block decision feedback equalizer), which can be regarded as a turbo equalizer implemented in the frequency domain [Benvenuto et al. 2010]. For this reason, it is desirable to employ IB-DFE schemes with SC-FDE with hierarchical constellations.

## 4.1.1 IB-DFE Receivers

In this section, we will review the basic IB-DFE structure described in Chapter 2 and apapt it to use symbol soft-decision values as feedback, and then generalize it for multi-resolution streams.

### 4.1.1.1 Basic IB-DFE Structure

Let us consider an SC-DFE scheme where the signal associated with a given block is

$$s(t) = \sum_{n=-N_G}^{N-1} s_n h_T (t - nT_S) \tag{4.1}$$

with $T_S$ denoting the symbol duration, $N_G$ denoting the number of samples at the cyclic prefix, and $h_T (t)$ denoting the adopted pulse shape. The transmitted symbols $s_n$ belong to a given alphabet $\mathfrak{S}$ (i.e., a given constellation) with dimension $M = \#\mathfrak{S}$ and are selected according to the corresponding bits $\beta_n^{(m)}$, $m = 1,2,...,\mu$ ($\mu = \log_2(M)$), that is, $s_n = f(b_n^{(1)}, b_n^{(2)},...,b_n^{(\mu)}$, with $b_n^{(m)} = 2\beta_n^{(m)} - 1$ (throughout this chapter, we assume that $\beta_n^{(m)}$ is the $m$th bit associated with the $n$th symbol and $b_n^{(m)}$ is the corresponding polar representation, that is, $\beta_n^{(m)} = 0$ or 1 and $b_n^{(m)} = -1$ or $+1$, respectively). As with other cyclic-prefix-assisted block transmission schemes, it is assumed that the time-domain block is periodic, with period $N$, that is, $s_{-n}^{(m)} = s_{N-n}^{(m)}$.

If we discard the samples associated with the cyclic prefix at the receiver, then there is no interference between blocks, provided that the length of the cyclic prefix is higher than the length of the overall channel impulse response. Moreover, the linear convolution associated with the channel is equivalent to a cyclic convolution relatively to the $N$-length, useful part of the received block, $\{y_n; n = 0,1,...,N-1\}$. This means that the corresponding frequency-domain block (i.e., the length-$N$ DFT (discrete Fourier transform) of the block $\{y_n; n = 0,1,...,N-1\}$) is $\{Y_k; k = 0,1,...,N-1\}$, where

$$Y_k = S_k H_k + N_k, \tag{4.2}$$

with $H_k$ denoting the channel frequency response for the $k$th subcarrier and $N_k$ the corresponding channel noise. Clearly, the impact of a time-dispersive channel reduces to a scaling factor for each frequency.

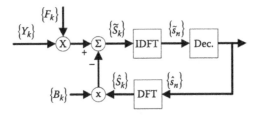

**Figure 4.1    Basic structure of an IB-DFE.**

To cope with these channel effects, we can employ a linear FDE. However, the performance can be substantially improved if the linear FDE is replaced by an IB-DFE, whose structure is depicted in Figure 4.1. For a given iteration, the output samples are given by

$$\tilde{S}_k = F_k Y_k - B_k \hat{S}_k, \tag{4.3}$$

where $\{F_k; k = 0,1,\dots,N-1\}$ and $\{B_k; k = 0,1,\dots,N-1\}$ denote the feedforward and the feedback coefficients, respectively. $\{\hat{S}_k; k = 0,1,\dots,N-1\}$, is the DFT of the block $\{\hat{s}_n; n = 0,1,\dots,N-1\}$, with $\hat{s}_n$ denoting the "hard" estimate of $\hat{s}_n$ from the previous FDE iteration.

The optimum coefficients $F_k$ and $B_k$ that maximize the overall SNR in the samples $\tilde{S}_k$ are given by [Dinis et al. 2003]*

$$F_k = \frac{\kappa H_k^*}{\alpha + (1 - \rho^2)|H_k|^2}, \tag{4.4}$$

and

$$B_k = \rho(F_k H_k - 1), \tag{4.5}$$

respectively, where

$$\alpha = E\left[|N_k|^2\right] / E\left[|S_k|^2\right] \tag{4.6}$$

and $\kappa$ is selected so as to ensure that

$$\sum_{k=0}^{N-1} F_k H_k / N = 1. \tag{4.7}$$

The correlation coefficient $\rho$, which can be regarded as the blockwise reliability of the decisions used in the feedback loop (from the previous iteration), is given by

$$\rho = \frac{E[\hat{S}_k S_k^*]}{E\left[|S_k|^2\right]} = \frac{E[\hat{s}_n s_n^*]}{E\left[|s_n|^2\right]}. \tag{4.8}$$

---

* It should be noted that, unlike [24], we are considering a normalized feedforward filter.

## 4.1.1.2 IB-DFE with Soft Decisions

The IB-DFE structure described above is usually denoted as "IB-DFE with hard decisions," although "IB-DFE with blockwise soft decisions" would probably be more adequate, as we will see in the following. In fact, (4.3) could be written as

$$\tilde{S}_k = F_k Y_k - B'_k \overline{S}_k^{Block},\qquad(4.9)$$

with $\rho B'_k = B_k$ and $\overline{S}_k^{Block}$ denoting the average of the block of overall time-domain bits associated with a given iteration, given by $\overline{S}_k^{Block} = \rho \hat{S}_k$ (as mentioned above, $\rho$ can be regarded as the blockwise reliability of the estimates $\{\hat{S}_k; n = 0, 1, \ldots, M-1\}$).

To improve the performance, we could replace the "blockwise averages" by "symbol averages," leading to what is usually denoted as "IB-DFE with soft decisions" [Benvenuto and Tomasin 2005; Gusmao et al. 2006, 2007]. A simple way of achieving this is to replace the feedback input $\{\overline{S}_k^{Block}; k = 0, 1, \ldots, N-1\}$ by $\{\overline{S}_k^{Symbol} = \overline{S}_k; k = 0, 1, \ldots, N-1\} = \text{DFT}\ \{\overline{s}_n^{Symbol}; n = 0, 1, \ldots, N-1\}$, with $\overline{s}_n^{Symbol}$ denoting the average symbol values conditioned to the FDE output of the previous iteration $\tilde{s}_n$, where $\{\overline{s}_n =; n = 0, 1, \ldots, N-1\}$ denotes the IDFT of the frequency-domain block $\{\overline{S}_k; k = 0, 1, \ldots, N-1\}$. To simplify the notation, we will use $\overline{s}_n$ (and $\overline{S}_k$) instead of $\overline{s}_n^{Symbol}$ (and $\overline{S}_k^{Symbol}$) in the remaining part of this section.

For normalized QPSK constellations (i.e., $s_n = b_n^{(I)} + j b_n^{(Q)} = \pm 1 \pm j$, with $b_n^{(I)}$ and $b_n^{(Q)}$ denoting the "in-phase" and "quadrature" bits of the $n$th symbol $s_n$, respectively) with Gray mapping it is easy to show that [Gusmao et al. 2006, 2007]

$$\overline{s}_n = \tanh\left(\frac{\lambda_n^I}{2}\right) + j\tanh\left(\frac{\lambda_n^Q}{2}\right) = \rho_n^{(I)}\hat{b}_n^{(I)} + j\rho_n^{(Q)}\hat{b}_n^{(Q)}\qquad(4.10)$$

with the LLRs (log likelihood ratios) of the "in-phase bit" and the "quadrature bit," associated with $b_n^{(I)} = Re\{s_n\}$ and $b_n^{(Q)} = Im\{s_n\}$, respectively, given by $\lambda_n^{(I)} = 2Re\{\tilde{s}_n\}/\sigma^2$ and $\lambda_n^{(Q)} = 2Im\{\tilde{s}_n\}/\sigma^2$, respectively. $\sigma^2$ is the variance of the noise at the FDE output; that is,

$$\sigma^2 = \frac{1}{2}E\left[|s_n - \tilde{s}_n|^2\right] \approx \frac{1}{2N}\sum_{n=0}^{N-1}E\left[|s_n - \tilde{s}_n|^2\right],\qquad(4.11)$$

where $\{\tilde{s}_n; n = 0, 1, \ldots, N-1\}$ denotes the IDFT of $\{\tilde{S}_k; k = 0, 1, \ldots, N-1\}$, the $\tilde{s}_n$ are the time-domain samples at the FDE output.

The hard decisions $\hat{s}_n = \hat{b}_n^{(I)} + j\hat{b}_n^{(Q)} = \pm 1 + \pm j$ are defined according to the signs of $\lambda_n^{(I)}$ and $\lambda_n^{(Q)}$, respectively, and $\rho_n^{(I)}$ and $\rho_n^{(Q)}$ can be regarded as the reliabilities associated with the in-phase and quadrature bits of the $n$th symbol, given by

$$\rho_n^{(I)} = E\left[b_n^{(I)}\hat{b}_n^{(I)}\right]/E\left[|b_n^{(I)}|^2\right] = \tanh\left(|\lambda_n^{(I)}|/2\right)\qquad(4.12)$$

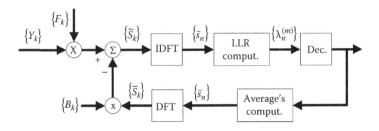

**Figure 4.2  IB-DFE with soft decisions.**

and

$$\rho_n^{(Q)} = E[b_n^{(Q)}\hat{b}_n^{(Q)}] / E[|b_n^{(Q)}|^2] = \tanh\left(|\lambda_n^{(Q)}|/2\right) \tag{4.13}$$

(for the first iteration, $\rho_n^{(I)} = \rho_n^{(Q)} = 0$ and $\bar{s}_n = 0$).

The feedforward coefficients are still obtained from (4.4), with the blockwise reliability given by

$$\rho = \frac{1}{2N}\sum_{n=0}^{N-1}(\rho_n^I + \rho_n^Q). \tag{4.14}$$

Therefore, the receiver with "blockwise reliabilities" (hard decisions) and the receiver with "symbol reliabilities" (soft decisions) employ the same feedforward coefficients; however, in the first, the feedback loop uses the hard decisions on each data block, weighted by a common reliability factor, while in the second the reliability factor changes from bit to bit. The receiver structure when we have soft decisions is depicted in Figure 4.2, which is closely related to the IB-DFE with hard decisions (Figure 4.1).

We can also define a frequency-domain turbo equalizer that employs the channel decoder outputs instead of the uncoded soft decisions in the feedback loop. The receiver structure is similar to the IB-DFE with soft decisions, but with a Soft-In, Soft-Out (SISO) channel decoder employed in the feedback loop. The SISO block, which can be implemented as defined in Vucetic and Yuan [2002], provides the LLRs of both the "information bits" and the "coded bits." The inputs of the SISO block are LLRs of the coded bits at the FDE. Once again, the feedforward coefficients are obtained from (4.4), with the blockwise reliability given by (4.12).

### 4.1.1.3 Multi-Resolution Systems

For multi-resolution systems, we have $\mu$ streams, each one associated with a different resolution and with a suitable error protection. The basic structure of the transmitter is depicted in Figure 4.3. The data stream associated with the *m*th

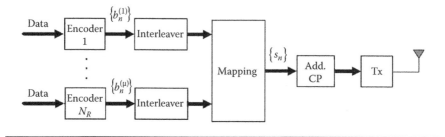

**Figure 4.3    Transmitter structure for SC-FDE with multi-resolution.**

resolution is encoded by a different channel encoder and the corresponding bits are interleaved, leading to the block $\{b_n^{(m)}, n = 0,1,\ldots,N-1\}$. The different blocks $\{b_n^{(m)}, n = 0,1,\ldots,N-1\}$, $m = 1,2,\ldots,\mu$, are mapped into the bock of time-domain symbols $\{s_n, n = 0,1,\ldots,N-1\}$, and the rest of the transmitter is similar to the transmitter for conventional constellations.[*]

The receiver can be the one depicted in Figure 4.4. Essentially, we have an IB-DFE receiver where the demapping block provides the LLR of the bits associated with each error protection and the operations of detection/decoding and computation of average bit values are performed separately for each resolution.

## *4.1.2 Analytical Characterization of Mapping Rules*

It can be shown [Montezuma and Gusmão 2001] that the constellation symbols can be expressed as the function of the corresponding bits as follows:[†]

$$s_n = g_0 + g_1 b_n^{(1)} + g_2 b_n^{(2)} + g_3 b_n^{(1)} b_n^{(2)} + g_4 b_n^{(3)} + \cdots$$

$$= \sum_{i=0}^{M-1} g_i \prod_{m=1}^{\mu} \left( b_n^{(m)} \right)^{\gamma_{m,i}}, \tag{4.15}$$

for each $s_n \in \mathfrak{S}$, where $(\gamma_{\mu,i} \ \gamma_{\mu-1,i} \ \cdots \ \gamma_{2,i} \ \gamma_{1,i}$ is the binary representation of $i$. Since we have $M$ constellation symbols in $\mathfrak{S}$ and $M$ coefficients $g_i$, (4.15) is a system of $M$ equations that can be used to obtain the coefficients $g_i$, $i = 0,1,\ldots,\mu-1$. Writing (4.15) in matrix format, we have

$$\mathbf{s} = \mathbf{W}\mathbf{g}, \tag{4.16}$$

---

[*] Without loss of generality, we are assuming that the bit rate associated with each resolution is identical. The extension to the case where we have different bit rates for different resolutions is straightforward.

[†] It should be noted that in this section $s_n$ denotes the $n$th constellation point but in the previous section $s_n$ denotes the $n$th transmitted symbol; the same applies to $b_n^{(m)}$ (or $\beta_n^{(m)}$), which here denotes the $m$th bit of the $n$th constellation point (instead of the $m$th bit of the $n$th transmitted symbol).

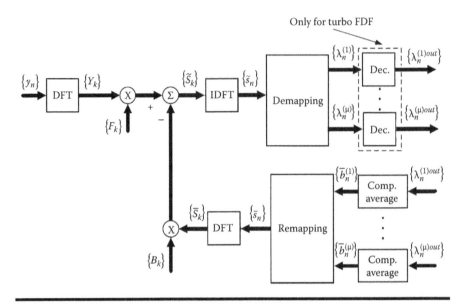

**Figure 4.4** **Iterative FDE receiver structure for SC-FDE with multi-resolution (the dashed part corresponds to the SISO decoders and is only required for a turbo FDE).**

with

$$s = [s_1 \; s_2 \; ... \; s_M]^T, \tag{4.17}$$

$$g = [g_0 \; g_1 \; ... \; g_{\mu-1}]^T \tag{4.18}$$

and $W$ is a Hadamard matrix with dimensions $M \times M$. Clearly, the vector of constellation points s is the Hadamard transform of the vector of coefficients g. Therefore, for a given constellation, we can obtain the corresponding coefficients $g_i$ from the inverse Hadamard transform of the vector of constellation points.

### 4.1.2.1 Special Modulation Mapping Cases

We will now present some special cases where the mapping is particularly simple, not requiring the computation of Hadamard transforms.

#### 4.1.2.1.1 PAM Constellations

For a uniform pulse amplitude modulation (M-PAM) constellation, we have $\mathfrak{S} = \{\pm 1, \pm 3, ..., \pm(M-1)\}$. If we have a natural binary mapping, the only nonzero coefficients are $g_1, g_2, g_4, ..., g_{M/2}$ (i.e., the coefficients $g_{2^i}, i = 0, 1, .., \mu - 1$). Moreover, $g_{2^i} = 2^{\mu-i}$, which means that

$$
\begin{aligned}
s_n &= g_1 b_n^{(1)} + g_2 b_n^{(2)} + g_4 b_n^{(3)} + \cdots \\
&= b_n^{(1)} + 2 b_n^{(2)} + 4 b_n^{(4)} + \cdots \\
&= \sum_{m=1}^{\mu} 2^{m-1} b_n^{(m)}
\end{aligned}
\tag{4.19}
$$

For a Gray mapping, the only nonzero coefficients are the ones with binary representations $(0...001), (0...011), ..., (111...1)$, (i.e., the coefficients $g_{2^i-1}, i = 0, 1, .., \mu - 1$). Moreover, $g_{2^i-1} = 2^{\mu-i-1}$, which means that

$$
\begin{aligned}
s_n &= 2^{\mu-1} b_n^{(1)} + 2^{\mu-2} b_n^{(1)} b_n^{(2)} + 2^{\mu-3} b_n^{(1)} b_n^{(2)} b_n^{(3)} + \cdots \\
&= \sum_{m=1}^{\mu} 2^{\mu-m} \prod_{m'=1}^{m} b_n^{(m')}.
\end{aligned}
\tag{4.20}
$$

For uniform 4-PAM constellations, we have

$$
s_n = 2 b_n^{(1)} + b_n^{(2)},
\tag{4.21}
$$

for a natural binary mapping and

$$
s_n = 2 b_n^{(1)} + b_n^{(1)} b_n^{(2)},
\tag{4.22}
$$

for a Gray mapping. For uniform 8-PAM constellations, we have

$$
s_n = 4 b_n^{(1)} + 2 b_n^{(2)} + b_n^{(3}
\tag{4.23}
$$

for a natural binary mapping and

$$
s_n = 4 b_n^{(1)} + 2 b_n^{(1)} b_n^{(2)} + b_n^{(1)} b_n^{(2)} b_n^{(3)}
\tag{4.24}
$$

for a Gray mapping.

The same approach can be employed for nonuniform hierarchical constellations such as the ones adopted in multi-resolution schemes [Jiang and Wilford 2005]. In fact, from (4.19) and (4.20) an M-PAM constellation with either natural binary mapping or Gray mapping can be regarded as the sum $\mu$ of binary sub-constellations, each one with twice the amplitude of the previous one. By reducing the amplitude of successive sub-constellations, we obtain hierarchical constellations with different error protections.

### 4.1.2.1.2  QAM Constellations

Usually, an quadrature amplitude modulation (M-QAM) constellation can be written as the sum of two PAMs each with dimension $\sqrt{M}$, one for the in-phase (real) component and the other for the quadrature (imaginary) component. Therefore, the corresponding mapping is straightforward: half the bits are used to define the in-phase component (as in the previous case, for Gray mapping or natural binary mapping) and the other half is used to define the quadrature component. For instance, for 16-QAM with Gray mapping, we have

$$s_n = 2b_n^{(1)} + b_n^{(1)}b_n^{(2)} + 2jb_n^{(3)} + jb_n^{(3)}b_n^{(4)} \tag{4.25}$$

and for 64-QAM with Gray mapping, we have

$$s_n = 4b_n^{(1)} + 2b_n^{(1)}b_n^{(2)} + b_n^{(1)}b_n^{(2)}b_n^{(3)} + 4jb_n^{(4)} + 2jb_n^{(4)}b_n^{(5)} + jb_n^{(4)}b_n^{(5)}b_n^{(6)} \tag{4.26}$$

The extension to other mapping rules and/or nonuniform QAM constellations is straightforward.

### 4.1.2.1.3  M-PSK Constellations

Clearly, the characterization of a bi-phase shift keying (BPSK) constellation is trivial. QPSK constellations with Gray mapping are also easily characterized since they are a special case within the QAM class. However, to characterize analytically a given M-PSK constellation is in general complex,[*] and we need to directly employ (4.15).

In the following, we present the analytical characterization of 8-PSK constellations (it is better to employ the Hadamard transform as described above when we want to characterize larger PSK constellations). Let us assume that the constellation is defined by two points, $s_1$ and $s_2$, as well as their reflections in the real and imaginary axis; that is, $\mathfrak{S} = \{s_1, s_2, s_2^*, s_1^*, -s_1, -s_2, -s_2^*, -s_1^*\}$, corresponding to the "tri-bits" 011, 111, 110, 000, 100, 101, 001, respectively (Gray mapping). Clearly, for a regular 8-PSK constellation $s_1 = \exp(j3\pi/8)$ and $s_2 = \exp(j\pi/8)$. By using other values of $s_1$ and $s_2$, we can define 8-PSK constellations with two or three different error protections; we can also define some 8-APSK (amplitude phase shift keying) constellations (e.g., for $s_1 = \exp(j\pi/4)$ and $s_1 = 2\exp(j\pi/4)$).

---

[*] Naturally, we could define a PSK constellation as a complex exponential of a suitable PAM constellation, but this does not help us in the receiver design since the constellation symbols are not a linear function of the corresponding bits.

If we define $\bar{s} = (s_1 + s_2)/2 = \bar{s}^R + j\bar{s}^I$ ($\bar{s}^R = Re\{\bar{s}\}$ and $\bar{s}^I = Im\{\bar{s}\}$) and $s_\Delta = s_1 - \bar{s} = s_\Delta^R + js_\Delta^I$ ($s_\Delta^R = Re\{s_\Delta\}$ and $s_\Delta^I = Im\{s_\Delta\}$), then the constellation point associated to the bits of $b_n^{(1)}$, $b_n^{(2)}$, and $b_n^{(3)}$ is

$$s_n = \bar{s}^R b_n^{(1)} + j\bar{s}^R b_n^{(2)} + s_\Delta^R b_n^{(1)} b_n^{(3)} + js_\Delta^I b_n^{(2)} b_n^{(3)}. \tag{4.27}$$

This means that $g_1 = \bar{s}^R$, $g_2 = j\bar{s}^R b_n^{(2)}$, $g_5 = s_\Delta^R$ and $g_6 = js_\Delta^I$, with the remaining equal to zero.

### 4.1.3 Computation of Receiver Parameters

An IB-DFE receiver with soft decisions (as described in Section 4.1) has the following constellation-dependent tasks (see Figure 4.2):

■ De-mapping the time-domain samples at the output of the FDE, $\tilde{s}_n$, into the corresponding bits. This can be implemented by computing the log likelihood ratios associated with each bit of each transmitted symbol.
■ Computation of the average symbol values conditioned to the FDE output of the previous iteration $\tilde{s}_n$, denoted $\bar{s}_n$.
■ Computation of the blockwise reliability $\rho$, required for obtaining the feedforward coefficients (see (4.4)).

The log likelihood ratio of the $m$th bit for the $n$th transmitted symbol is given by

$$\lambda_n^{(m)} = \log\left(\frac{Pr(\beta_n^{(m)} = 1|\tilde{s}_n)}{Pr(\beta_n^{(m)} = 0|\tilde{s}_n)}\right)$$

$$= \log\left(\frac{\displaystyle\sum_{s \in \Psi_1^{(m)}} \exp\left(-\frac{|\tilde{s}_n - s|^2}{2\sigma^2}\right)}{\displaystyle\sum_{s \in \Psi_0^{(m)}} \exp\left(-\frac{|\tilde{s}_n - s|^2}{2\sigma^2}\right)}\right), \tag{4.28}$$

where $\Psi_1^{(m)}$ and $\Psi_0^{(m)}$ are the subsets of $\mathfrak{S}$ where $\beta_n^{(m)} = 1$ or 0, respectively (clearly, $\Psi_1^{(m)} \bigcup \Psi_0^{(m)} = \mathfrak{S}$ and $\Psi_1^{(m)} \bigcap \Psi_0^{(m)} = \varnothing$). As an example, Figure 4.5 shows the regions associated with $\Psi_0^{(m)}$ and $\Psi_1^{(m)}$ ($m = 1, 2, 3$) for a uniform 8-PAM constellation with Gray mapping.

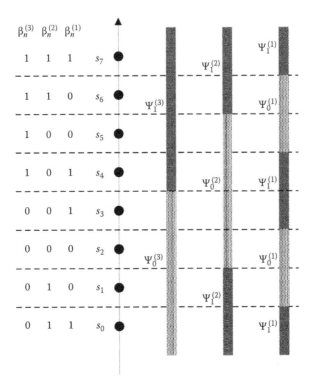

**Figure 4.5** Regions associated with $\Psi_0^{(m)}$ and $\Psi_1^{(m)}$ ($m = 1, 2, 3$) for a uniform 8 PAM constellation with Gray mapping.

To obtain the average symbol values conditioned to the FDE output, $\bar{s}_n$, we need to compute the average bit values conditioned to the FDE output, $\bar{b}_n^{(m)}$. These are related to the corresponding log likelihood ratio as follows:

$$\bar{b}_n^{(m)} = \tanh\left(\frac{\lambda_n^{(m)}}{2}\right). \tag{4.29}$$

By taking advantage of the analytical characterization of the mapping rules (or, equivalently, the specific formulas of Section 3.2) and assuming uncorrelated bits (e.g., thanks to of suitable interleaving), we have

$$\bar{s}_n = \sum_{i=0}^{M-1} g_i \prod_{m=1}^{\mu} \left(\bar{b}_n^{(m)}\right)^{\gamma_{m,i}}. \tag{4.30}$$

Finally, the reliability of the estimates to be used in the feedback loop is given by

$$\rho = \frac{E[\hat{s}_n s_n^*]}{E\left[|s_n|^2\right]} = \frac{\sum\limits_{i=0}^{M-1}|g_i|^2 \prod\limits_{m=1}^{\mu}\left(\rho_n^{(m)}\right)^{\gamma_{m,i}}}{\sum\limits_{i=0}^{M-1}|g_i|^2}, \tag{4.31}$$

where $\rho_n^{(m)}$ is the reliability of the $m$th bit of the $n$th transmitted symbol, given by

$$\rho_n^{(m)} = \left|\overline{b}_n^{(m)}\right|. \tag{4.32}$$

As an example, let us consider a uniform 4-PAM constellation with Gray mapping (i.e., the symbols are characterized by (4.21)). Figure 4.6 shows the LLR values of the different bits, $\lambda_n^{(m)}$, as a function of the output of the FDE, $\tilde{s}_n$, for different SNR values, and Figure 4.7 shows the average value of each bit conditioned to the FDE output, $\overline{b}_n^{(m)}$, under the same conditions. The regions where each bit is 0 or 1 are clear when we have high SNR, but for SNR these regions are not so evident.

Figure 4.8 shows the average symbol value conditioned to the FDE output, $\overline{s}_n$, in the conditions of Figures 4.6 and 4.7. Once again, the four levels are clear for high

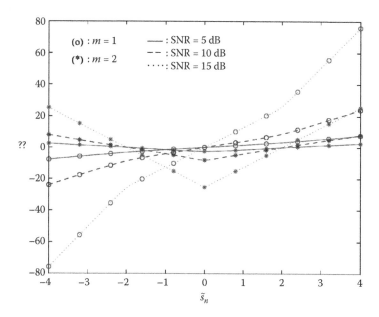

**Figure 4.6** Evolution of the LLR of the different bits, $\lambda_n^{(m)}$, for an uniform 4-PAM constellation with Gray mapping.

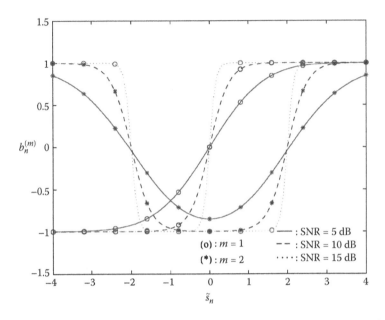

**Figure 4.7** Evolution of the average value of the different different bits conditioned to the FDE output, $\bar{b}_n^{(m)}$, for a uniform 4-PAM constellation with Gray mapping.

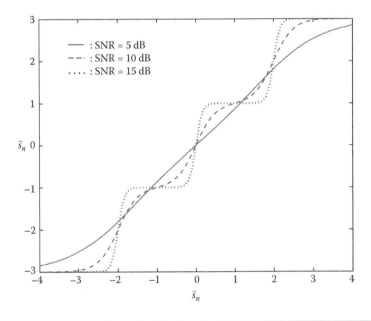

**Figure 4.8** Evolution of $\bar{s}_n$ for a uniform 4-PAM constellation with Gray mapping.

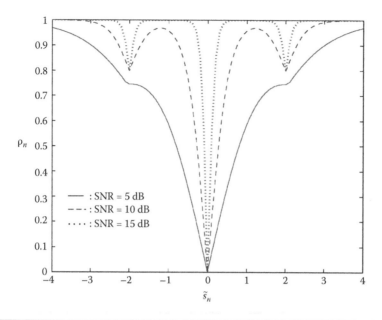

**Figure 4.9** **Evolution of $\rho_n$ for a uniform 4-PAM constellation with Gray mapping.**

SNR, and the transition between levels becomes smoother as we reduce the SNR. The corresponding symbol reliability $\rho_n$ is depicted in Figure 4.9. As expected, the reliability is lower between levels, becoming 0 for $\tilde{s}_n = 0$; for $\tilde{s}_n \approx 0$ or $\pm 3$ the reliability is close to 1, unless the SNR is very small.

## 4.1.4 Performance Results

In this section, we present a set of performance results concerning IB-DFE receivers with soft decisions for generalized constellations. The blocks have $N = 256$ symbols, plus an appropriate cyclic prefix, and we consider a severely time-dispersive channel with perfect synchronization and channel estimation at the receiver.

Let us first consider a uniform 64-QAM constellation with Gray mapping based on two separate 8-PAM constellations characterized by $g_7/g_3 = g_3/g_1 = 0.5$.

Figure 4.10 shows the BER performance for the described IB-DFE receivers. Clearly, the performance improves significantly with the iterations: when compared with a conventional linear FDE, we have about 7 dB gain for BER $=10^{-4}$ after four iterations.

Let us consider now a nonuniform 64-QAM constellation based on two 8-PAM constellations characterized by $g_7/g_3 = g_3/g_1 = 0.4$ (Gray mapping). These constellations allow bits with three different error protections, denoted by least protected bits (LPB), intermediate protected bits (IPB), and most protected bits (MPB), respectively.

Figure 4.11 shows the uncoded BER performance for the different type of bits when we have a conventional IB-DFE receiver. These results are expressed as a

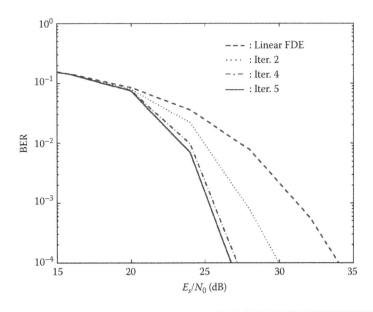

**Figure 4.10    BER for uniform 64-QAM with Gray mapping.**

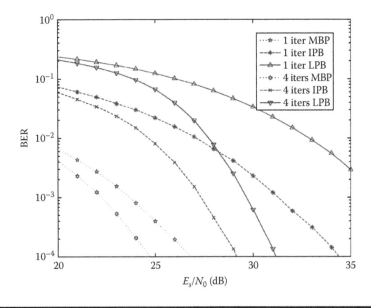

**Figure 4.11    Uncoded BER for the different bits of a nonuniform 64-QAM modulation with Gray mapping, with different IB-DFE iterations at the receiver.**

function of $E_s/N_0$, with $E_s$ denoting the average symbol energy and $N_0$ the noise power spectral density. Clearly, the performance improves significantly with the number of iterations, outperforming significantly the linear FDE. This improvement is higher for LPB (more than 10 dB for BER $= 10^{-4}$), since they suffer more from the residual ISI that is inherent to a linear FDE optimized under the MMSE criterion.

Let us consider now the impact of channel coding for both conventional IB-DFE and a turbo FDE (i.e., an IB-DFE where the channel decoder is involved in the feedback loop). We consider a rate-1/2 turbo code [Berrou et al. 1993] based on two identical recursive convolutional codes with two constituent codes characterized by $G(D) = [1(1 + D^2 + D^3) / (1 + D + D^3)]$ and interleaving depth corresponding to a single FFT block. The coded setting presents three types of simulations; one using 1 IB-DFE iteration followed by 12 iterations of the turbo code (IB-DFE+TC 1 iter), another using 4 IB-DFE iterations followed by 12 iterations of the turbo code (IB-DFE + TC 4 iters), and a final one using a joint decoding method consisting of 4 iterations of a composite decoding operation (joint IB-DFE-TC 4 iters). This composite decoding operation consists of 1 IB-DFE iteration and 3 turbo code iterations. This means that the total number of turbo code iterations is the same as the other setups ($4 \times 3 = 12$), with a total of 4 IB-DFE iterations.

Figure 4.12 shows the BER performance for the different types of bits. Analyzing the results for this setup, interesting conclusions can be drawn. Starting off by comparing solely the IB-DFE + TC with one and four iterations, it is seen that, as expected, the "IB-DFE + TC 4 iters" outperforms the "IB-DFE + TC 1 iter" for the LPB and IPB. However, for the MPB, the situation is inverted; although marginally,

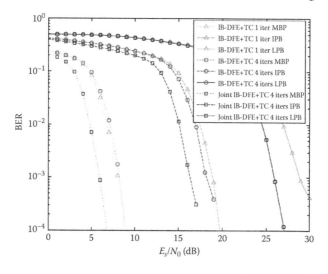

**Figure 4.12 Coded BER performance for the different bits of a nonuniform 64-QAM modulation with Gray mapping, with different error protections and decoding schemes.**

the performance of the "IB-DFE+TC 1 iter" is better. This is due to the uncoded performance of the "IB-DFE 1 iter" being also marginally better than "IB-DFE 4 iters" for very low $E_s/N_0$ values (its these low $E_s/N_0$ values that the turbo code takes as input before decoding, with an offset of −3 dB due to the coding rate of 1/2). The joint "IB-DFE+TC 4 iters" comes to yield the best performance with similar complexity to the other decoding schemes (the main algorithm in terms of complexity is the turbo decoding maximum a posteriori procedure). The greatest gains are obtained for the MPBs, decreasing progressively until the LPB. This is due to the great performance of the uncoded performance of the IB-DFE 1 iter at very low $E_s/N_0$; the first iteration of the joint "IB-DFE-TC 4 iters" is largely influenced by this, since only one IB-DFE iteration is performed at first, followed by three turbo-decoding operations. The second iteration has as input the turbo decoder log likelihood ratio (LLR) results, with a much lower error level than at the first iteration, and thus a gain of dB can be achieved. The case for the LPB is different; the performance is the same as the "IB-DFE + TC 4 iters." In fact, for higher $E_s/N_0$ values, the uncoded 4 IB-DFE iterations are more beneficial than solely one IB-DFE iteration, and thus the joint "IB-DFE+TC 4 iters" starts off poorly, but regains performance equivalence with the "IB-DFE+TC 4 iters." Notice that although the joint algorithm is recursive, the complexity of both algorithms is exactly the same. Overall, it can be concluded that the joint algorithm yields the best performance overall.

## 4.2 Channel Capacity for Multi-Antenna Systems

The capacity limit of any (SISO) telecommunications system is taken to be the resulting throughput obtained through the full usage of the allowed spectrum. However, if multiple transmit and receive antennas are employed, the capacity may be raised due to code reuse across transmit antennas. If there is a sufficient number of receive antennas, it is possible to resolve all messages, as long as the channel correlation between antennas is not too high. The pioneering work of Telatar [1995] and Foschini [Foschini and Gans 1998] established the mathematical roots from the information theory field that with multipath propagation, multiple antennas at both the transmitter and the receiver can establish essentially multiple parallel channels that operate simultaneously on the same frequency band and at the same time.

Claude Shannon [1948] derived the following capacity formula in 1948 for an AWGN channel:

$$C = W \log_2\left(1 + \frac{S}{N}\right) \text{[bits/s]}, \qquad (4.33)$$

where $W$ is the bandwidth of the channel (in Hz), $S$ is the signal power, and $N$ is the total noise power of the channel (both in watts). In a more practical case of a time variable and randomly fading wireless channel, the capacity is written as

$$C = W \log_2\left[1 + \frac{S}{N} \cdot |H^2|\right] \text{[bits/s]}, \qquad (4.34)$$

where $H^2$ is the normalized channel power transfer function (for the SISO case $H$ is a $1 \times 1$ unit power complex Gaussian amplitude of the channel).

The channel coding theorem (CCT) states two properties:

■ The direct part states that for rate $R < C$ there exists a coding system with arbitrarily low block and bit error rates as the code length increases $N \rightarrow \infty$.
■ The converse part states that for $R \geq C$ the bit and block error rates are strictly higher than zero for any coding system.

The CCT therefore establishes rigid limits on the maximal supportable transmission rate of an AWGN channel in terms of power and bandwidth.

The bandwidth/spectral efficiency (in bits/s/Hz) characterizes how efficiently a system uses its allotted bandwidth and is defined as

$$\eta = \frac{\text{Transmission Rate}}{\text{Channel bandwidth}} \tag{4.35}$$

From it, the Shannon limit is calculated as

$$\eta_{MAX} = \log_2\left(1 + \frac{S}{N}\right). \tag{4.36}$$

The average signal power $S$ can be expressed as

$$S = \frac{k \cdot E_b}{T} = R \cdot E_b, \tag{4.37}$$

where $E_b$ is the energy per bit, $k$ is the number of bits transmitted per symbol, $T$ is the duration of a symbol, and $R = k/T$ is the transmission rate of the system in bits/s. The total noise power is given by

$$N = N_0 W, \tag{4.38}$$

where $N_0$ is the one-sided noise power spectral density. The Shannon limit is obtained in terms of the bit energy and noise power spectral density, given by

$$\eta_{MAX} = \log_2\left(1 + \frac{R E_B}{N_0 W}\right). \tag{4.39}$$

This can be resolved to obtain the minimum bit energy required for reliable transmission, called the Shannon bound:

$$\frac{E_b}{N_o} \geq \frac{2^{\eta_{MAX}} - 1}{\eta_{MAX}}. \tag{4.40}$$

The fundamental limit states that, for infinite amounts of bandwidth, that is, $\eta_{MAX} \to 0$, we obtain

$$\frac{E_b}{N_o} \geq \lim_{\eta_{MAX} \to 0} \frac{2^{\eta_{MAX}} - 1}{\eta_{MAX}} = \ln(2) = -1.59\text{dB}. \tag{4.41}$$

This is the absolute minimum signal energy to noise power spectral density ratio required to reliably transmit one bit of information.

As more $N_{RX}$ receive antennas are deployed, the statistics of capacity improve (single input, multiple output (SIMO) case) to

$$C = W \log_2 \left( 1 + \frac{S}{N} \sum_{i=rx}^{N_{RX}} |h_{rx}|^2 \right), \tag{4.42}$$

where $h_{rx}$ is the gain for receive antenna *rx*. Note the crucial feature of (4.42) in that increasing the value of $N_{RX}$ only results in a logarithmic increase in average capacity. Similarly, if transmit diversity is opted for, in the common case where the transmitter does not have channel knowledge, we have a multiple-input single-output (MISO) system with $N_{TX}$ antennas, and the capacity is given by [Foschini and Gans 1998]

$$C = W \log_2 \left( 1 + \frac{S}{N \cdot N_{TX}} \sum_{rx=1}^{N_{TX}} |h_{rx}|^2 \right), \tag{4.43}$$

where the normalization by $N_{TX}$ ensures a fixed total transmitter power and shows the absence of array gain in that case (compared to the case in (4.42), where the channel energy can be combined coherently). Again, note that capacity has a logarithmic relationship with $N_{TX}$. Now, considering the use of diversity at both transmitter and receiver, a MIMO system. For $N_{TX}$ and $N_{RX}$ antennas, the general capacity equation [Foschini and Gans 1998; Telatar 1995] is obtained as

$$C_{EP} = W \log_2 \left( \det \left( I_{N_{RX}} + \frac{S}{N \cdot N_{TX}} HH' \right) \right), \tag{4.44}$$

where $H$ is the channel matrix and $H'$ is the transpose-conjugate of $H$. Note that both (4.43) and (4.44) are based on equal power (EP) uncorrelated sources; hence, the subscript in (4.44). Foschini [Foschini and Gans 1998] and Telatar [Telatar 1995] both demonstrated that the capacity in (4.44) grows linearly with $m = \min (N_{TX}, N_{RX})$, rather than logarithmically (as in (4.43)). This result can be intuited as follows: the determinant operator yields a product of $\min (N_{TX}, N_{RX})$ nonzero eigenvalues of its (channel-dependent) matrix argument, each eigenvalue characterizing the SNR over a so-called channel eigenmode. An eigenmode corresponds to the transmission using a pair of right and left singular vectors of the channel matrix

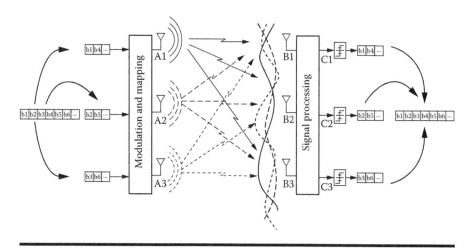

**Figure 4.13**  **Spatial multiplexing with three transmit and three receive antennas.**

as transmit and receive antenna weights, respectively. Thanks to the properties of the log, the overall capacity is the sum of capacities of each of these modes, hence the effect of capacity multiplication. Note that the linear growth predicted by the theory coincides with the transmission example of Figure 4.13.

Clearly, this growth is dependent on properties of the eigenvalues. If they decayed away rapidly, then linear growth would not occur. However (for simple channels), the eigenvalues have a known limiting distribution [Silverstein 1995] and tend to be spaced out along the range of this distribution. Hence, it is unlikely that most eigenvalues are very small, and the linear growth is indeed achieved.

The fundamental consideration is that it is possible to augment the capacity/throughput, depending on the number of transmit and receive antennas. The downside of this scheme is the receiver complexity, sensitivity to interference, and correlation between antennas, which is more significant as the antennas are closer together.

With the capacity defined by (4.44) as a random variable, the issue arises as to how best to characterize it. Two simple methods are commonly used: the mean (or ergodic) capacity [Telatar 1995; Rapajic and Popescu 2000] and capacity outage [Foschini and Gans 1998; Lozano et al. 2001; Driessen and Foschini 1999; Shiu 1999]. Capacity outage measures (usually based on simulation) are often denoted $C_{0.1}$ or $C_{0.01}$, that is, those capacity values supported 90% or 99% of the time and that indicate the system reliability. A full description of the capacity would require the probability density function or equivalent. Some results are available in Sengupta and Mitra [2000].

Some care is necessary in interpreting the previous equations. Capacity, as discussed here and in most MIMO work (Foschini and Gans [1998] and Telatar [1995]), is based on a "quasi-static" analysis where the channel varies randomly from burst to burst. Within a burst, the channel is assumed fixed, and it is also assumed that sufficient bits are transmitted for the standard infinite time horizon of information theory to be meaningful. A second note is that this discussion will concentrate on single-user MIMO systems, but many results also apply to multi-user systems with receive diversity. Finally, the linear capacity growth is only valid under certain channel conditions. It was originally derived for the independent and identically distributed (i.i.d.) flat Rayleigh fading channel and does not hold true for all cases. For example, if large numbers of antennas are packed into small volumes, then the gains in $H$ may become highly correlated and the linear relationship will plateau out due to the effects of antenna correlation [Chiurtu et al. 2001; Gesbert et al. 2002; Wei et al. 2002]. In contrast, other propagation effects not captured in (4.44) may serve to reinforce the capacity gains of MIMO such as multipath delay spread. This was shown in particular in the case when the transmit channel is known [Raleigh and Cioffi 1998] but also in the case when it is unknown [Bolcskei et al. 2002].

More generally, the effect of the channel model is critical. Environments can easily be chosen that give channels where the MIMO capacities do not increase linearly with the numbers of antennas. However, most measurements and models available to date do give rise to channel capacities that are of the same order of magnitude as the promised theory. Also, the linear growth is usually a reasonable model for moderate numbers of antennas that are not extremely close-packed.

Note that, unlike in CDMA, where users' signatures are quasi-orthogonal by design, the separation of the MIMO channel relies on the presence of rich multipath, which is needed to make the channel spatially selective. Therefore, MIMO can be said to effectively exploit multipath; this also implies that MIMO systems might not be a good radio solution for every scenario, namely, for line-of-sight-directed radio links.

## 4.3 Antenna Correlation

The antenna correlation effect is a serious aspect to be taken into account by the system designer. In fact, the spacing of the antennas at both the transmitter and receiver is essential to guarantee a low correlation value between antennas. A low correlation value implies that the channel associated with each antenna is independent of the other antennas, thus providing a greater degree of diversity.

Good antenna spacing is easy to obtain at the base station, but it is much harder to achieve a sufficient spacing at the UE. This section analyzes the potential performance decrease associated with antenna correlation.

Antenna correlation deals with the correlation between channels established between adjacent pairs of transmitting/receiving antennas, according to various parameters such as antenna spacing, angle spread, and angle of arrival. The correlation between antennas is commonly defined as the envelope correlation coefficient between signals received at two antenna elements. The received baseband signals are modeled as two complex random processes $X$ and $Y$ with an envelope correlation coefficient of

$$\rho_{env} = \left| \frac{E\left\{(X - E(X))(Y - E(Y))^*\right\}}{\sqrt{E\left\{|X - E(X)|^2\right\} E\left\{|Y - E(Y)|^2\right\}}} \right| \quad (4.45)$$

(note that this is not equal to the correlation coefficient of the envelopes (magnitude) of two signals, a measure commonly used in cases where no complex data is available).

Since the correlation coefficient is independent of the mean, only the random parts of the channel are of interest. In order to simplify notation, only the random channel components will be considered in the remainder of this section. Considering the general case of frequency selective propagation, the channel is modeled as a tapped-delay line:

$$g(t, \tau) = \sum_{l=1}^{L} g_l(t) \delta(\tau - \tau_l). \quad (4.46)$$

where $L$ is the number of taps, $g_l(t)$ are the time-varying tap coefficients, and $\tau_l$ are the tap delays. Considering two channels with channel impulse responses $g_1(t,\tau)$ and $g_2(t,\tau)$ and assuming that equivalent taps in both channels have equal power ($\sigma_{1,l}^2 = \sigma_{2,l}^2 = \sigma_l^2$), and that taps with different delays are uncorrelated within a channel as well as between channels,

$$E\left\{g_{ik}(t)g_{jl}^*(t)\right\} = 0, \ \forall k \neq l, \quad (4.47)$$

the antenna correlation coefficient becomes

$$\rho_{env} = \left| \frac{\rho_1\sigma_1^2 + \rho_2\sigma_2^2 + \cdots + \rho_L\sigma_L^2}{\sigma_1^2 + \sigma_2^2 + \cdots + \sigma_L^2} \right| \quad (4.48)$$

where $\rho_l$ are the correlation coefficients between each pair of taps $g_{1l}(t)$ and $g_{2l}(t)$, that is,

$$\rho_l = \frac{E\left\{g_{1l}(t)g_{2l}^*(t)\right\}}{\sigma_{1l}^2\sigma_{2l}^2}. \quad (4.49)$$

From (4.48), it can be seen that the antenna correlation can be related to the individual tap correlations, where $\sigma_l^2$ are the individual tap gains. To obtain a simple solution for setting these tap correlations depending on the required antenna correlation, the condition can be imposed that all tap correlations should be equal. Therefore, (4.48) simply states that all tap correlations have to be set to the antenna correlation.

To generate a sequence of random state vectors with specified first-order statistics (mean vector μ and correlation matrix R), the following transformation can be used:

$$\tilde{v} = \sqrt{R} \cdot v + \mu, \qquad (4.50)$$

where $v$ is a vector of independent sequences of circularly symmetric complex Gaussian-distributed random numbers with zero mean and unit variance. The correlation matrix $R$ is defined and factored as

$$R = E\left\{\tilde{v}\tilde{v}^H\right\} = \begin{bmatrix} 1 & \rho_{12} & .. \\ \rho_{12}^* & 1 & .. \\ .. & .. & . \end{bmatrix} = R^{1/2} R^{H/2} = R^{1/2} R^{1/2}, \qquad (4.51)$$

with $\rho_{12}$ being the correlation between channel 1 and channel 2. In Shiu et al. [2000], the correlation between different MIMO channel elements is modeled under the assumption that the correlation among receive antennas is independent of the correlation between transmit antennas (and vice versa). The underlying justification for this approach is to assume that only immediate surroundings of the antenna array impose the correlation between array elements and have no impact on correlations observed between the elements of the array at the other end of the link, which is a reasonable assumption for indoor [Yu et al. 2001], city center environments (such as Pedestrian A), and Vehicular A. A way to include this type of antenna signal correlation in the MIMO channel model for Rayleigh flat-fading-like channels, using real-valued correlation coefficients, is given by

$$R = R_{RX} \otimes R_{TX}, \qquad (4.52)$$

where ⊗ denotes the Kronecker product and $R_{TX}(M_{TX} \times M_{TX})$ and $R_{RX}(N_{RX} \times N_{RX})$ denote the correlation observed on the transmitter and receiver, respectively. For complex correlation coefficients, the lower-diagonal part of matrix $R$ must have its coefficients turned into the respective conjugate values. A simple model for deriving the correlation values that uses an approximation for the fading correlation between two adjacent antenna elements averaged over all possible orientations of the two antennas in a given wave-field [Durgin and Rappaporto 1999] will be used as a basis, for the sake of simplicity:

$$\rho(d) = e^{-23\Lambda^2 d^2}. \qquad (4.53)$$

In this equation, $d$ is the distance in wavelengths between two antennas, and $\Lambda$ is the angular spread according to Durgin and Rappaporto [1999]. Note that this definition of $\Lambda$ is very general; that is, it is defined for any distribution of power in the azimuth plan, and values close to 0.0 denote completely directional scenarios, whereas those at 1.0 represent more uniform spreading of energies in space. Based on this approximation, a correlation model is proposed for linear arrays at both the transmitter and the receiver with equidistant antenna spacing $d_{TX}$ and $d_{RX}$, respectively, resulting in the following Toeplitz structure correlation matrices

$$
R_{RX} = \begin{bmatrix}
1 & \rho_r & \rho_r^4 & \cdots & \rho_r^{(n_r-1)^2} \\
\rho_r & 1 & \rho_r & \ddots & \vdots \\
\left(\rho_r^4\right) & \rho_r & 1 & \ddots & \rho_r^4 \\
\vdots & \ddots & \ddots & \ddots & \rho_r \\
\left(\rho_r^{(n_r-1)^2}\right) & \cdots & \left(\rho_r^4\right) & \rho_r & 1
\end{bmatrix}
\tag{4.54}
$$

$$
R_{TX} = \begin{bmatrix}
1 & \rho_t & \rho_t^4 & \cdots & \rho_t^{(n_t-1)^2} \\
\rho_t & 1 & \rho_t & \ddots & \vdots \\
\left(\rho_t^4\right) & \rho_t & 1 & \ddots & \rho_t^4 \\
\vdots & \ddots & \ddots & \ddots & \rho_t \\
\left(\rho_t^{(n_t-1)^2}\right) & \cdots & \left(\rho_t^4\right) & \rho_t & 1
\end{bmatrix},
$$

where $\rho_r$ and $\rho_t$ represent $\rho(d_{RX})$ and $\rho(d_{TX})$, respectively. Note that this model can range from the totally uncorrelated scenario ($\rho_r = \rho_t = 0$) to the fully correlated scenario ($\rho_r = \rho_t = 1$). For small $\rho_r$ and $\rho_t$ (much smaller than 1), the terms with powers higher than 4 can be discarded, which translates the Toeplitz matrices to a tri-diagonal structure. This tri-diagonal correlation model can be viewed as approximating a scenario in which it is assumed that the fading correlation has a certain given value for any pair of adjacent antenna elements, and that any other pair of antenna elements exhibits independent fading on both sides of the communication link. For further simplification of the spatial correlation, the following relation can be set:

$$
\rho_r = \rho_t,
\tag{4.55}
$$

which leads to a single-parameter MIMO correlation model. For nonlinear antenna arrangements, such as triangular or squared dispositions (Figure 4.14), different matrix structures are obtained, according to the dependence on the distance between antennas.

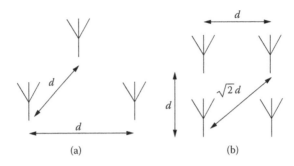

**Figure 4.14   Triangular (a) and square (b) antenna dispositions.**

The square disposition was assumed for the $4 \times 4$ setting, since it seems the most obvious in terms of space savings. Therefore, for the cases of $2 \times 2$ and $4 \times 4$, the correlation matrices are as follows:

$$2 \times 2 : R_{RX} = \begin{bmatrix} 1 & \rho_r \\ \rho_r & 1 \end{bmatrix} \quad R_{TX} = \begin{bmatrix} 1 & \rho_t \\ \rho_t & 1 \end{bmatrix}, \qquad (4.56)$$

$$4 \times 4 : R_{RX} = \begin{bmatrix} 1 & \rho_r & \rho_r & \rho_r^2 \\ \rho_r & 1 & \rho_r^2 & \rho_r \\ \rho_r & \rho_r^2 & 1 & \rho_r \\ \rho_r^2 & \rho_r & \rho_r & 1 \end{bmatrix} \quad R_{TX} = \begin{bmatrix} 1 & \rho_t & \rho_t & \rho_t^2 \\ \rho_t & 1 & \rho_t^2 & \rho_t \\ \rho_t & \rho_t^2 & 1 & \rho_t \\ \rho_t^2 & \rho_t & \rho_t & 1 \end{bmatrix}.$$

$$(4.57)$$

## 4.4  Spatial Multiplexing—MIMO for OFDM/SC-FDE

In this section, schemes for reducing transmit power levels via use of diversity and enhanced decoding schemes alongside capacity augmentation techniques via the use of MIMO will be discussed for FDE schemes.

Let us consider the uplink of a broadband wireless system where an SC modulation is employed by each MT. We can define an space-division multiple access (SDMA) architecture corresponding to a MIMO system with $P$ users (MTs), transmitting independent data blocks, and $N$ receive antennas at the BS, as depicted in Figure 4.15. It is assumed that each MT has a single transmit antenna (the generalization to the case where we have multiple antennas at the MTs is straightforward). The time-domain block transmitted by the $p$th user is $\{s_{m,p}; m = 0,1,...,M-1\}$, with $s_{m,p}$

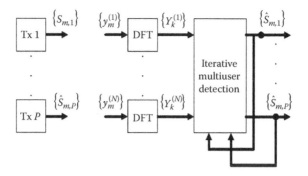

**Figure 4.15   System model.**

denoting the $m$th data symbol of the $p$th user, which is selected from a given constellation (e.g., a QPSK), under an appropriate mapping rule. A cyclic prefix, preceding each block, is used to avoid interblock interference and to make the linear convolution associated with the channel equivalent to a cyclic convolution with respect to the useful, $M$-length part of the block. At the receiver, the cyclic prefix is discarded.

The time-domain block at the $n$ th receive antenna is $\{y_m^{(n)}; m = 0,1,\ldots, J \cdot M - 1\}$, where an oversampling factor $J$ is assumed. The corresponding frequency-domain block, obtained after an appropriate size-$J \cdot M$ DFT operation (discrete Fourier transform), is $\{Y_k^{(n)}; k = 0,1,\ldots, JM - 1\}$, where

$$Y_k^{(n)} = \sum_{p=1}^{P} S_{k,p} H_{k,p}^{(n)} + N_k^{(n)},  \qquad (4.58)$$

with $H_{k,p}^{(n)}$ denoting the overall channel frequency response from the $p$th transmitting antenna to the $n$th receiver antenna, for the $k$th frequency, and $N_k^{(n)}$ denoting the corresponding channel noise. The block $\{S_{k,p}; k = 0,1,\ldots, M-1\}$ is the size-$M$ DFT of the $p$th user's data block $\{s_{m,p}; m = 0,1,\ldots, M - 1\}$, and it is assumed that $S_{k,p}$ is periodic, with period $M$ (i.e., $S_{k,p} = S_{k+M,p}$ for any $k$).

## 4.4.1 Receiver Structures

We consider a frequency-domain iterative multiuser detection that combines IB-DFE principles with LST (layered space-time) interference cancellation. Each iteration consists of $P$ detection stages, one for each user.

When detecting a given user, the interference from previously detected users is canceled, as with conventional LST receivers. However, unlike with conventional LST receivers, we also cancel the residual ISI from the user that is being detected. Moreover, these interference and residual ISI cancellations take into account the reliability of each of the previously detected users. For a given iteration, the receiver

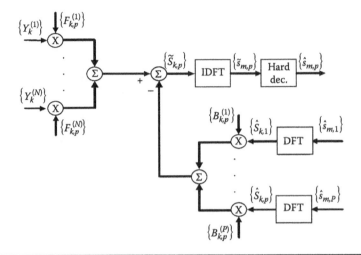

**Figure 4.16    Detection of the *p*th user, for a given iteration.**

structure for the detection of *p*th user is illustrated in Figure 4.16. We have $N$ frequency-domain feedforward filters (one for each receive antenna) and $P$ frequency-domain feedback filters (one for each user). The feedforward filters are designed to minimize both the ISI and the multiuser interference that cannot be canceled by the feedback filters, due to decision errors in the previous detection steps. This structure can be regarded as an equalizer with multiuser interference suppression properties. After an IDFT operation, the corresponding time-domain outputs are passed through a hard-decision device so as to provide an estimate of the data block transmitted by the *p*th user.

We consider in this section two alternative detection schemes:

■ Iterative Detection with Successive Interference Cancellation (SIC)
The SIC architecture is closely related to the LST receivers proposed in Dinis and Gusmão [2004]: for the first iteration and the detection of a given user, the interference from previously detected users are canceled; for the remaining iterations, we cancel the interference from all users (using the most updated version of each user), as well as the residual ISI for the user that is being detected. The detection procedure can be summarized as follows:

First iteration:
 (i)    Detect user 1.
 (ii)   Detect user 2 by removing the interference from user 1.
 (iii)  Detect user 3 by removing the interference from users 1 and 2.
 (iv)   Proceed until the detection of user $P$.

Remaining iterations:

(i) Repeat the detection of user 1, now removing the interference from all users ($p \neq 1$) and the residual ISI ($p = 1$).

(ii) Repeat the detection of user 2, now removing the interference from all users ($p \neq 2$) and the residual ISI ($p = 2$).

(iii) Proceed until the detection of user $P$.

■ **Iterative Detection with Parallel Interference Cancellation (PIC)**

This second architecture combines IB-FDE principles with parallel interference cancellation. Each iteration consists of $P$ parallel detection stages. At the first iteration, all of the users are detected simultaneously using linear processing. For the remaining iterations, we repeat the detection of each user ($p = 1, 2, ..., P$), now using the users' estimate from the previous iteration to remove the interference from the other users ($p' \neq p$) and the residual ISI ($p' = p$). Although the complexity of this structure is almost the same as the previous one, a parallel design can be advantageous from the implementation point of view.

## 4.4.2 Derivation of the Equalizer Coefficients

Throughout this section, it is assumed that the average received powers for the different users are sorted in descending order, that is,

$$\sum_{n=0}^{N-1}\sum_{k} E[|H_{k,p}^{(n)}|^2]E_{S,p} < \sum_{n=0}^{N-1}\sum_{k} E\left[|H_{k,p-1}^{(n)}|^2\right]E_{S,p-1}, \quad p = 1,2,...,P, \quad (4.59)$$

with

$$E_{S,p} = E\left[|s_{m,p}|^2\right] = \frac{1}{M} E\left[|S_{k,p}|^2\right] \quad (4.60)$$

denoting the average symbol energy for the $p$th user.

Let us first assume that there is no oversampling at the receiver (i.e., $J = 1$). In this case, the frequency-domain samples associated with the $p$th user at the output of the equalizer/multiuser detector are given by

$$\tilde{S}_{k,p} = \sum_{n=1}^{N} F_{k,p}^{(n)}Y_k^{(n)} - \sum_{p'=1}^{P} B_{k,p}^{(p')}S_{k,p'}$$

$$= \sum_{n=1}^{N} F_{k,p}^{(n)}Y_k^{(n)} - B_{k,p}^{(p)}S_{k,p} - \sum_{p'\neq p} B_{k,p}^{(p')}S_{k,p'} \quad (4.61)$$

where $F_{k,p}^{(n)}$ $(k = 0,1,\ldots,M-1; n = 1,2,\ldots,N)$ denote the feedforward coefficients and $(B_{k,p}^{(p')} k = 0,1,\ldots,M-1; p = 1,2,\ldots,P)$ denote the feedback coefficients. The coefficients $\{B_{k,p}^{(p)}; k = 0,1,\ldots,M-1\}$ are used for residual ISI cancelation and the coefficients $(\{B_{k,p}^{(p')}; k = 0,1,\ldots,N-1\}$ $p' \neq p)$ are used for interference cancelation. The block $\{S_{k,p'}; k = 0,1,\ldots,M-1\}$ is the DFT of the block $\{s_{m,p'}; m = 0,1,\ldots,M-1\}$, where the time-domain samples $\hat{s}_{m,p'}, m = 0,1,\ldots,M-1$, are the latest estimates for the $p'$th user transmitted symbols, that is, the hard decisions associated with the block of time-domain samples $\{\tilde{s}_{m,p'}; m = 0,1,\ldots,M-1\} = \text{IDFT} \{\tilde{S}_{k,p'}; k = 0,1,\ldots,N-1\}$. For the $i$th iteration of a SIC receiver, $\hat{s}_{m,p'}$ is associated with the $i$th iteration for $p' < p$ and with the $(i-1)$th iteration for $p' \geq p$ (in the first iteration, we do not have any information for $p' \geq p$ and $\hat{s}_{m,p'} = 0$); for the PIC receiver, $\hat{s}_{m,p'}$ is always associated with the previous iteration (for the first iteration $\hat{s}_{m,p'} = 0$).

Due to decision errors, we have $\hat{s}_{m,p} \neq s_{m,t}$ for some symbols. Consequently, $\hat{S}_{k,p} \neq S_{k,p}$. However, the frequency-domain estimates, $\hat{S}_{k,p}$, can be written as

$$\hat{S}_{k,p} = \rho_p S_{k,p} + \Delta_{k,p} \tag{4.62}$$

where the correlation coefficient $\rho_p$ is given by

$$\rho_p = \frac{E[\hat{s}_{m,p} s_{m,p}^*]}{E_{S,p}} = \frac{E[\hat{S}_{k,p} S_{k,p}^*]}{M E_{S,p}} \tag{4.63}$$

and $\Delta_{k,p}$ denotes a zero-mean error term. It is assumed that $E[\Delta_{k,p} S_{k',p}] \approx 0$, regardless of $k$ and $k'$. Therefore,

$$E[|\Delta_{k,p}|^2] = (1 - \rho_p^2) M E_{S,p}. \tag{4.64}$$

By combining (4.58), (4.61), and (4.62), we obtain

$$\tilde{S}_{k,p} = \sum_{l=1}^{N} F_{k,p}^{(n)} \left( \sum_{p'=1}^{P} H_{k,p'}^{(n)} S_{k,p'} + N_k^{(n)} \right)$$

$$- B_{k,p}^{(p)} \left( \rho_p S_{k,p} + \Delta_{k,p} \right) - \sum_{p' \neq p} B_{k,p}^{(p')} \left( \rho_{p'} S_{k,p'} + \Delta_{k,p'} \right)$$

$$= \gamma_p S_{k,p} + \left( \sum_{n=1}^{N} F_{k,p}^{(n)} H_{k,p}^{(n)} - \gamma_p - \rho_p B_{k,p}^{(p)} \right) S_{k,p} \tag{4.65}$$

$$+ \sum_{p' \neq p} \left( \sum_{n=1}^{N} F_{k,p}^{(n)} H_{k,p'}^{(n)} - \rho_{p'} B_{k,p}^{(p')} \right) S_{k,p'}$$

$$- \sum_{p'=1}^{P} B_{k,p}^{(p')} \Delta_{k,p'} + \sum_{n=1}^{N} F_{k,p}^{(n)} N_k^{(n)}$$

with

$$\gamma_p = \frac{1}{M} \sum_{k=0}^{M-1} \sum_{n=1}^{N} F_{k,p}^{(n)} H_{k,p}^{(n)}.$$

(4.66)

Clearly, $\gamma_p$ can be regarded as the average overall channel frequency response for the $p$th user, after combining the outputs of the $N$ feedforward filters.

This means that $\tilde{S}_{k,p}$ contains a "signal" component, $\gamma_p S_{k,p}$, and four "noise" components:

■ $\left( \sum_{n=1}^{N} F_{k,p}^{(n)} H_{k,p}^{(n)} - \gamma_p - \rho_p B_{k,p}^{(p)} \right) S_{k,p}$ is the residual ISI;

■ $\sum_{p' \neq p} \left( \sum_{n=1}^{N} F_{k,p}^{(n)} H_{k,p'}^{(n)} - \rho_{p'} B_{k,p}^{(p')} \right) S_{k,p'}$ is the residual multiuser interference;

■ $\sum_{p'=1}^{P} B_{k,p}^{(p')} \Delta_{k,p'}$ accounts for the errors in $\hat{s}_{n,p'}$ ($p' = 1, 2, \ldots, P$);

■ $\sum_{n=1}^{N} F_{k,p}^{(n)} N_k^{(n)}$ comes from the channel noise.

The forward and backward coefficients, $\{F_{k,p}^{(n)}; k = 0, 1, \ldots, N-1\}$, $n = 1, 2, \ldots, N$, and $\{B_{k,p}^{(p')}; k = 0, 1, \ldots, N-1\}$, $p' = 1, 2, \ldots, P$, respectively, are chosen so as to maximize the signal-to-noise plus interference ratio (SNIR) for the $p$th user, defined as

$$SNIR_p = \frac{|\gamma_p|^2 E_S}{E\left[|\varepsilon_{m,p}^{eq}|^2\right]},$$

(4.67)

with

$$\varepsilon_m^{eq} = \tilde{s}_{m,p} - \gamma_p s_{m,p}$$

(4.68)

denoting the overall noise. The SNIR can also be written as

$$SNIR_p = \frac{1}{M} \sum_{k=0}^{M-1} SNIR_{k,p}^F,$$

(4.69)

where $SNIR_{k,p}^F$ denotes the SNIR associated with the corresponding frequency-domain samples, defined as

$$SNIR_{k,p}^F = \frac{|\gamma_p|^2 M E_S}{E[|\varepsilon_{k,p}^{Eq}|^2]},$$

(4.70)

with the block $\{\varepsilon_{k,p}^{Eq}; k = 0,1,\ldots,M-1\}$ denoting the DFT of the block $\{\varepsilon_m^{eq}; m = 0,1,\ldots,N-1\}$. Clearly, the maximization of the SNIR in the time-domain samples $\tilde{s}_{m,p}$ is equivalent to the maximization of the SNIR in the corresponding frequency-domain samples, $SNIR_{k,p}^F$.

We can write

$$E\left[\left|\varepsilon_{k,p}^{Eq}\right|^2\right] = \left|\sum_{n=1}^{N} F_{k,p}^{(n)} H_{k,p}^{(n)} - \gamma_p - \rho_p B_{k,p}^{(p)}\right|^2 ME_S$$

$$+ \sum_{p' \neq p}\left|\sum_{n=1}^{N} F_{k,p}^{(n)} H_{k,p'}^{(n)} - \rho_{p'} B_{k,p}^{(p')}\right|^2 ME_S$$

$$+ \sum_{p'=1}^{P}\left|B_{k,p}^{(p')}\right|^2 (1-\rho_{p'}^2) ME_S + \sum_{n=1}^{N}\left|F_{k,p}^{(n)}\right|^2 2M\sigma_N^2,$$

(4.71)

with

$$\sigma_N^2 = \frac{1}{2M} E\left[|N_k^{(n)}|^2\right]$$

(4.72)

denoting the variance of both the in-phase and quadrature components of the channel noise at the input of each antenna.

The maximization of $SNIR_{k,p}^F$ is equivalent to solving the following set of $N + P$ equations for each frequency

$$\frac{\partial E\left[\left|\varepsilon_{k,p}^{Eq}\right|^2\right]}{\partial F_{k,p}^{(n)}}$$

$$= 2ME_S H_{k,p}^{(n)*}\left(\sum_{n'=1}^{N} F_{k,p}^{(n')} H_{k,p}^{(n')} - \gamma_p - \rho_p B_{k,p}^{(p)}\right)$$

(4.73)

$$+ 2ME_S \sum_{p' \neq p} H_{k,p'}^{(n)*}\left(\sum_{n'=1}^{N} F_{k,p}^{(n')} H_{k,p'}^{(n')} - \rho_{p'} B_{k,p}^{(p')}\right)$$

$$+ 4M\sigma_N^2 F_{k,p}^{(n)} = 0, \quad n = 1,2,\ldots,N,$$

$$\frac{\partial E\left[\left|\varepsilon_{k,p}^{Eq}\right|^2\right]^{(p)}}{\partial B_{k,p}}$$

$$= -2ME_S\rho_p\left(\sum_{n'=1}^{N}F_{k,p}^{(n')}H_{k,p}^{(n')} - \gamma_p - \rho_p B_{k,p}^{(p)}\right)$$

$$+ 2ME_S(1-\rho_p^2)B_{k,p}^{(p)} = 0$$

$(4.74)$

and

$$\frac{\partial E\left[\left|\varepsilon_{k,p'}^{Eq}\right|^2\right]}{\partial B_{k,p}^{(p')}}$$

$$= -2ME_S\rho_{p'}\left(\sum_{n'=1}^{N}F_{k,p}^{(n')}H_{k,p'}^{(n')} - \rho_{p'}B_{k,p}^{(p')}\right)$$

$$+ 2ME_S(1-\rho_{p'}^2)B_{k,p}^{(p')} = 0, \quad p' \neq p.$$

$(4.75)$

These equations can be rewritten in the form

$$H_{k,p}^{(n)*}\left(\sum_{n'=1}^{N}F_{k,p}^{(n')}H_{k,p}^{(n')} - \gamma_p - \rho_p B_{k,p}^{(p)}\right)$$

$$+ \sum_{p'\neq p}H_{k,p'}^{(n)*}\left(\sum_{n'=1}^{N}F_{k,p}^{(n')}H_{k,p'}^{(n')} - \rho_{p'}B_{k,p}^{(p')}\right)$$

$$+ \frac{F_{k,p}^{(n)}}{SNR_p} = 0, \quad n = 1,2,\ldots,N,$$

$(4.76)$

$$\rho_p\left(\sum_{n'=1}^{N}F_{k,p}^{(n')}H_{k,p}^{(n')} - \gamma_p - \rho_p B_{k,p}^{(p)}\right) = (1-\rho_p^2)B_{k,p}^{(p)}$$

$(4.77)$

and

$$\rho_{p'}\left(\sum_{n'=1}^{N}F_{k,p}^{(n')}H_{k,p'}^{(n')} - \rho_{p'}B_{k,p}^{(p')}\right) = (1-\rho_{p'}^2)B_{k,p}^{(p')}, \quad p' \neq p,$$

$(4.78)$

where $SNR_p = \dfrac{E_{S,p}}{2\sigma_N^2}$.

From (4.77), the optimum values of $B_{k,p}^{(p)}$ are

$$B_{k,p}^{(p)} = \rho_p \left( \sum_{n'=1}^{N} F_{k,p}^{(n',i)} H_{k,p}^{(n')} - \gamma_p \right); \tag{4.79}$$

from (4.78), the optimum values of $B_{k,p}^{(p')}$, $p' \neq p$, are

$$B_{k,p}^{(p')} = \rho_{p'} \sum_{n'=1}^{N} F_{k,p}^{(n',i)} H_{k,p'}^{(n')}. \tag{4.80}$$

By replacing (4.79) and (4.80) in (4.76), we get the set of $N$ equations

$$(1-\rho_p^2) H_{k,p}^{(n)*} \sum_{n'=1}^{N} F_{k,p}^{(n')} H_{k,p}^{(n')}$$

$$+ \sum_{p' \neq p} (1-\rho_{p'}^2) H_{k,p'}^{(n)*} \sum_{n'=1}^{N} F_{k,p}^{(n')} H_{k,p'}^{(n')} + \frac{F_{k,p}^{(n)}}{SNR_p} \tag{4.81}$$

$$= \gamma_p (1-\rho_p^2) H_{k,p}^{(n)*}, \quad n = 1, 2, \ldots, N.$$

These feedforward coefficients can be used in (4.79) and (4.80) for obtaining the feedback coefficients $B_{k,p}^{(p)}$ and $B_{k,p}^{(p')}$, $p' \neq p$, respectively.

It can be shown that the solution of the system of (4.81) can be written in the form

$$F_{k,p}^{(n)} = \sum_{p'=1}^{P} H_{k,p'}^{(n)*} C_{k,p}^{(p')}, \tag{4.82}$$

with the set of coefficients $\{C_k^{(p')}; p' = 1, 2, \ldots, P\}$ satisfying the set of $P$ equations

$$\sum_{p''=1}^{P} C_{k,p}^{(p'')} \cdot \left( (1-\rho_{p'}^2) \sum_{n'=1}^{N} H_{k,p'}^{(n')*} H_{k,p'}^{(n')} + \frac{1}{SNR_p} \delta_{p',p''} \right) \tag{4.83}$$

$$= \delta_{p,p'}, \quad p' = 1, 2, \ldots, P$$

($\delta_{p,p'} = 1$ if $p = p'$ and 0 otherwise).

The computation of the feedforward coefficients from (4.82) is simpler than the direct computation, from (4.81), especially when $P < N$.

For the special case where $P = 1$ (and $p = 1$), it can be shown that

$$F_{k,1}^{(n)} = \frac{SNR_1 \cdot H_{k,1}^{(n)^*}}{1 + SNR_1(1 - \rho_1^2)\displaystyle\sum_{n'=1}^{N} |H_{k,1}^{(n')}|^2}, l = 1, 2, \ldots, N, \qquad (4.84)$$

which corresponds to feedforward coefficients of an IB-DFE with $N$-branch space diversity [Dinis et al. 2003]. For the first iteration,

$$F_{k,1}^{(n)} = \frac{SNR_1 \cdot H_{k,1}^{(n)^*}}{1 + SNR_1 \displaystyle\sum_{n'=1}^{N} |H_{k,1}^{(n')}|^2}, l = 1, 2, \ldots, N, \qquad (4.85)$$

corresponding to a linear FDE with $N$-branch space diversity [Gusmao et al. 2003].

It should be noted that, for the first iteration ($i = 0$), we do not have any information about $S_{k,p'}$ for $p' \leq p$, for the SIC receiver, or all $S_{k,p'}$, for the PIC receiver. Therefore, the corresponding correlation coefficients are zero, leading to $B_{k,p}^{(p')} = 0$. After the first iteration, if the residual BER is not too high, $\hat{s}_{m,p'} = s_{m,p'}$ for most of the data symbols, and $\hat{S}_{m,p'} \approx S_{k,p'}$; this means that we can use the feedback coefficients to eliminate a significant part of the residual ISI, as well as the residual multiuser interference. Naturally, when $\hat{s}_{m,p'} = s_{m,p'}$ for all data symbols, $\hat{S}_{k,p'} = S_{k,p'}$ and $\rho_{p'} = 1$, leading to

$$F_{k,p}^{(n)} = SNR_p \cdot H_{k,p}^{(n)^*}, n = 1, 2, \ldots, N, \qquad (4.86)$$

$$B_{k,p}^{(p)} = SNR_p \left( \sum_{n'=1}^{N} |H_{k,p}^{(n')}|^2 - \frac{1}{M} \sum_{k=0}^{M-1} \sum_{n'=1}^{N} |H_{k,p}^{(n')}|^2 \right) \qquad (4.87)$$

and

$$B_{k,p}^{(p')} = SNR_p \sum_{n'=1}^{N} |H_{k,p}^{(n')}|^2, \qquad (4.88)$$

which corresponds to the total elimination of the ISI and total multiuser interference cancellation.

Let us assume now that we have an oversampling factor $J$. In that case, since $S_{k,p} = S_{k+lM,p}$ for any $l$, there is an implicit diversity effect, unless the transmitted signals have the Nyquist bandwidth (i.e., for a square-root raised-cosine shaping with roll-off zero). Therefore, for a given iteration, the $k$th frequency-domain sample associated with the $p$th user is given by

$$\tilde{S}_{k,p} = \sum_{l}\sum_{n=1}^{N} F_{k+lM,p}^{(n)} Y_{k+lM,p}^{(n)} - \sum_{p'=1}^{P} B_{k,p}^{(p')} \hat{S}_{k,p'} \tag{4.89}$$

where $S_{k,p'}$ is composed of the most updated estimates, for the SIC receiver, or the estimated from the previous iteration, for the PIC receiver.

From (4.79) and (4.80), the optimum feedback coefficients are

$$B_{k,p}^{(p)} = \rho_p \left( \sum_{l}\sum_{n'=1}^{N} F_{k+lM,p}^{(n')} H_{k+lM,p}^{(n')} - \gamma_p \right), \tag{4.90}$$

and

$$B_{k,p}^{(p')} = \rho_{p'} \sum_{l}\sum_{n'=1}^{N} F_{k+lM,p'}^{(n')} H_{k+lM,p}^{(n')}, \tag{4.91}$$

for $p' \neq p$, with

$$\gamma_p = \frac{1}{M} \sum_{l}\sum_{n=1}^{N}\sum_{k=0}^{M-1} F_{k+lM,p}^{(n)} H_{k+lM,p}^{(n)}. \tag{4.92}$$

And, from (4.82), the feedforward coefficients can be written in the form

$$F_{k,p}^{(n)} = \sum_{l}\sum_{p'=1}^{P} H_{k+lM,p'}^{(n)*} C_{k,p}^{(p')}, \tag{4.93}$$

with the set of coefficients $\{C_{k,p}^{(p')}; p' = 1,2,\ldots,P\}$ satisfying the set of $P$ equations

$$\sum_{p''=1}^{P} C_{k,p}^{(p'')} \cdot \left( (1-\rho_{p'}^2) \sum_{l}\sum_{n'=1}^{N} H_{k+lM,p''}^{(n')*} H_{k+lM,p'}^{(n')} + \frac{1}{SNR_p} \delta_{p',p''} \right) \tag{4.94}$$

$$= \delta_{p,p'}, \quad p' = 1,2,\ldots,P.$$

### 4.4.3 Correlation Coefficient

The correlation factors, $\rho_p$, $p = 1,2,\ldots,P$, can be regarded as a blockwise measure of the reliability of the decisions used in the feedback loops. Therefore, they are key parameters in the proposed receiver structures since, by weighting the decided blocks appropriately (see (4.90) and (4.91)), we can have a "turbo-like" behavior without error propagation.

In the following, we will show how to calculate the correlation coefficient $\rho_p$ when the transmitted symbols $s_m$ belong to a QPSK constellation (for the sake of simplicity, we omit the dependence on the user number $p$). In this case,

$$s_m = s_m^I + js_m^Q = \pm d \pm jd \tag{4.95}$$

where $s_m^I = Re\{s_m\}$ and $s_m^Q = Im\{s_m\}$ are the in-phase and quadrature components of $s_m$, respectively. The minimum Euclidian distance within the constellation is $D = 2d$ and the average symbol energy is

$$E_S = E\left[|s_m|^2\right] = \frac{D^2}{4}. \tag{4.96}$$

By assuming proper normalization, the time-domain samples at the output of the equalizer/multiuser detector can be written as

$$\tilde{s}_m = \tilde{s}_m^I + j\tilde{s}_m^Q = s_m + v_m, \tag{4.97}$$

where $\tilde{s}_m^I = Re\{\tilde{s}_m\}$, $\tilde{s}_m^Q = Im\{\tilde{s}_m\}$ and $v_m$ is the overall noise component. We will assume that $v_m$ is approximately Gaussian distributed, with zero mean.

The symbol estimates can be written as

$$\hat{s}_m = s_m + \varepsilon_m^I + j\varepsilon_m^Q, \tag{4.98}$$

where the error coefficients $\varepsilon_m^I$ (or $\varepsilon_m^Q$) are zero if there is no error in $s_m^I$ (or $s_m^Q$) and $\pm D$ otherwise. This means that $\varepsilon_m^I$ and $\varepsilon_m^Q$ are random variables with

$$\varepsilon_m^I = \begin{cases} 0, & \text{probability } P_{\varepsilon_m^I} \\ \pm D, & \text{probability } 1 - P_{\varepsilon_m^I} \end{cases} \tag{4.99}$$

and

$$\varepsilon_m^Q = \begin{cases} 0, & \text{probability } P_{\varepsilon_m^Q} \\ \pm D, & \text{probability } 1 - P_{\varepsilon_m^Q} \end{cases} \tag{4.100}$$

where $P_{\varepsilon_m^I}$ and $P_{\varepsilon_m^Q}$ denote the error probability in $s_m^I$ and $s_m^Q$, respectively.

We have

$$E[s_m \hat{s}_m^*] = E_S + E[s_m^I \varepsilon_m^I] + E[s_m^Q \varepsilon_m^Q], \tag{4.101}$$

where

$$E[s_m^I \varepsilon_m^I] = -\frac{D^2}{2} P_{\varepsilon_m^I} = -E_S P_{\varepsilon_m^I} \qquad (4.102)$$

and

$$E[s_m^Q \varepsilon_m^Q] = -\frac{D^2}{2} P_{\varepsilon_m^Q} = -E_S P_{\varepsilon_m^Q}. \qquad (4.103)$$

By replacing (4.102) and (4.103) in (4.101) and using (4.63), we get the correlation factor for the $p$th user

$$\rho = 1 - 2P_e \qquad (4.104)$$

where $P_e = P_{\varepsilon_m^I} = P_{\varepsilon_m^Q}$ denotes the average BER.

### 4.4.4 Complexity Analysis

Both SIC and PIC receivers require $N$ size-$J \cdot M$ DFT operations, one for each receiver antenna, and a pair of DFT/IDFT operations (with size-$M$) for the detection of each user, at each iteration. For the equalization/interference cancellation, we need $J \cdot N \cdot P \cdot M$ multiplications for the first iteration of a PIC receiver and $J \cdot N \cdot P \cdot M + P(P-1)M/2$ multiplications for the first iteration of a SIC receiver; for the remaining iterations, we need $(JN + P) \cdot P \cdot M$ multiplications for both SIC and PIC receivers.

The most complex part of the algorithm is the computation of the feedforward coefficients, since we need to solve the $M$ systems of $P$ equations (one for each of the $M$ frequencies), for each iteration and each user. Naturally, for slow-varying channels, this operation is not required for all blocks.

### 4.4.5 Performance Results

In this section, we consider the use of the proposed receiver in an SDMA system where each user has one transmit antenna and the base station has $N$ receive antennas. The data block consists of $M = 64$ QPSK data symbols, plus an appropriate cyclic prefix (similar results were obtained for other values of $M$, provided that $M \gg 1$). We consider uncoded BER performances under perfect synchronization and channel estimation conditions. A square-root raised-cosine filtering with rolloff zero is assumed so that there is no oversampling at the receiver (i.e., $J = 1$).

First, we evaluate the performance of the proposed receivers for a channel that has independent Rayleigh-distributed fading at each frequency. This channel model can be regarded as a severely time-dispersive scenario with rich multipath propagation. Figure 4.17 shows the BER performances for different users and different

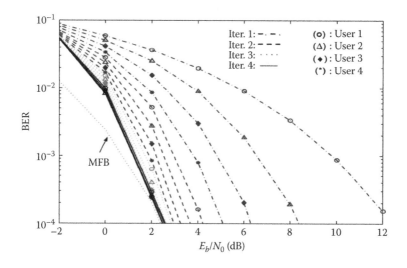

**Figure 4.17 BER for the different users and iterations, along corresponding MFB for $N = P = 4$.**

iterations, when we have $P = 4$ users with the same average received power and $N = 4$ receive antennas at the BS. An SIC receiver is assumed. For the sake of comparison, we also include the matched filter bound (MFB) performance, defined as

$$P_{MFB} = P_{MFB,p} = E\left[ Q\left( \sqrt{\frac{SNR_p}{M} \sum_{l} \sum_{n=1}^{N} \sum_{k=0}^{M-1} |H_{k+lM,p}^{(n)}|^2} \right) \right] \quad (4.105)$$

(identical for all users, in this case).

From Figure 4.17, we can observe that, for the first iteration, the users have very different performances: more than 6 dB from user 1 to user 4, at BER $= 10^{-4}$. This difference decreases as we increase the number of iterations, with all users having almost the same performance after three iterations. Moreover, the resulting performance can be very close to the MFB after four iterations: the required $E_b/N_0$ for an average BER $= 10^{-4}$ are 13.8, 6.8, 5.7, 5.7, and 5.4 dB, for iterations 1 to 4 and for the MFB, respectively. This shows that the proposed receiver is able to eliminate a significant part of the ISI and multiuser interference.

Figure 4.18 shows the average BER performances (averaged over all the users) for SIC and PIC receivers, for the different iterations. After the first iteration, the performance of the PIC receiver is almost 2 dB worse than the performance of the SIC receiver. After iteration 3, this gap reduces to less than 1 dB and after iteration 4 to 0.2 dB. Once again, the BER performance after four iterations is very close to the corresponding MFB for both structures. It should be mentioned that, for the PIC receiver, all users have the same average BER.

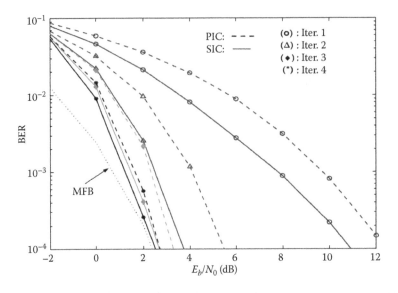

**Figure 4.18   Average BER of an SIC / PIC receiver, along MFB for N = P = 4.**

Figure 4.19 shows the average BER performance after four iterations, for different $N$ and $P(N \geq P)$. Once again, the BER performance after four iterations is close to the corresponding MFBs, regardless of $N$ and $P$.

Let us consider now a scenario where the received powers are not the same for all users (e.g., due to a wrong power control or different uncoded BER requirements

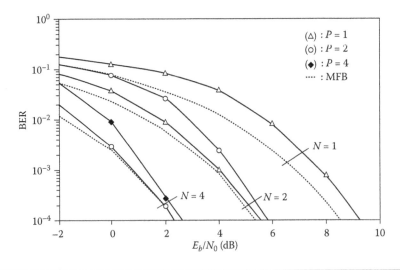

**Figure 4.19   Average BER performance for different values of N and P, after four iterations.**

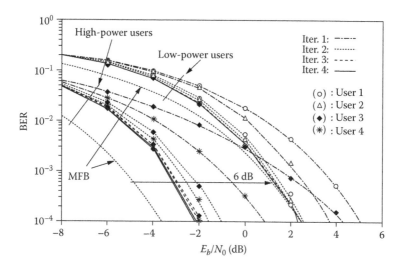

**Figure 4.20  BER performance for low-power and high-power users, as a function of the $E_b/N_0$ of the low-power users.**

for the different users). We consider $P = 4$ users and an SIC receiver with receive antennas (similar conclusions could have been drawn for PIC receivers and other values of $N$ and $P$, provided that $N = P$). The average received power for users 1 and 2 is 6 dB larger than the received powers for users 3 and 4. For a given iteration, the receiver detects first the high-power and then the low-power users. Figure 4.20 shows the BER performances for the different users. Once again, the iterative detection procedure allows significant performance gains, and after four iterations the BER performances are similar for users with the same average power. Looking at Figure 4.20, we note that by increasing the iterations, the performance of the low-power users asymptotically approaches the MFB. However, for the high-power users, the BER at $10^{-4}$ is 2 dB from MFB. This can be explained by the fact that the BER is much lower for the high-power users, allowing an almost perfect interference cancellation and performances close to the MFB. The higher BERs for the low-power users preclude an appropriate interference cancellation on the high-power users (see also Figure 4.21).

Figure 4.21 concerns the situation where $N = 4$ and $P = 6$, that is, an overloaded scenario where the number of users is larger than the number of receive antennas at the BS. In this case, a perfect multiuser separation is not possible, since we have $N = 4$ degrees of freedom to separate $P = 6$ users (see (4.93)). However, the iterative receiver structure presented here has an acceptable performance, with significant interference cancellation, although, even after 10 iterations, the achievable performance is still about 4 dB from the MFB.

The previous results showed that our receiver structures have a good performance for the channel where we have uncorrelated fading on the different

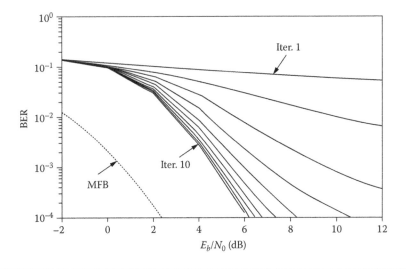

**Figure 4.21   Average BER for *N* = 4 and *P* = 6**

frequencies considered previously. In the following, we will evaluate the performance for a more realistic channel. We consider the HYPERLAN/2 power delay profile type C [ETSI 1998b], where we have a maximum delay 1050 ns. The symbol rate is assumed to be 62.5 ns, and as usual it is assumed that the cyclic extension is longer than the overall channel impulse response. Figure 4.22 illustrates the BER performance for the different users when an SIC receiver is considered. Once again, the BER performances for the different users are substantially different for the first iteration (almost 6 dB); however, after four iterations the BER performances for all users become similar. For this channel, the BER performances after four iterations is still about 1.5 dB from the MFB.

Figure 4.23 shows the average BER for both SIC and PIC receivers. Once again, the PIC receiver has worse performance after the first iteration, but the performance of the two are similar after four iterations.

Since this channel is also severely time-dispersive, the main difference relative to a channel with uncorrelated Rayleigh fading on the different frequencies is the diversity effect inherent to the multipath propagation. To evaluate the impact of this multipath diversity on the proposed receivers, we considered a reference channel where we have $L$ rays with uncorrelated Rayleigh fading and delays uniformly distributed in the interval $[0;T_G]$, with $T_G$ denoting the duration of the cyclic extension (it is assumed that $T_G = 0.2T$, with $T$ denoting the duration of the useful part of the blocks). Figure 4.24 shows the impact of the number of rays $L$ on the required $E_b / N_0$ for BER $= 10^{-4}$. Clearly, this difference decreases with $L$ (as expected, the MFB improves also with $L$). This means that the richer the multipath propagation, the closer we approach the MFB.

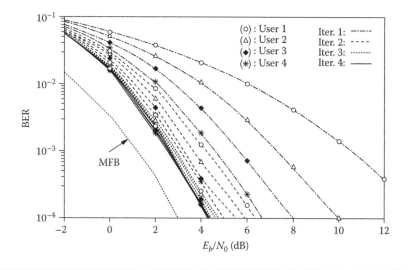

**Figure 4.22 BER for the different users and iterations, along MFB for *N* = *P* = 4.**

## 4.5 Channel Estimation

Both multiplexed and implicit pilot transmition strategies already introduced in Chapter 3 will be considered in this section. Though the use of multiplexed pilots is straightforward (pilots send multiplexed with the data symbols on predefined positions, so that the receiver knows which symbols are pilots and simply computes how the channel affected them), the use of implicit pilots in MIMO schemes requires some clarification.

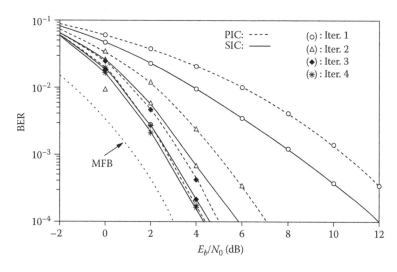

**Figure 4.23 Average BER of an SIC/PIC receiver, along MFB, for *N* = *P* = 4.**

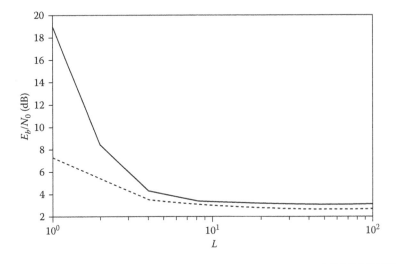

**Figure 4.24**  Required $E_b/N_0$ for BER = $10^{-4}$ after four iterations (solid) and for the corresponding MFB (dashed).

### 4.5.1 Use of Implicit Pilots

After passing the signal to the frequency domain, the implicit pilots can be added. It is better to add them in the frequency domain, since most of the processing is done there.

The transmitted sequences have the form (uppercase $S$ being the signal in the frequency domain)

$$X_k^{ntx} = S_k^{ntx} + S_k^{pilot} \qquad (4.106)$$

where $S_k^{ntx}$ is the data symbol transmitted by the $k$th subcarrier (out of a total of $N$) and $S_k^{pilot}$ is the corresponding implicit pilot. In Figure 4.25, we show a transmitter chain that incorporates for the SC-FDE scheme. The pilot symbols are added in the frequency domain (for simplicity at the receiver, though they

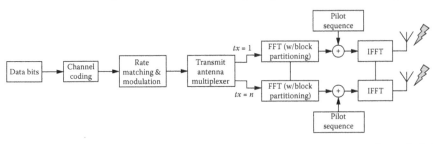

**Figure 4.25**  Transmitter scheme for the SC-FDE scheme.

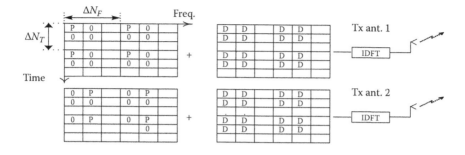

**Figure 4.26 Proposed frame structure for a MIMO-SC-FDE transmission with implicit pilots (P—pilot symbol, D—data symbol).**

could also be added in the time domain), and the data is then passed through the IFFT for transmission.

Assuming only one user, the data bits are passed through a turbo coder, after which they are submitted to rate matching (taking into account the use of FFTs for faster processing, the antenna multiplexer and block partitioning). All of the antennas will transmit a part of the message. The data bits are partitioned into blocks and the cyclic prefix is added to each block, so that the total size of the block is a power of 2, for efficient use of the FFT.

We will consider the frame structure of Figure 4.26 for an SC-FDE system with $N$ carriers. According to this structure, the pilot grid is generated using a spacing of $\Delta N_T$ symbols in the time domain (number of blocks) and $\Delta N_F$ symbols in the frequency domain (in the case of more than one transmitter, the minimum $\Delta N_F$ should be equal to the number of transmit antennas, so that there is no interference between pilot symbols on different antennas).

### 4.5.2 Iterative Receiver for Implicit Pilots

The transmission of pilot symbols superimposed on data will clearly result in interference between them. To reduce the mutual interference and achieve reliable channel estimation and data detection, we follow the same approach of Chapter 3 and employ a receiver capable of jointly performing these tasks through iterative processing. The structure of the proposed iterative receiver is shown in Figure 4.27. According to the figure, the signal, which is considered to be sampled and with the cyclic prefix removed, is converted to the frequency domain after an appropriate size-$N$ FFT operation. If the cyclic prefix is longer than the overall channel impulse response, the *nrx* receive antenna is given as

$$R_{k,l,nrx} = \sum_{ntx=1}^{Ntx} \left( \left( S_{k,l,ntx} + S_{k,l,ntx}^{Pilot} \right) H_{k,l,ntx,nrx} + N_{k,l,nrx} \right) \tag{4.107}$$

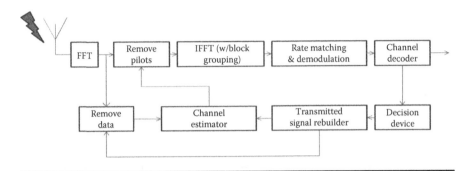

**Figure 4.27   Iterative receiver structure.**

with $H_{k,l,ntx,nrx}$ denoting the overall channel frequency response for the $k$th frequency of the $l$th time block between the $ntx$ transmit and $nrx$ receive antenna, and $N_{k,l,nrx}$ denoting the corresponding channel noise.

Before entering the equalization block, the pilot symbols are removed from the sequence resulting

$$(Y_{k,l,nrx})^{(q)} = R_{k,l,nrx} - \sum_{ntx=1}^{Ntx}\left( S_{k,l,ntx}^{Pilot}\left(\hat{H}_{k,l,ntx,nrx}\right)^{(q)} \right) \tag{4.108}$$

where $(H_{k,l,ntx,nrx})^{(q)}$ are the channel frequency response estimates and $q$ is the current iteration. Note that, in the case of a known channel without any pilots for estimation, $Y_{k,l,ntx,nrx} = R_{k,l,nrx}$.

The equalized samples are then simply computed as

$$(S_{k,l,ntx})^{(q)} = \frac{(H_{k,l,ntx,nrx})^{(q)^*}(Y_{k,l,ntx})^{(q)}}{\left|H_{k,l,ntx,nrx}^{(q)}\right|^2}, \tag{4.109}$$

where

$$(Y_{k,l,ntx})^{(q)} = \sum_{nrx}\left( (Y_{k,l,nrx})^{(q)} - \sum_{ntx1 \neq ntx}\left( S_{k,l,ntx1}^{Pilot}\left(H_{k,l,ntx1,nrx}\right)^q \right) \right). \tag{4.110}$$

The sequences of the equalized samples are then passed through the IFFT and block grouping, and then demodulated and delivered to the channel decoder. Each channel decoder has two outputs. One is the estimated information sequence and the other is the sequence of LLR estimates of the code symbols. These LLRs are passed through the decision device, which outputs either soft-decision or hard-decision estimates of the code symbols. These estimates enter the transmitted signal rebuilder, which performs the same operations of the transmitter (coding, modulation), and the FFT operation. The reconstructed symbol sequence can then be used for improving the channel estimates for the subsequent iterations, as will be explained next.

### 4.5.3 Channel Estimation Procedure

Let us first assume that $S_{k,l} = 0$; that is, there is no data overlapping the training block, as in conventional schemes. In that case, the channel frequency response is, for a SISO (single-input, single-output) scheme:

$$H_{k,l} = \frac{Y_{k,l}}{S_{k,l}^{TS}} = H_{k,l} + \frac{N_{k,l}}{S_{k,l}^{TS}} = H_{k,l} + \epsilon_{k,l}^{H} \tag{4.111}$$

The channel estimation error $\epsilon_{k,l}^{H}$ is Gaussian distributed, with zero mean and

$$E\left[\left|\epsilon_{k,l}^{H}\right|^2 S_{k,l}\right] = E\left[\left|N_{k,l}\right|^2\right]E\left[\frac{1}{\left|S_{k,l}^{TS}\right|^2}\right]. \tag{4.112}$$

Since the power assigned to the training block is proportional to $E\left[\dfrac{1}{\left|S_{k,l}^{TS}\right|^2}\right]$ and $E\left[\left|S_{k,l}^{TS}\right|^2\right]$, the training blocks should be constant and equal to $\left|S_{k,l}^{TS}\right|^2 = 2\sigma_t^2$ for all $k$. On the other hand, if we want to minimize the envelope fluctuations of the transmitted signal, $\left|S_{n,l}^{TS}\right|$ should also be constant. This can be achieved by employing Chu sequences, which have both $\left|S_{n,l}^{TS}\right|$ and $\left|S_{k,l}^{TS}\right|$ constant [Benvenuto et al. 2010]. The $n$th element of a length-$N$ Chu sequence is

$$s_{n,l}^{TS} = \begin{cases} \exp\left(j\pi rn^2/N\right), & N \text{ even} \\ \\ \exp\left(j\pi rn(n-1)/N\right), & N \text{ odd} \end{cases}, \tag{4.113}$$

where $r$ is relatively prime to $N$.

### 4.5.3.1 Use of Implicit Pilots

Let us consider now the use of implicit pilots, that is, $S_{k,l} \neq 0$ for the training blocks. In the following, we will assume that

$$E\left[\left|S_{k,l}\right|^2\right] = NE\left[\left|s_n\right|^2\right] = 2\sigma_D^2 \tag{4.114}$$

and, for the frequencies that have pilots,

$$E\left[\left|S_{k,l}^{TS}\right|^2\right] = NE\left[\left|s_n^{TS}\right|^2\right] = 2\sigma_T^2. \tag{4.115}$$

Clearly, we will have interference between data symbols and pilots. This leads to performance degradation for two reasons:

1. The data symbols produce interference on pilots, which might lead to inaccurate channel estimates. To reduce this effect, we should have

$$\sigma_D^2 \ll \sigma_T^2. \tag{4.116}$$

2. The pilots produce interference on data symbols, which might lead to performance degradation (even if the channel estimation was perfect). To reduce this effect, we should have

$$\sigma_T^2 \ll \sigma_D^2. \tag{4.117}$$

Conditions (4.116) and (4.117) are mutually exclusive. Moreover, the use of implicit pilots leads to increased envelope fluctuations on the transmitted signals [Wu et al. 2008]. To overcome these problems, we can employ pilots with relatively low power (i.e., $\sigma_T^2 \ll \sigma_D^2$) and average the pilots over a large number of blocks so as to obtain accurate channel estimates. This is very effective since the data symbols have usually zero mean and different data blocks are uncorrelated. Naturally, there are limitations on the length of this averaging window, since the channel should be constant within it (not to mention the associated delays). Once we have an accurate channel estimate, we can detect the data symbols, eventually removing first the signal associated with the pilots.

Let us assume a frame with $N_T$ time-domain blocks, each with $N$ subcarriers. If the cyclic prefix of each FFT block has $N_G = NT_g/T$ samples, we will need $N_G$ equally spaced frequency-domain pilots for the channel estimation. For pilot spacing in time and frequency $\Delta N_T$ and $\Delta N_F$, respectively, the total number of pilots in the frame is given by

$$N_P^{Frame} = \frac{N}{\Delta N_F} \cdot \frac{N_T}{\Delta N_T}. \tag{4.118}$$

This means that we have a pilot multiplicity or redundancy of

$$N_R = \frac{N_P^{Frame}}{N_G} = \frac{N}{N_G \Delta N_F} \cdot \frac{N_T}{\Delta N_T}, \tag{4.119}$$

and the SNR associated with the channel estimation procedure is

$$SNR_{est} = \frac{N_R \sigma_T^2}{\sigma_N^2 + \sigma_D^2} = N_R \frac{\sigma_T^2}{\sigma_D^2} \cdot \frac{SNR_{data}}{1 + SNR_{data}}, \tag{4.120}$$

where $\sigma_N^2 = \frac{1}{2}E\left[|N_{k,l}|^2\right]$ and the SNR associated with data symbols is given by $SNR_{data} = \sigma_d^2 / \sigma_N^2$. For moderate and high SNR values,

$$SNR_{est} \approx N_R \frac{\sigma_T^2}{\sigma_D^2}, \tag{4.121}$$

that is, we have an irreducible noise floor of $N_R \dfrac{\sigma_T^2}{\sigma_D^2}$.

To avoid significant performance degradation due to channel estimation errors, $SNR_{est}$ should be much higher than $SNR_{data}$. This could be achieved with $\sigma_T^2 \ll \sigma_D^2$, provided that $N_R \ll 1$.

To obtain the frequency channel response estimates, the receiver applies the following steps:

Data symbol estimates are removed from the pilots. The resulting sequence becomes

$$(\tilde{R}_{k,l,nrx})^{(q)} = R_{k,l,nrx} - \sum_{ntx=1}^{Ntx}((\hat{S}_{k,l,ntx})^{(q-1)}(\hat{H}_{k,l,ntx,nrx})^{(q-1)}), \tag{4.122}$$

where $(\hat{S}_{k,l,ntx})^{(q-1)}$ and $(\hat{H}_{k,l,ntx,nrx})^{(q-1)}$ are the data and channel response estimates of the previous iteration. This step can only be applied after the first iteration. In the first iteration, we set $(\tilde{R}_{k,l,nrx})^{(1)} = R_{k,l,nrx}$ (note the receiver structure in Figure 4.27). The channel frequency response estimates is computed using a moving average with size $W$, while at the same time removing the pilots, as follows (data is considered to be zero mean):

$$(H_{k,l,ntx,nrx})^{(q)} = \frac{1}{W} \sum_{l'=l-\lfloor W/2 \rfloor}^{l+\lfloor W/2 \rfloor} \frac{(\tilde{R}_{k,l'nrx})^{(q-1)}}{s_{k,l'ntx}^{Pilot}}. \tag{4.123}$$

After the first iteration, the data estimates can also be used as pilots for channel estimation refinement. This is especially useful if the spacing of pilot symbols in the time domain is $\Delta N_T > 1$. The respective channel estimates are computed as

$$(\tilde{H}_{k,l,ntx,nrx})^{(q)} = \frac{(Y_{kl\,nrx})^{(q-1)}(\hat{S}_{kl\,ntx})^{(q-1)*}}{\left|(\hat{S}_{kl\,ntx})^{(q-1)}\right|^2}. \tag{4.124}$$

These channel estimates are enhanced by ensuring that the corresponding impulse response has a duration $N_G$. This is accomplished by computing the time domain impulse response of (4.123) and (4.124) through $\{(\tilde{h}_{i,l})^{(q)}; i = 0, 1, \ldots, N-1\} = \text{IDFT}\{(\tilde{H}_{k,l})^{(q)}; k = 0, 1, \ldots, N-1\}$ (zeros can be used for the missing carriers if $\Delta N_F > 1$, in order to perform an "FFT-interpolation"), followed by the truncation of this sequence according to $\{(\hat{h}_{i,l})^{(q)} = w_i (\tilde{h}_{i,l})^{(q)}; i = 0, 1, \ldots, N-1\}$ with if $w_i = 1$ the $i$th time domain sample is inside the cyclic prefix duration and $w_i = 0$ otherwise.

The final frequency response estimates are then simply computed using $\{(\hat{H}_{k,l})^{(q)}; k = 0, 1, \ldots, N-1\} = \text{DFT}\{(\hat{h}_{i,l})^{(q)}; i = 0, 1, \ldots, N-1\}^* \Delta N_F$.

### 4.5.3.2 Data-Aided Estimation Algorithm with Multiplexed Pilots

Data-aided estimation can also be employed for algorithms with multiplexed pilots, in order to use fewer pilots and promote bandwidth efficiency. To obtain the frequency channel response estimates, the receiver applies the following steps:

■ In the first iteration, the channel estimates are simply computed from the pilots. Assuming that only the first and last blocks are used for multiplexed pilots, the remaining blocks' estimates may be found via a linear interpolation.

■ In the subsequent iterations, the process of getting the new channel estimates is simple, admitting now that the estimated data bits are our new pilots. Of course, this procedure will not yield good results if the estimates are wrong. Before doing this, however, it is necessary to remove the transmit signal interference for each receive antenna, in order to obtain a channel estimate for a pair of transmit-receive antennas. This is done using the channel and data estimates of the previous iteration through

$$(\ddot{H}_{k,l,ntx,nrx}) = \frac{(R_{k,l,nrx} - \sum_{n=1,\neq ntx}^{Ntx}((\hat{s}_{k,l,n})^{(q-1)}(\hat{H}_{k,l,nrx})^{(q-1)}).(\hat{s}_{k,l,ntx})^{(q-1)*}}{\left|(\hat{S}_{kl\,ntx})^{(q-1)^2}\right|}. \quad (4.125)$$

The channel estimates are then processed through the DFT and IDFT in order to guarantee that the corresponding impulse response has a duration $N_G$.

In the second and successive iterations, the resulting channel estimate is blended with the estimate from the previous iteration, with a specific weight for the iteration at hand (scalar $IW$ weights of 10%, 20%, and 30% are suggested).

$$(\tilde{H}_{k,l,ntx,nrx})^{(q)} = IW^{(q)}\ddot{H}_{k,l,ntx,nrx} + (1 - IW^{(q)})(\hat{H}_{k,l,ntx,nrx})^{(q-1)} \quad (4.126)$$

The resulting estimates are passed through a moving average filter to cancel out some of the noise effect and provide continuity:

$$(\hat{H}_{k,l,ntx,nrx})^{(q)} = \frac{1}{W}\sum_{l'=1-[W/2]}^{1+[W/2]}(\tilde{H}_{k,l',ntx,nrx})^{(q)} \quad (4.127)$$

For all interpolations, the channel estimates are enhanced by ensuring that the corresponding impulse response has a duration $N_G$.

## 4.6 Diversity and Spatial-Multiplexing Schemes with Channel Estimation

Diversity and spatial-multiplexing schemes for SC-FDE and OFDM using channel estimation will be discussed in this section.

## 4.6.1 Diversity Schemes for SC-FDE with IB-DFE

Let us consider block transmission schemes where the $l$th transmitted block has the form

$$s_l(t) = \sum_{n=-N_G}^{N-1} s_{n,l} h_T(t - nT_S), \qquad (4.128)$$

with $T_S$ denoting the symbol duration, $N_G$ denoting the number of samples at the cyclic prefix, and $h_t(t)$ being the adopted pulse shaping filter. Throughout this section, we assume square-root raised cosine filtering. The signal $s_l(t)$ is transmitted over a time-dispersive channel. The received signal is sampled, and the cyclic prefix is removed. The resulting time-domain block is

$$\{y_{n,l}; n = 0,1,...,N-1\}, \qquad (4.129)$$

which is then subject to the frequency domain equalization. For SC-FDE schemes, the $l$th time-domain length-$N$ data block to be transmitted is denoted as

$$\{s_{n,l}^{Tx}; n = 0,1,...,N-1\}. \qquad (4.130)$$

The same length is applicable to other variables using index $n$ (i.e., variables of type $x_{n,l}$) throughout this section. $s_{n,l}^{Tx}$ corresponds to

$$s_{n,l}^{Tx(m)} = s_{n,l}^{(m)} + s_{n,l}^{P(m)}, \qquad (4.131)$$

where $s_{n,l}^{(m)}$ is the $n$th data symbol transmitted in the $l$th time domain block by the $m$th transmit antenna ($1 \le m \le M$, with $M = 1$ for SISO, $M = 2$ for STBC2, and $M = 4$ for STBC4, where $M$ stands for the number of transmit antennas). Moreover, $s_{n,l}^{(m)}$ is selected from a given constellation (e.g., a QPSK constellation) under an appropriate mapping rule. $s_{n,l}^{P(m)}$ corresponds to the block of superimposed pilots (it is assumed that $s_{n,l} = s_{N+n,l}$, $n = -N_G, -N_G+1,...,-1$). The transmitter frequency-domain block is

$$\{S_{k,l}^{Tx(m)}; k = 0,1,...,N-1\}, \qquad (4.132)$$

where $S_{k,l}^{Tx(m)} = S_{k,l}^{(m)} + S_{k,l}^{P(m)}$. The same length is applicable to other variables with index $k$ (i.e., variables of type $X_{k,l}$) throughout this section.

Lower- and uppercase signal variables correspond to time- and frequency-domain variables, respectively. The mapping between one and the other is achieved through DFT and IDFT operations (i.e., $X_{k,l} = \text{DFT}\{x_{n,l}\}$ and). $x_{n,l} = \text{IDFT}\{X_{k,l}\}$

$\mathrm{DFT}[x]$ denotes discrete Fourier transform of $x$, and $k$ stands for the subcarrier index (in the frequency domain, $N$ stands for the number of subcarriers per block).

Assuming that the cyclic prefix is longer than the overall channel impulse response of each channel, the $l$th frequency-domain block before the FDE block (i.e., the DFT of the $l$th received time-domain block, after removing the cyclic prefix) is

$$Y_{k,l} = \sum_{m=1}^{M} S_{k,l}^{Tx(m)} H_{k,l}^{(m)} + N_{k,l}$$

$$= \sum_{m=1}^{M} \left( S_{k,l}^{(m)} + S_{k,l}^{P(m)} \right) H_{k,l}^{(m)} + N_{k,l}. \tag{4.133}$$

In the above expression, $H_{k,l}^{(m)}$ denotes the channel frequency response for the $k$th subcarrier and the $l$th time domain block (the channel is assumed invariant in the frame) and the $m$th transmit antenna. $N_{k,l}$ is the frequency-domain block channel noise for that subcarrier and the $l$th block. We assume a frame structure with $N$ subcarriers per block and $N_T$ time-domain blocks, each one corresponding to an FFT block. For OFDM schemes, the data symbols are transmitted in the frequency domain; that is, $S_{k,l}^{(m)}$ are selected according to an appropriate constellation.

For SC-FDE signals and SISO, the output of the linear equalizer comes as

$$\tilde{A}_{k,l} = Y_{k,l} H_{k,l}^* \Big/ \left( \alpha + \left| H_{k,l} \right|^2 \right). \tag{4.134}$$

Since using $\alpha = 0$ might lead to noise enhancement effects in the channel estimates when $\left| \bar{S}_{k,l}^{(i)} \right|^2$ is small, we will consider

$$\alpha = \mathrm{E}\left\{ \left| N_{k,l} \right|^2 \right\} / \mathrm{E}\left\{ \left| S_{k,l} \right|^2 \right\}. \tag{4.135}$$

In the SISO OFDM case, we simply need to invert the channel effects [Gusmao et al. 2003], which means that the equalization process would lead to $\tilde{A}_{k,l} = Y_{k,l} / H_{k,l}$, corresponding to (4.134) with $\alpha = 0$. However, for PSK constellations, we just need the phase, which means that the equalization could be accomplished by $\tilde{A}_{k,l} = Y_{k,l} H_{k,l}^*$. This avoids numerical problems for subcarriers with frequency response close to zero (it also has the advantage of properly weighting the subcarriers for the channel decoding).

### 4.6.1.1 Space Time Block Coding for Two Antennas

If we employ Alamouti's transmit diversity, we need some processing at the transmitter. Alamouti's coding can be implemented either in the time domain or in the frequency domain. In this section, we consider time-domain coding (the extension to frequency domain coding is straightforward). By considering space time block coding with two transmit antennas, the time-domain blocks to be transmitted by the $m$th antenna ($m = 1$ or $2$) are $s_{n,l}^{(m)}$, with [Marques da Silva and Correia 2002; Marques da Silva et al. 2009b]

$$s_{n,2l-1}^{(1)} = a_{n,2l-1}$$

$$s_{n,2l-1}^{(2)} = -a_{n,2l}^{*}$$

$$s_{n,2l}^{(1)} = a_{n,2l} \qquad\qquad (4.136)$$

$$s_{n,2l}^{(2)} = a_{n,2l-1}^{*}.$$

Considering the matrix-vector representation, let us define $s_{n,l}^{[1,2]}$ as

$$s_{n,l}^{[1,2]} = \begin{bmatrix} a_{n,2l-1} & a_{n,2l} \\ -a_{n,2l}^{*} & a_{n,2l-1}^{*} \end{bmatrix}, \qquad (4.137)$$

where different rows of the matrix refer to transmit antenna order and different columns refer to symbol period orders. $a_{n,l}$ refers to the $n$th sample of the symbol selected from a given constellation, to be transmitted in the $l$th time domain block.

Assuming for now the conventional linear FDE for SC schemes, Alamouti's post-processing for two antennas (denoted in this section as STBC2) becomes [Marques da Silva et al. 2009b],

$$\tilde{A}_{k,2l-1} = \left[ Y_{k,2l-1} H_{k,l}^{(1)*} + Y_{k,2l}^{*} H_{k,l}^{(2)} \right] \beta_{k}$$

$$\tilde{A}_{k,2l} = \left[ Y_{k,2l} H_{k,l}^{(1)*} - Y_{k,2l-1}^{*} H_{k,l}^{(2)} \right] \beta_{k}. \qquad (4.138)$$

Defining

$$Y_{k,l}^{[1,2]} = \begin{bmatrix} Y_{k,2l-1} & Y_{k,2l}^{*} \\ Y_{k,2l} & -Y_{k,2l-1}^{*} \end{bmatrix} \qquad (4.139)$$

and

$$H_{k,l}^{[1,2]} = \begin{bmatrix} H_{k,l}^{(1)*} & H_{k,l}^{(2)} \end{bmatrix}^{T}, \qquad (4.140)$$

(4.138) can be expressed in the matrix-vector representation as

$$\tilde{A}_{k,l}^{[1,2]} = \left[ Y_{k,l}^{[1,2]} \times H_{k,l}^{[1,2]} \right] \times \beta_{k}, \qquad (4.141)$$

where

$$\tilde{\mathbf{A}}_{k,l}^{[1,2]} = \begin{bmatrix} \tilde{A}_{k,2l-1} & \tilde{A}_{k,2l} \end{bmatrix}^T \qquad (4.142)$$

$$\tilde{A}_{k,2l-j} = A_{k,2l-j} \overbrace{\sum_{m=1}^{M} \left| H_{k,l}^{(m)} \right|^2 \beta_k}^{\text{Desired Symbol}} + \overbrace{N_{k,2l-j}^{eq}}^{\text{Noise Component}} \qquad j = 0,1 \qquad (4.143)$$

where

$$\beta_k = \left( \alpha + \sum_{m=1}^{M} \left| H_{k,l}^{(m)} \right|^2 \right)^{-1}. \qquad (4.144)$$

$N_{k,l}^{eq}$ denotes the equivalent noise for detection purposes, with

$$E\left\{ \left| N_{k,l}^{eq} \right|^2 \right\} = \left[ 2\sigma_N^2 \left( \sum_{m=1}^{M} \left| H_{k,l}^{(m)} \right|^2 \right) \right] \beta_k^2, \qquad (4.145)$$

and with

$$\sigma_N^2 = E\left\{ \left| N_{k,l} \right|^2 \right\} \Big/ 2. \qquad (4.146)$$

Alamouti's postprocessing for OFDM signals is the same as that defined in (4.138) but assuming

$$\beta_k^{-1} = \sum_{m=1}^{M} \left| H_{k,l}^{(m)} \right|^2. \qquad (4.147)$$

This corresponds to (4.144) with $\alpha = 0$. For PSK constellations, we could simply remove the $\beta_k$ coefficient because it does not affect the phase.

### 4.6.1.2 Space Time Block Coding for Four Antennas

Using unspecified complex valued modulation, such an improvement is possible only for the two-antenna scheme. Higher schemes with four and eight antennas with code rate one exists only in the case of binary transmission [Hochwald et al. 2001]. The proposed STBC4 scheme has $M = 4$ transmit antennas, presenting a code rate one. The symbol construction can be generally written as [Marques da Silva and Correia 2002]

$$
s_{n,l}^{[1,4]} = \begin{bmatrix} s_{n,l}^{[3,4]} & s_{n,l}^{[1,2]} \\ s_{n,l}^{[1,2]*} & -s_{n,l}^{[3,4]*} \end{bmatrix},
\tag{4.148}
$$

where $s_{n,l}^{[3,4]}$ is the same as $s_{n,l}^{[1,2]}$, by replacing the subscripts 1 by 3 and 2 by 4, as well as by replacing $2l$ by $4l$ (e.g., $a_{n,2l-1} \rightarrow a_{n,4l-3}$).

Similarly to (4.136) and (4.137), and considering the space time block coding with four transmit antennas as in (4.148), the time-domain blocks to be transmitted by the $m$th antenna ($m = 1, 2, 3,$ or 4) are $s_{n,l}^{(m)}$, which can be expressed in the matrix-vector representation as

$$
s_{n,l}^{[1,4]} = \begin{bmatrix}
a_{n,4l-3} & a_{n,4l-2} & a_{n,4l-1} & a_{n,4l} \\
-a_{n,4l-2}^* & a_{n,4l-3}^* & -a_{n,4l}^* & a_{n,4l-1}^* \\
a_{n,4l-1}^* & a_{n,4l}^* & -a_{n,4l-3}^* & -a_{n,4l-2}^* \\
-a_{n,4l} & a_{n,4l-1} & a_{n,4l-2} & -a_{n,4l-3}
\end{bmatrix}.
\tag{4.149}
$$

Note that, in (4.149), different rows of the matrix refer to different transmit antennas and different columns refer to different symbol periods.

The $l$th frequency-domain block before the FDE block (i.e., the DFT of the $l$th received time-domain block, after removing the cyclic prefix) is $y_{n,l} = \text{IDFT}\{Y_{k,l}\}$, with $Y_{k,l}$ as defined by (4.133).

Let us define

$$
\tilde{A}_{k,l}^{[1,4]} = \begin{bmatrix} \tilde{A}_{k,4l-3} & \tilde{A}_{k,4l-2} & \tilde{A}_{k,4l-1} & \tilde{A}_{k,4l} \end{bmatrix}^T,
\tag{4.150}
$$

$$
Y_{k,l}^{[1,4]} = \begin{bmatrix} Y_{k,l}^{[3,4]} & -Y_{k,l}^{[1,2]*} \\ Y_{k,l}^{[1,2]} & Y_{k,l}^{[3,4]*} \end{bmatrix}
\tag{4.151}
$$

and

$$
H_{k,l}^{[1,4]} = \begin{bmatrix} H_{k,l}^{(1)*} & H_{k,l}^{(2)} & H_{k,l}^{(3)} & H_{k,l}^{(4)*} \end{bmatrix}^T,
\tag{4.152}
$$

with $Y_{k,l}^{[3,4]}$ as defined for $Y_{k,l}^{[1,2]}$ by replacing the subscripts 1 by 3 and 2 by 4, as well as by replacing $2l$ by $4l$ (e.g., $Y_{k,2l-1} \rightarrow Y_{k,4l-3}$).

Assuming, for now, the conventional SC-FDE decoding (no IB-DFE receiver), the postprocessing STBC for four antennas ($M = 4$) becomes [Marques da Silva and Correia 2002],

$$\tilde{A}_{k,l}^{[1,4]} = \left[ Y_{k,l}^{[1,4]} \times H_{k,l}^{[1,4]} \right] \times \beta_k \qquad (4.153)$$

If we do not use implicit pilots, then the decoded symbols become

$$\tilde{A}_{k,4l-j} = \overbrace{A_{k,4l-j} \sum_{m=1}^{M} \left| H_{k,l}^{(m)} \right|^2}^{\text{Desired Symbol}} - \overbrace{C_k A_{k,4l-p}}^{\text{Residual Interference}} + \overbrace{N_{k,4l-j}^{eq}}^{\text{Noise Component}} \qquad (4.154)$$

with $j = 0,1,2,3$, and with $p = 3 - j$. We also define

$$C_k = 2\operatorname{Re}\left\{ H_{k,l}^{(1)*} H_{k,l}^{(4)} - H_{k,l}^{(2)} H_{k,l}^{(3)*} \right\} \Big/ \left\{ \left( \sum_{m=1}^{M} \left| H_{k,l}^{(m)} \right|^2 \right) \right\}, \qquad (4.155)$$

which stands for the residual interference coefficient generated in the STBC decoding process of order four.

In the following, we will show how we can remove this residual interference.

### 4.6.1.3 Receiver Design for STBC Detection and Channel Estimation

In this section, we present a receiver with STBC detection and channel estimation for SC-FDE with superimposed pilots. Without loss of generality, it is assumed that there is a pilot for each subcarrier of each block of the frame (and for each transmit antenna), that is, $\Delta N_F = \Delta N_T = 1$, leading to $N_P^{Frame} = NN_T$ and a pilot multiplicity or redundancy of

$$N_R = N_P^{Frame} / N_G = NN_T / N_G . \qquad (4.156)$$

The principles behind this receiver are the following:

We first obtain the preliminary channel frequency response estimate

$$\tilde{H}_k^{(1)(m)} = \frac{1}{N_T} \sum_{l=1}^{N_T} \frac{Y_{k,l}^{Rx} S_{k,l}^{P(m)*}}{2\sigma_P^2}, \qquad (4.157)$$

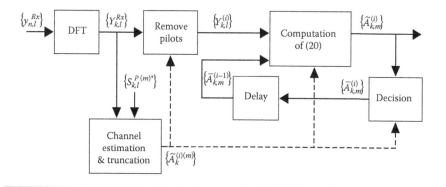

**Figure 4.28 Block diagram of the initial phase of STBC detection and channel estimation.**

where $Y_{k,l}^{Rx}$ denotes the $l$th received frequency-domain block $(l = 1, 2, .., N_T)$. This initial phase of STBC detection and channel estimation is depicted in Figure 4.28, where solid lines refer to the main detection chain and dashed lines represent the auxiliary chain. Note that $\tilde{H}_k^{(i)(m)}$ denotes the channel estimate for the $m$th transmit antenna and the $i$th iteration.

The channel estimate is enhanced through the truncation $\hat{H}_{k,l}^{(m)} = \text{DFT}\left\{ h_{n,l}^{(m)} w_n \right\}$. The pilots are removed from the received frequency-domain blocks, leading to [Marques da Silva et al. 2009]

$$Y_{k,l}^{(1)} = Y_{k,l}^{Rx} - \sum_{m=1}^{M} \hat{H}_k^{(1)(m)} S_{k,l}^{P(m)} \tag{4.158}$$

and the $N_T$ blocks of detected samples $\tilde{A}_{k,l}^{(1)}$ are generated using the IB-DFE receiver. Thus, considering space time block coding with four antennas, and SC-FDE signals with IB-DFE in the receiver, the decoded symbols $\tilde{A}_{k,l}^{(1)}$ becomes

$$\tilde{A}_{k,l}^{[1,M](i)} = \overbrace{\left[ \mathbf{Y}_{k,l}^{[1,M]} \times \mathbf{F}_{k,l}^{[1,M](i)} \right]}^{\text{STBC decoding with feedforward of IB-DFE}}$$

$$+ \overbrace{C_k \left\{ \left[ \hat{\mathbf{A}}_{k,l}^{[1,M](i)} \right]^T \right\}^T}^{\substack{\text{Cancellation of residual interference STBC4} \\ \text{(not applicable to STBC2)}}} - \overbrace{B_{k,l}^{(i)} \bar{\mathbf{A}}_{k,l}^{[1,M](i-1)}}^{\text{Feedback of IB-DFE}} \tag{4.159}$$

with $M = 2$ for STBC2 and $M = 4$ for STBC4. Note that the component that corresponds to the cancellation of the residual interference is not applicable to STBC2. In (4.159) $C_k$ is defined by (4.155). The vector

$$\mathbf{F}_{k,l}^{[1,M]} = \left[ F_{k,l}^{(1)} \quad \cdots \quad F_{k,l}^{(M)} \right]^T \tag{4.160}$$

is composed of the feedforward coefficients

$$\left\{ F_{k,l}^{(i)(m)}; m = 1,2,...,M \right\} \tag{4.161}$$

and the feedback coefficients are $B_{k,l}^{(i)}$, both defined in the next section. $\hat{A}_{k,l}^{[1,M]}$ and $\tilde{A}_{k,l}^{[1,M]}$ are vectors defined using the same structure as for $\tilde{A}_{k,l}^{(1-M)}$ (both defined in the next section). The block $\hat{A}_{n,4l-j}^{(i-1)}$ denotes the DFT of the data estimates associated to the previous iteration; that is, the hard decisions associated with the time-domain block at the output of $\hat{a}_{n,4l-j}^{(i-1)}$. In addition, $\left\{ \overline{A}_{k,4l-j}^{(i-1)}; j = 0,1,2,3 \right\}$ denotes the average signal conditioned to the IB-DFE output for the previous iteration $\tilde{a}_{n,4l-j}^{(1)}$ from (4.174). In the case of a SISO system, (4.159) takes the form

$$\tilde{A}_{k,l}^{(i)} = Y_{k,l}F_{k,l}^{(i)} - B_{k,l}^{(i)}\overline{A}_{k,l}^{(i-1)}. \tag{4.162}$$

That is, there is a single branch (there is no STBC decoding), and there is no cancellation of the residual interference. In the case of STBC2 (two transmit antennas), there is no cancellation of the residual interference component. Furthermore, for OFDM schemes with STBC, (4.159) also applies with the difference that there is no feedback component, and the feedforward component only has a numerator of (4.167).

The decoded STBC blocks are submitted to a decision device, and the average values of the data symbols $\overline{A}_{k,4l-j}^{(1)}$ are obtained using (4.174). These average data values are then used to generate the average values of the transmitted symbols $\overline{S}_{k,l}^{(1)}$ that will be used in the next iteration.

For the second and further iterations, the pilots are removed from the received blocks and the average values of the data symbols will be used as pilots for obtaining the channel frequency response estimate

$$\tilde{H}_k^{(i)(m)} = \frac{\displaystyle\sum_{l=1}^{N_T} Y_{k,l}^{(i-1)}\overline{S}_{k,l}^{(i-1)*}}{\displaystyle\sum_{l=1}^{N_T} \left| \overline{S}_{k,l}^{(i-1)} \right|^2 + \alpha}. \tag{4.163}$$

Since using $\alpha = 0$ might lead to noise enhancement effects in the channel estimates when $\displaystyle\sum_{l=1}^{N_T} \left| \overline{S}_{k,l}^{(i)} \right|^2$ is small, we will consider

$$\alpha = \mathrm{E}\left\{ |N_{k,l}|^2 \right\} / \mathrm{E}\left\{ |S_{k,l}|^2 \right\}. \tag{4.164}$$

If we have moderate to high SNR and/or $N_T \gg 1$ then

$$\sum_{l=1}^{N_T} \left| \overline{S}_{k,l}^{(i)} \right|^2 \approx N_T 2\sigma_P^2 \tag{4.165}$$

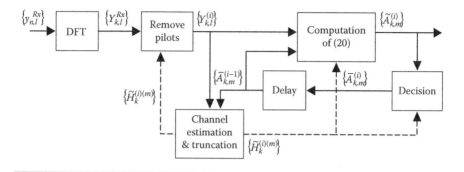

**Figure 4.29   Block diagram of the second and other phases of STBC detection and channel estimation.**

and we could use $\alpha = 0$. The next iterations (after the first one depicted in the previous figure) are depicted in Figure 4.29.

It is worth noting that, in the case of STBC4, since $\tilde{A}^{(i)}_{k,4l-j}$ presents (initially) residual interference, the detection of $A^{(i)}_{k,4l-j}$ should be accompanied by the detection of $A^{(i)}_{k,4l-p}$ (with $p = 3\text{-}j$) to allow the cancellation of the estimated residual interference generated in the decoding process.

To further improve performance with STBC4 the residual interference to be subtracted (which is a function of the estimate of the symbol that generates interference), we consider an iterative interference cancellation (IIC) that can be implemented as follows:

i.   Compute $\hat{A}^{(i)(q)}_{k,4l-j}$ using (4.159) without canceling the residual interference.

ii.  Based on $\hat{A}^{(i)(q)}_{k,4l-j}$, compute $\hat{A}^{(i)(q)}_{k,4l-p}$ after canceling the corresponding residual interference.

iii. Based on $\hat{A}^{(i)(q)}_{k,4l-p}$, compute $\hat{A}^{(i)(q+1)}_{k,4l-j}$ after canceling the residual interference $(C_k\hat{A}^{(i)}_{k,4l-p})$.

Repeat steps 0 and 0 iteratively to improve the accuracy of $\hat{A}^{(i)}_{k,4l-p}$ (cancellation of the residual interference), which will finally be used to improve the accuracy of $\hat{A}^{(i)}_{k,4l-j}$.

## 4.6.1.4 Computation of the IB-DFE Coefficients

It can be shown that the optimum feedback coefficients are [Benvenuto and Tomasin 2002b; Dinis et al. 2003]

$$B^{(i)}_{k,l} = \sum_{m=1}^{M} F^{(i)(m)}_{k,l} H^{(m)}_{k,l} - 1 \tag{4.166}$$

and the feedforward coefficients are given by

$$F^{(i)(m)}_{k,l} = \frac{\kappa^{(i)}_l Q^{(m)}_{k,l}}{\alpha + \left(1 - \left(\rho^{(i-1)}_l\right)^2\right)\sum_{m=1}^{M}\left|H^{(m)}_{k,l}\right|^2} \tag{4.167}$$

with $Q_{k,l}^{(m)} = H_{k,l}^{(m)*}$ for $m = 1$ or $4$ and $Q_{k,l}^{(m)} = H_{k,l}^{(m)}$ for $m = 2$ or $3$. In the particular case of SISO, we only have $m = 1$ (with $M = 1$) and $Q_{k,l} = H_{k,l}^*$. In case of STBC of order two (i.e., STBC2), we have $Q_{k,l}^{(m)} = H_{k,l}^{(m)*}$ for $m = 1$ and $Q_{k,l}^{(m)} = H_{k,l}^{(m)}$ for $m = 2$. For OFDM signals, (4.167) becomes $F_{k,l}^{(i)(m)} = Q_{k,l}^{(m)}$ (i.e., the denominator of (4.167) $\kappa_l^{(i)}$ does not exist). In (4.167) a constant is chosen to ensure

$$\gamma_l^{(i)} = \frac{1}{N} \sum_{m=1}^{M} \sum_{k=0}^{N-1} F_{k,l}^{(i)(m)} H_{k,l}^{(m)} \qquad (4.168)$$

and the correlation factor $\rho_{4l-j}^{(i-1)}$ is defined as

$$\rho_{4l-j}^{(i-1)} = \frac{E\left\{ \hat{a}_{n,4l-j}^{(i-1)} a_{n,4l-j}^* \right\}}{E\left\{ \left| a_{n,4l-j} \right|^2 \right\}}. \qquad (4.169)$$

The computation of the correlation factor (as well as other related parameters) depends on the adopted constellation.* For QPSK constellations, its computation is relatively simple, and it can be shown that the correlation coefficient is given by [Gusmao et al. 2006]

$$\rho_{4l-j}^{(i)} = \frac{1}{2N} \sum_{n=0}^{N-1} \left( \rho_{n,4l-j}^{I(i)} + \rho_{n,4l-j}^{Q(i)} \right). \qquad (4.170)$$

($\rho_{4l-j}^{(i)}$ is almost independent of $l$ for large values of $N$, provided that $H_{k,l}^{(m)}$ remains constant for the frame duration), where

$$\rho_{n,4l-j}^{I(i)} = \left| \tanh\left( \frac{L_n^{I(i)}}{2} \right) \right|$$

$$\rho_{n,4l-j}^{Q(i)} = \left| \tanh\left( \frac{L_n^{Q(i)}}{2} \right) \right|. \qquad (4.171)$$

The LLRs of the in-phase and quadrature bits associated with $a_{n,4l-j}^{I(i)}$ and $a_{n,4l-j}^{Q(i)}$, respectively, are given by

$$L_n^{I(i)} = \frac{2}{\sigma_i^2} \tilde{a}_{n,4l-j}^{I(i)}$$

$$L_n^{Q(i)} = \frac{2}{\sigma_i^2} \tilde{a}_{n,4l-j}^{Q(i)} \qquad (4.172)$$

---

* Moreover, the performance with larger constellations can be rather poor due to the residual interference levels.

respectively, with

$$\sigma^2_{i,4l-j} = \frac{1}{2}\mathrm{E}\left\{\left|a_{n,4l-j} - a^{(i)}_{n,4l-j}\right|^2\right\} \approx \frac{1}{2N}\sum_{n=0}^{N-1}\left|\hat{a}^{(i)}_{n,4l-j} - \bar{a}^{(i)}_{n,4l-j}\right|^2 \qquad (4.173)$$

(as with $\rho^{(i)}_{4l-j}$, $\sigma^2_{i,4l-j}$ is almost independent of $l$ for large values of $N$, provided that $H^{(m)}_{k,l}$ remains constant for the frame duration).

The conditional average values associated with the data symbols are given by

$$\bar{a}^{(i)}_{n,4l-j} = \tanh\left(\frac{L^{I(i)}_{n,4l-j}}{2}\right) + j\tanh\left(\frac{L^{Q(i)}_{n,4l-j}}{2}\right), \qquad (4.174)$$

Therefore, the four symbols ($j=0,1,2,3$) that comprise the STBC4 block need to be decoded independently by the IB-DFE receiver, with the exception of the symbol estimates that originate the residual interference generated in the STBC4 decoding process, as shown in (4.159).

The IB-DFE with soft decisions does not need to perform the channel decoding in the feedback loop. As an alternative, we can define a Turbo FDE that employs the channel decoder outputs, instead of the uncoded soft decisions in the feedback loop of the IB-DFE. The main difference between IB-DFE with soft decisions and the turbo FDE is in the decision device: in the first case, the decision device is a symbol-by-symbol soft-decision (for QPSK constellations, this corresponds to the hyperbolic tangent, as in (4.174)); for the Turbo FDE a soft-in, soft-out channel decoder is employed in the feedback loop. The soft-in soft-out block, which can be implemented as defined in Vucetic and Yuan [2002], provides the LLRs of both the "information bits" and the "coded bits." The inputs of the soft-in, soft-out block are the LLRs of the "coded bits" at the FDE output, given by (4.172) and (4.173). The receiver for OFDM schemes with STBC2 is straightforward [Dinis et al. 2003]. It is worth noting that these STBC schemes can easily be extended to multiple receive antennas by employing receive diversity.

### 4.6.1.5 Performance Results

In this section, we present a set of performance results concerning the proposed receivers for SC-FDE with two- and four-antenna STBC schemes. We consider coded bit error rate (BER) and block error rate (BLER) performances, which are expressed as a function of $E_b / N_0$, where $N_0$ is the one-sided power spectral density of the noise and $E_b$ is the energy of the transmitted bits (i.e., the degradation due to the useless power spent on the cyclic prefix is not included).

Each block has $N = 256$ symbols selected from a QPSK constellation under a Gray mapping rule (similar results were observed for other values of $N$, provided that $N \gg 1$).

With respect to the channel effects, the performance is essentially dictated by the number of relevant multipath propagation components. We can identify three major scenarios:

- If we have a single component the channel is flat in the frequency, but the fading effects can be very strong due to lack of diversity effects. In this case, we have significant gains when we employ transmit diversity schemes, but there is no need to employ sophisticated equalization schemes.
- If we have a large number of components, the channel is severely time-dispersive but we have significant multipath diversity. Since SC-FDE schemes (and to a weaker degree, coded OFDM schemes) can cope with strong interference levels and take full advantage of multipath diversity, there are only negligible improvements when we employ transmit diversity (there can even be some degradation due to residual interference).
- If we have a small number of components, the channel is dispersive and the fading effects can still be relevant. In this scenario, the gain obtained with transmit diversity is potentially high.

Clearly, our schemes are recommendable for the latter case. For this reason, we will consider a channel with two uncorrelated paths, with the same average power and a separation of 0.125 μs. We assumed uncorrelated Rayleigh fading between different transmit and receive antennas. The channel is also assumed to be invariant during the frame. The duration of the useful part of the blocks ($N$ symbols) is 1 μs and the cyclic prefix has a duration of 0.125 μs. For SC-FDE systems, we considered the linear FDE (i.e., just the first iteration of the IB-DFE receiver) and turbo FDE with five iterations (which employs the channel decoder output in each of the IB-DFE iterations). Beyond this number, the performance improvement was almost negligible.

Linear power amplification is considered at the transmitter, and perfect synchronization is assumed at the receiver. The square-root raised cosine filtering was used with a roll-off factor of 0.2. The channel encoder is a convolutional code with generators $1 + D^2 + D^3 + D^5 + D^6$ and $1 + D + D^2 + D^3 + D^6$, and the coded bits associated with a given block are interleaved and mapped into the constellation points.

Figure 4.30 shows coded BER results for the SC-FDE with various transmit diversity schemes, using both linear FDE receiver and turbo FDE (i.e., IB-DFE with soft decisions, considering the channel decoding in the feedback loop). Clearly, the increased diversity due to STBC schemes leads to significant performance improvements relatively to the SISO case. From this figure, it is also clear that the turbo FDE receiver always performs better than the linear FDE receiver. It can also be observed that the STBC4 with the linear FDE receiver does not perform better than STBC2, due to the residual interference (inter-symbol interference generated in the STBC4 decoding process).

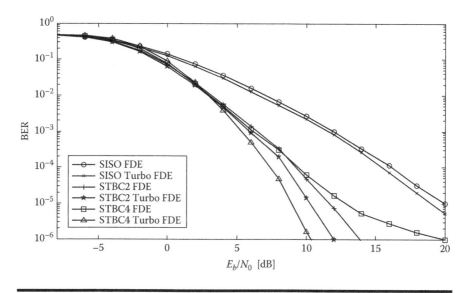

**Figure 4.30  BER results with ideal channel estimation (various transmit diversity schemes).**

However, when we add the turbo FDE to the STBC4 receiver, we have significant performance improvement, with STBC4 outperforming the others, due to the increased diversity (as expected, this advantage is higher when we do not have significant multipath diversity). It is worth noting that the receiver includes the IIC algorithm previously described, which cancels the residual interference generated in the STBC4 decoding process.

Figure 4.31 shows coded BLER results for the SC-FDE with various transmit diversity schemes, using both linear FDE receiver and the turbo FDE receiver. As in the case of the BER curves, the increased diversity due to STBC schemes leads to significant performance improvements relative to the SISO case. As before, the turbo FDE always performs better than the linear FDE receiver, but the relative gain is higher than in the case of the BER curves. Note that the BER can be significantly affected by a channel realization with a deep fade, while for the BLER a channel realization with a deep fade leads simply to a single block (or frame) error. Moreover, the STBC4 with the linear FDE receiver does not perform better than the STBC2, due to the residual interference. However, for the turbo FDE receiver, the performance tends to improve when we increase the diversity order.

Figure 4.32 shows a performance comparison between SC-FDE and OFDM when channel coding is considered (it is well known that uncoded performance is very poor for OFDM schemes). Note that the OFDM receiver for the STBC4 does not include cancellation of the residual interference. Clearly, there is a visible advantage of SC-FDE with the turbo FDE over the OFDM system for all schemes. This advantage is visibly higher for higher diversity order, especially for STBC4,

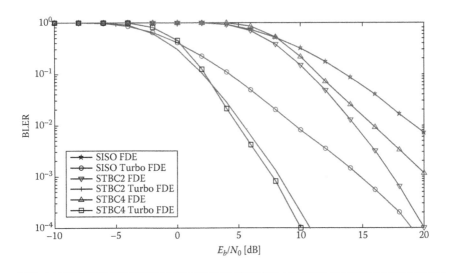

**Figure 4.31  BLER results with ideal channel estimation (various transmit diversity schemes).**

namely, due to the cancellation of the residual interference, which can be included in the turbo FDE (IB-DFE) receiver without significant increase of its complexity.

Figure 4.33 shows the coded results for the STBC4 using the turbo FDE with conventional pilots. Clearly, conventional pilots with powers 3 dB or less below the power of the data present approximately the same performance. However, there is a noticeable degradation when the pilots' power is 6 dB below the power of data.

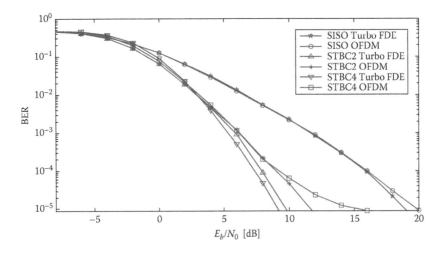

**Figure 4.32  BER results with ideal channel estimation (turbo FDE vs. OFDM with various transmit diversity schemes).**

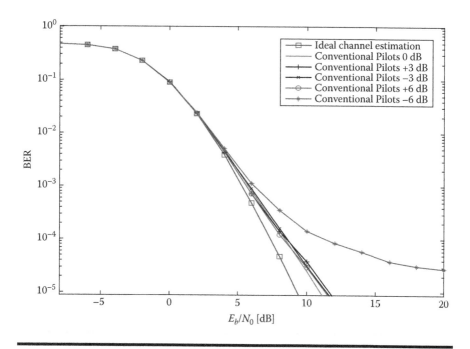

**Figure 4.33** **BER results for the STBC4 with turbo FDE receiver, using conventional pilots vs. ideal channel estimation.**

Figure 4.34 shows the performance comparison between implicit pilots (with several power levels and number of blocks) and the ideal channel estimation. Clearly, by considering a power relation 0 dB or –3 dB (between the implicit and superimposed pilots power related to the data power) and 15 blocks, the performance degradation is very low, when compared to the ideal channel estimation. By reducing the window size to 5 blocks, it a performance degradation is noticeable. When we compare the results obtained with the implicit pilots (Figure 4.34) against those obtained with conventional pilots (Figure 4.33), it is clear that better performance is achieved with the latter, with the additional advantage of its increased spectrum efficiency. The residual interference generated in the STBC4 decoding process is overcome by the design of the proposed iterativ Figure 4.35 presents coded BLER results for the STBC4 with the turbo FDE receiver, using implicit pilots and conventional pilots versus ideal channel estimation. Similar results from those observed in the two previous BER figures can be observed. Implicit pilots with a relative power of –3 dB and using 15 blocks achieve a BLER performance very close to the ideal channel estimation. Furthermore, it is clear the worst performance is achieved with conventional pilots with 0 dB, as compared to the implicit pilots. The typical noise floor inherent in the use of implicit pilots (due to the

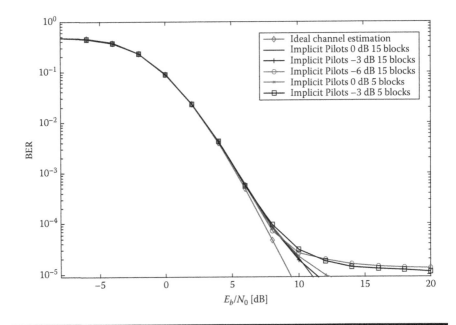

**Figure 4.34   BER results for the STBC4 with turbo FDE receiver, using implicit pilots vs. ideal channel estimation.**

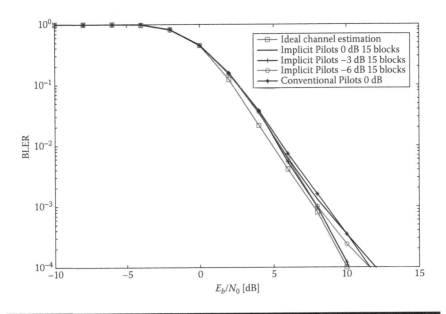

**Figure 4.35   BLER results for the STBC4 with turbo FDE receiver, using implicit pilots and conventional pilots vs. ideal channel estimation**

interference between training and data symbols) was overcome by employing the IB-DFE receiver with (coded) soft decisions, by adopting pilots with relatively low power (i.e., $\sigma_P^2 \ll \sigma_D^2$), and by averaging the pilots over a large number of blocks.

## 4.6.2 CRM for Multi-Antenna Systems

As explained in detail in Chapter 3, CRM is a technique for achieving signal space diversity (SSD) in SISO and MIMO OFDM/OFDMA systems and can be easily combined with turbo or LDPC codes to improve the system performance without a substantial reduction of the spectral efficiency. In this section, we provide the details of using CRM in MIMO schemes.

### 4.6.2.1 Multi-Antenna Transmitter for CRM

CRM can be easily incorporated into OFDM systems. Figure 4.36 shows the block diagram of an OFDM transmitter with CRM and multiple transmitting antennas.

An information block is encoded, interleaved, and mapped according to the constellation symbols. A rotation matrix (RM) is then applied by grouping the symbols into $M_{CRM}$-tuples and multiplying them by the RM $A_{M_{CRM}}$.

The resulting sequence is split into $M_{tx}$ parallel streams, which are interleaved in the symbol interleaver. The objective of the symbol interleaver is to explore the characteristics of OFDM transmissions in severe time-dispersive environments whose channel frequency response can change significantly between different subcarriers. The interleaver ensures that samples of a super-symbol are mapped to distant subcarriers, thus taking advantage of the diversity in the frequency domain. Finally, pilot symbols are inserted into the resulting sequence before it is converted to the time domain using a size-$N$ Inverse discrete Fourier transform (IDFT) and transmitted as a conventional OFDM transmission.

### 4.6.2.2 Multi-Antenna Receiver for CRM

Figure 4.37 represents the receiver block diagram for OFDM transmissions with CRM assuming the use of $N_{rx}$ receiving antennas.

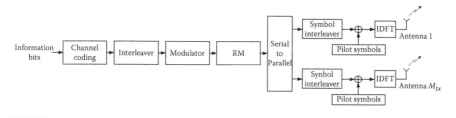

**Figure 4.36   Transmitter block diagram for MIMO-OFDM transmissions using CRM.**

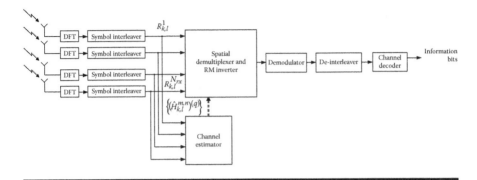

**Figure 4.37   Receiver structure for OFDM transmissions with CRM.**

At the receiver, the signal is sampled, the cyclic prefix removed, and the resulting signal is converted to the frequency domain with an appropriate size-$N$ DFT operation. The sequence of symbols is then de-interleaved. If the cyclic prefix is longer than the overall channel impulse response, each received $M_{CRM}$-sized super-symbol can be expressed using matrix notation as

$$R = H \cdot X + N, \qquad (4.175)$$

where $\mathbf{H}$ is the frequency response channel matrix. Matrix $\mathbf{H}$ can be defined as a block-wise diagonal matrix

$$\mathbf{H} = \begin{bmatrix} H_1 & & 0 \\ & \ddots & \\ 0 & & H_{M_{CRM}/M_{tx}} \end{bmatrix}, \qquad (4.176)$$

with

$$\mathbf{H}_k = \begin{bmatrix} H_k^{1,1} & \cdots & H_k^{1,M_{tx}} \\ \vdots & \ddots & \vdots \\ H_k^{N_{rx},1} & \cdots & H_k^{N_{rx},M_{tx}} \end{bmatrix}, \quad k = 1,\ldots,M_{CRM}/M_{tx}. \qquad (4.177)$$

Index $k$ represents a subcarrier position. It is important to note that due to the presence of the symbol interleaver, the different subcarriers denoted by index $k$ may not be necessarily adjacent. To simplify the following explanations, we will assume that $M_{CRM}$ is a multiple of the number of transmitting antennas $M_{tx}$. $N$ is an $(N_{rx} \cdot M_{CRM}/M_{tx}) \times 1$ vector containing additive white Gaussian noise (AWGN) samples.

The super-symbol's samples enter the spatial demultiplexer and CRM inverter block, which has the purpose of separating the streams transmitted simultaneously by the multiple antennas and inverting the rotation applied at the transmitter. Two alternative methods will be considered in the following: the minimum mean square error (MMSE) equalizer and the maximum likelihood-based soft output (MLSO) detector.

The MMSE criterion is applied to each individual subcarrier using

$$\hat{X}_k = (H_k)^H \cdot \left[ H_k (H_k)^H + \sigma^2 I \right]^{-1} R_k \tag{4.178}$$

where $\hat{X}_k$ is the $M_{tx} \times 1$ vector with the estimated subset of coordinates from the super-symbol mapped to subcarrier $k$, $R_k$ is the $N_{rx} \times 1$ received signal vector in subcarrier $k$ with one different receive antenna in each position, and $\sigma^2$ is the noise variance. Using the rotated super-symbol estimates, $\hat{X}_k$, the component symbol estimates are computed through

$$\hat{S} = \left( A_{MCRM} \right)^{-1} \cdot \hat{X}. \tag{4.179}$$

In the case of using the MLSO criterion, each symbol estimate is computed as

$$\hat{S}_l = E\left[ S_l | R \right]$$

$$= \sum_{s_i \in \Lambda} s_i \cdot P\left( S_l = s_i | R \right) \tag{4.180}$$

$$= \sum_{s_i \in \Lambda} s_i \cdot \frac{P\left( S_l = s_i \right)}{p(R)} p\left( R | S_l = s_i \right),$$

with $s_i$ representing a constellation symbol from the modulation alphabet $\Lambda$, $E[\cdot]$ denoting the expected value, $P(\cdot)$ a discrete probability, and $p(\cdot)$ a probability density function (PDF). Considering equiprobable symbols we have $P\left( S_l = s_i \right) = 1/M$, where $M$ is the constellation size. The PDF values required in (4.180) can be computed as

$$p\left( R | S_l = s_i \right) = \frac{1}{M^{MCRM - 1}} \sum_{S_l^{compl} \in \Lambda^{MCRM - 1}} p\left( R | S_l = s_i, S_l^{compl} \right), \tag{4.181}$$

with

$$p\left( R | S_l = s_i, S_l^{compl} \right) = \frac{1}{\left( 2\pi\sigma^2 \right)^{N_{rx} MCRM / M_{tx}}} \exp\left[ \sum_{n=1}^{N_{rx} MCRM / M_{tx}} - \frac{\left| R_n - H(n,:) \cdot A_{MCRM} \cdot s \right|^2}{2\sigma^2} \right], \tag{4.182}$$

where $S_l^{compl}$ is an $(M_{CRM} - 1) \times 1$ vector representing a possible combination of symbols transmitted together with $S_l$ in the same super-symbol, $\mathbf{s}$ is an $M_{CRM} \times 1$ vector comprising $S_l^{compl}$ and $s_i$, $R_n$ is the $n$th received sample in (4.175), and $\mathbf{H}(n,:)$ is the $n$th line of channel matrix $\mathbf{H}$.

Independent of which of the two methods was applied, the resulting symbol estimates are serialized, demodulated, and de-interleaved before entering the channel decoder block, which produces the final estimate of the information sequence.

As an example of the possible use of CRM, we present some simulation results considering an UTRA LTE-based system with transmission bandwidth of 10 MHz. Two hundred subcarriers are occupied with QPSK-modulated data in each FFT block. The channel impulse response is based on a Typical Urban (TypU) environment [3GPP 2011c] with Rayleigh fading assumed for the different paths. A velocity of 50 km/h was considered. The channel encoder was a rate-1/3 turbo code based on two parallel recursive convolutional codes characterized by $G(D) = [1 \ (1 + D2 + D3)/(1 + D + D3)]$. Puncturing is applied to the parity bits for achieving higher coding rates. At the receiver, a maximum of 12 turbo decoding iterations are applied. The results presented next will be shown as a function of $E_b/N_0$, where $E_b$ is the average information bit energy and $N_0$ is the single-sided noise power spectral density.

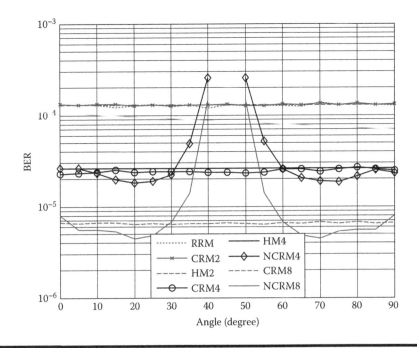

**Figure 4.38** **BER performance of uncoded transmission for MIMO $2 \times 2$ with different rotation matrices, $E_b/N_0 = 20$ dB.**

Figure 4.38 shows the BER performance of a MIMO 2 × 2 uncoded OFDM transmission with different rotation matrices, namely, CRM, real rotation matrix (RRM), and Hadamard matrices (HM), versus the angle for $E_b/N_0 = 20$ dB. In Chapter 3, the BER performance for SISO was analyzed also for $E_b/N_0 = 20$ dB, and thus a direct comparison between the two figures is possible. We conclude that with MIMO the BER performance is lower than with SISO and is almost independent of the chosen angle φ, which did not happen in the SISO case. Results for nonorthogonal CRM (NCRM) matrices are also presented in this figure. Once again, it is evident that the BER performance improves with the increase of the dimension of CRM matrices, allowing the reduction of the BER from $1.5 \times 10^{-4}$ (RRM,0°) to $4.5 \times 10^{-6}$ (NCRM8,20°). The BER reduction due to NCRM can achieve two orders of magnitude.

Figure 4.39 shows the BER performance of MIMO 2 × 2 OFDM transmissions employing turbo codes with coding rate ¾ and $E_b/N_0 = 10$ dB. The comparison between the matching figure in Chapter 3 for SISO results also indicates that except for NCRM, the BER performance does not depend on the angle. It should be noted that in this case, for dimension 8, the performance is worse than with dimension 4. This should be explained by the existence of an optimal combination of channel coding with MIMO and SSD. The BER reduction due to NCRM4 is one order of magnitude.

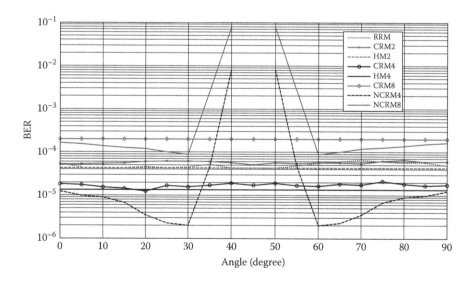

**Figure 4.39 BER performance of MIMO 2 × 2 transmission turbo coded with rate ¾ for different rotation matrices, $E_b/N_0 = 10$ dB.**

### 4.6.3 Capacity Augmentation with Channel Estimation

In this section, the MIMO schemes for spatial diversity and capacity augmentation described in Section 4.4 are simulated with channel estimation. For the more complex case of using implicit pilots, the general receiver depicted in Figure 4.27 can be explicitly drawn as Figure 4.40.

Therefore, in the case of using implicit pilots, the received sampled sequence can be expressed as (after conversion to the frequency domain)

$$R_{k,l}^n = \sum_{m=1}^{M_{tx}} \left( S_{k,l}^m + S_{k,l}^{m,Pilot} \right) H_{k,l}^{m,n} + N_{k,l}^n. \tag{4.183}$$

The pilot symbols are then removed from the sequence, resulting in

$$\left( Y_{k,l}^n \right)^{(q)} = R_{k,l}^n - \sum_{m=1}^{M_{tx}} S_{k,l}^{m,Pilot} \left( \hat{H}_{k,l}^{m,n} \right)^{(q)}, \tag{4.184}$$

where $\left( \hat{H}_{k,l}^{m,n} \right)^{(q)}$ is the channel frequency response estimate and $q$ is the current iteration. Since only one of the antennas can transmit a pilot over each carrier, the summation in (4.184) only has one term. The sequences of samples (4.184) enter the spatial demultiplexer block, which can apply any of the equalization methods discussed in the receiver structure, and follow the processing sequence already described in the previous sections.

#### 4.6.3.1 Main Simulation Parameters

To study and compare the behavior of the proposed schemes, several Monte Carlo simulations were performed for 16-HQAM (k1 = 0.4) and 64-HQAM (k1 = k2 = 0.4) constellations (other values of k1 and k2 could be used but the conclusions

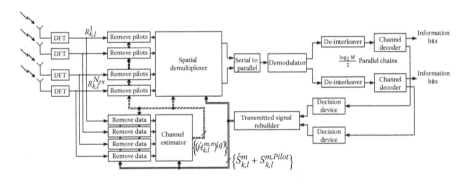

**Figure 4.40  Iterative receiver structure for implicit pilots' transmission.**

would be similar). Two streams of bits with different error protection levels were used in the case of 16-HQAM and three for 64-HQAM. Although this study is valid for any OFDM system, we performed the simulations using the parameters from UTRA LTE documents [3GPP 2006; 3GPP 2007a] for a 10 MHz bandwidth, which were already described in Chapter 3. The channel impulse response is based on the Vehicular A environment from [ETSI 1998c]. The channel encoders were rate-1/2 turbo codes based on two identical recursive convolutional codes characterized by $G(D) = [1 \ (1 + D^2 + D^3)/(1 + D + D^3)]$ [3GPP 2003]. A random interleaver was used within the turbo encoders. The results presented next will be shown as a function of $E_S/N_0$, where $E_S$ is the average symbol energy and $N_0$ is the single-sided noise power spectral density. When using data-multiplexed pilots, the spacing employed was $\Delta N_F = 5 + M_{tx}$ and $\Delta N_T = 4$ or 7 (the two possible configurations proposed in [3GPP 2007a]) and a sinc filter interpolation with length $W = 2$ was applied at the receiver.

In the graph legends, MPB designates most protected bits, IPB means intermediate protected bits, and LPB corresponds to least protected bits. The pilots powers will be denoted as $\beta_P$, which is defined as $\beta_P = E\left[\left|S_{k,l}^{m,Pilot}\right|^2\right] \Big/ E\left[\left|S_{k,l}^m\right|^2\right]$, where $E[\cdot]$ represents the expected value computed over all positions $(k,l)$ containing pilot symbols in the case of the numerator and over all positions containing data symbols in the case of the denominator. For data-multiplexed pilots, $\beta_P$ will always be 0 dB.

### 4.6.3.2 Simulation Results

Table 4.1 shows the individual block sizes employed for these simulations, which once again admit the use of a subframe composed of 7 OFDM blocks.

First, it was important to evaluate the impact of the use of an iterative receiver approach when multiple antennas are being employed. Figure 4.41 compares the performance of the different receiver methods of Table 4.2 for a MIMO 2 × 2 transmission employing a 16-HQAM ($k_l = 0.4$) constellation with data-multiplexed pilots (for $\Delta N_T = 7$). It is clear that, although the receiver with the MMSE equalizer alone performs worse than when using the MLSO equalizer, the performance

**Table 4.1 Individual Block Sizes (in Bits) Employed for Mimo 2 × 2 Simulations**

| Pilot Method | $\Delta N_T$ | Block Size |
|---|---|---|
| Data multiplexed | 4 | 3852 |
| Data multiplexed | 7 | 4024 |
| Implicit | 1 | 4196 |

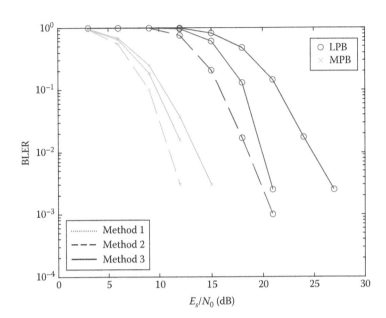

**Figure 4.41  Performance of a 16-HQAM (k1 = 0.4) hierarchical constellation with data-multiplexed pilots $\Delta N_T = 7$ using several receiver methods. MIMO 2 × 2.**

can be substantially improved and achieves lower BLERs when an IC is applied in the last iterations. This is due to the fact that the information feedback by the receiver in the first iterations affects the MMSE equalizer through the improvement of the channel estimates. After a certain number of iterations, the use of the MMSE may not be able to provide any more substantial improvements (if the channel estimates are already sufficiently accurate). Therefore, in the last iterations, it is preferable to employ an IC since the feedback information allows the subtraction of better estimates of interference and, thus, there is always the possibility of

**Table 4.2  Different Receiver Methods Applied for the Mimo 2 × 2 Simulations**

| Receiver Method | Inner Turbo Decoder Iterations | Receiver Iterations |
|:---:|:---:|:---|
| 1 | 12 | 1 with MMSE |
| 2 | 3 | 2 with MMSE + 2 with IC |
| 3 | 12 | 1 with MLSO |

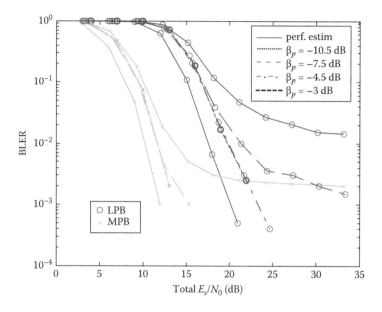

**Figure 4.42   Performance of a 16-HQAM (k1 = 0.4) hierarchical constellation for different $\beta_p$ values. MIMO 2 × 2.**

improving the performance with a high number of iterations even if the channel estimates are already accurate. Similar conclusions could be achieved for implicit pilots, as was done in [Souto et al. 2007b]. For the remainder of this section, the receiver configuration considered will be method 3.

When using multiple transmit antennas, the implicit pilot will be affected by higher levels of interference since there will be $M_{tx}$ data symbols being transmitted at the same time as the pilots. Therefore, before comparing the two pilot transmission methods, it was important to study how the overall performance of the implicit pilot method would be affected for different power levels applied for the pilots. Figure 4.42 shows some performance results. The first conclusion is that the curves for $\beta_p = -10.5$ dB, which correspond to a case where the percentage of power spent on the pilots is the same as in the data-multiplexed pilot transmission with $\Delta N_T = 7$, have considerably high irreducible BLER floors. This means that the implicit pilots are being severely affected by the interference of the multiple data streams. Only when the power level is raised to −4.5 dB did we start obtaining acceptable performances. If we keep increasing the power spent on the pilots, the improvement in the channel estimation quality does not result in a better performance as can be seen from the $\beta_p = -3$ dB curve. In fact, increasing $\beta_p$ even further would just result in similar curves dislocated to the right (worse performances) due to the higher power required for the pilots.

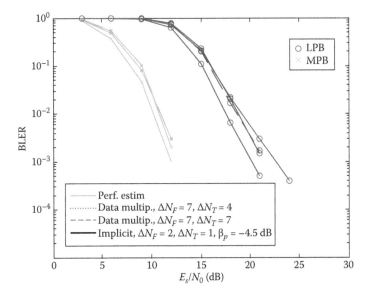

**Figure 4.43  Data-multiplexed pilots vs. implicit pilots. 16-HQAM (k1 = 0.4), MIMO 2 × 2.**

Therefore, for comparison of the implicit pilots against the data multiplexed pilots we used $\beta_p = -4.5$ dB. The results are presented in Figure 4.43. Once again both pilot spacing considered for data-multiplexed pilots have similar performances, which means that the longest time domain spacing can be used also for MIMO 2 × 2 transmissions. As for the implicit pilot results, they are very close to the data-multiplexed pilots' performance for the MPB and only show a small degradation in the LPB.

# Chapter 5

# Link and System-Level Evaluation*

## 5.1 Requirements and Scenarios

In this section, some important aspects related to LTE and E-MBMS requirements and scenarios are described. LTE is a mobile broadband technology, regarded as the common evolution for 3G, including WCDMA/HSPA. HSPA can be designed with either continuous coverage or discontinuous coverage.[†] From the first phase, most LTE radio network designs prefer conforming to the continuous networking principle. Continuous networking can improve cell edge user experience, reduce inter-radio access technologies (RAT) handover requirement, and facilitate future evolution from 3G to LTE [3GPP 2010a].

LTE continuous networking involves considering continuous coverage of the reference signal received power (RSRP). This consists of the average of the power of all resource elements that carry cell-specific reference signals over the entire bandwidth. The reference signal received quality (RSRQ) measurement provides additional information when RSRP is not sufficient to make a reliable handover or cell reselection decision. RSRQ is the ratio between the RSRP and the received signal strength indicator (RSSI) and, depending on the measurement bandwidth, represents the number of resource blocks. RSSI is the total received wideband power including all interference and thermal noise. Best service cells, traffic channel for

---

* The authors would like to acknowledge Paulo Sousa and José Seguro for the contribution to the simulation results presented in this chapter.
† Note that its fundamental network (WCDMA network) is continuous.

both uplink and downlink, physical uplink shared channel (PUSCH), and physical downlink shared channel (PDSCH) are also required.

As was previously described, E-MBMS is performed either in single-cell or multi-cell mode. In single-cell transmissions, E-MBMS traffic is mapped to the downlink shared channel (DL-SCH). In multi-cell mode, transmissions from cells are carefully synchronized to form a multicast/broadcast–single frequency network (MB-SFN) [3GPP 2010d; Marques Silva et al. 2010].

MBSFN is an elegant application of OFDM for cellular broadcast. The principle of operation is quite simple. Identical transmissions are broadcast from closely coordinated cells simultaneously on a common frequency. Signals from adjacent cells arrive at the receiver and are dealt with in the same manner as multipath delayed signals. In this manner, UE can combine the energy from multiple transmitters with no additional receiver complexity.

If the UE is at a cell boundary, the relative delay between the two signals is quite small. However, if the UE is close to one base station and relatively distant from a second base station, the amount of delay between the two signals can be quite large. For this reason, MBSFN transmissions might be supported using 7.5 kHz subcarrier spacing (instead of 15 kHz) and a longer CP [3GPP 2010d], as depicted in Figure 5.1. MBSFN networks also use a common reference signal from all transmitters within the network to facilitate channel estimation. As a consequence of the MBSFN transmission scheme, UE can roam between cells with no handover procedure required. Signals from various cells will vary in strength and in relative delay, but in the aggregate the received signal is still dealt with in the same manner as a conventional single channel OFDM transmission. Figure 5.2 illustrates the MBSFN transmission scheme, and includes an illustration of the single cell point-to-multipoint transmission, the other scenario to be evaluated later in this chapter.

According to 3GPP specifications [3GPP 2009a, 2009b, 2009c, 2007b], the deployment scenario where a dedicated carrier is used for broadcast only (MBSFN) is characterized by the following properties:

- E-UTRA MBMS is envisaged to achieve a cell edge spectrum efficiency of 1 bit/s/Hz, equivalent to the support of, at least, 16 mobile TV channels, at around 300 kbps per 5 MHz channel (in an urban or suburban environment).
- In the deployment scenario, where a carrier is shared between broadcast and unicast traffic, the target performance at cell edge for broadcast traffic should be in line with the existing target performance for the unicast traffic.

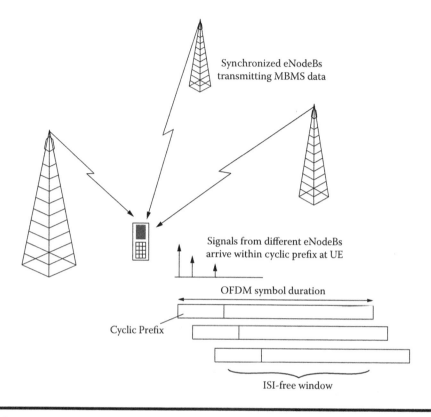

Synchronized eNodeBs transmitting MBMS data

Signals from different eNodeBs arrive within cyclic prefix at UE

OFDM symbol duration

Cyclic Prefix

ISI-free window

**Figure 5.1  MBSFN transmission with long cyclic prefix to avoid ISI.**

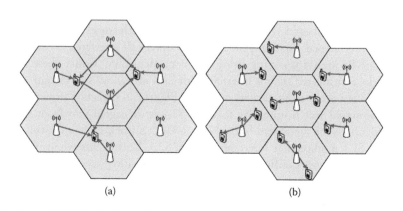

(a)

(b)

**Figure 5.2  MBSFN transmission (a) and single-cell point-to-multipoint SC-PMP (b).**

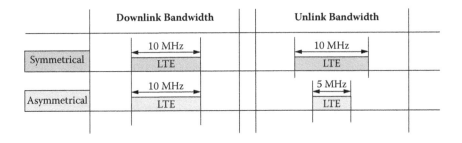

**Figure 5.3  Example of symmetrical bandwidths and asymmetrical bandwidths.**

LTE make use of various frequency bands such as 2.6 GHz, 1.8 GHz, advanced wireless services (AWS*) band, 700 MHz, digital divide band (800 MHz), 2.1 GHz, etc. These bands can be employed to deploy LTE networks in different countries. Thus, LTE involves both opportunities and challenges. On the one hand, no other systems have such rich frequency bands as LTE. Moreover, so many different frequency bands bring challenges to LTE production and networking as well as difficulties for international roaming.

LTE also supports very flexible bandwidths. LTE can support six kinds of bandwidths: 20, 15, 10, 5, 3, and 1.4 MHz [Correia et al. 2010a; 3GPP 2009a].

In most scenarios, LTE FDD will support symmetrical uplink and downlink bandwidths. Special scenarios (e.g., frequency bandwidth limited scenario, interference limited scenario, etc.) may need asymmetrical uplink and downlink bandwidths. The reader should refer to Figure 5.3 where an example of symmetrical bandwidths† and asymmetrical bandwidths‡ is depicted.

## 5.1.1 Inter-Cell Interference Coordination Schemes

Since the LTE system employs OFDMA [Marques da Silva 2012] in the downlink and SC-FDMA in the uplink, intra-cell interference is mitigated due to orthogonality between subcarriers. However, inter-cell interference remains the main source of interference, especially at cell edges. Inter-cell interference occurs when multiple eNBs are transmitting using the same frequency. Under these circumstances, the transmitted signals interfere with each other, collisions occur, and the UEs might not receive the packets correctly. Figure 5.4 depicts a situation where S1.1 is the signal sent from eNB1 to UE1, S2.1 is the signal sent from eNB2 to UE1, and S2.2 is the signal sent from eNB2 to UE2. Since UE1 is at the cell border, the signal S1.1 might suffer interference from S2.1. This occurs when there is no inter-cell interference coordination, such as in the SC-PMP scenario. However, in

---

* AWS band corresponds to 1.7 GHz in the uplink and 2.1 GHz in the downlink.
† Downlink bandwidth 10 MHz is the same as uplink bandwidth 10 MHz.
‡ Downlink bandwidth is 10 MHz while uplink bandwidth is 5 MHz.

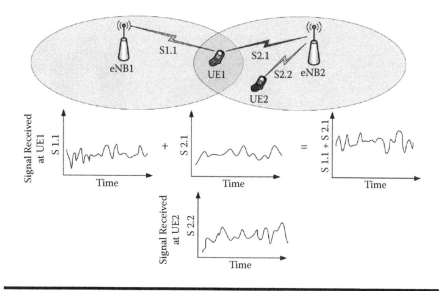

**Figure 5.4 Example of potential inter-cell interference.**

the MBSFN scenario, signals S1.1 and S2.1 are soft combined and therefore UE1 receives S1.1 + S2.1 correctly.

Coordinated multi-point transmission is a technique used to combat inter-cell interference and improve cell edge performance. It is used by MBSFN, thus allowing adjacent base stations to coordinate their transmissions with diversity gains. It relies on a high-speed backbone (fiber) transmission system to exchange information between adjacent base stations.

Inter-cell interference coordination (ICIC), another solution that can be used to overcome this limitation, consists of managing the radio resources (radio resource blocks). This chapter covers the traditional frequency reuse schemes and hybrid reuse (HR), also referred to as soft reuse schemes.

## *5.1.2 Frequency Reuse*

Frequency reuse or fractional frequency reuse is a technique traditionally applied in RAN planning and consists partitioning the available spectrum in fractions, where each of those fractions is assigned to different neighbor cells. The goal of this is to ensure that at least the cells around a given cell are using different subcarriers. This way direct inter-cell interference from neighbor cells is avoided, but inter-cell interference from other distant cells that reuse the same part of the spectrum still exists. The factor of reuse is given by

$$\text{reuse factor} = \frac{1}{\text{number of subcarrier partitions}} \tag{5.1}$$

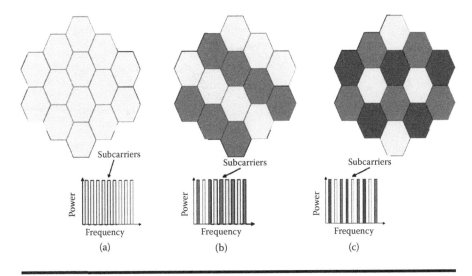

**Figure 5.5    Reuse 1 scheme (a), Reuse 1/2 scheme (b), Reuse 1/3 scheme (c).**

Figure 5.5 illustrates three particular cases of frequency reuse, namely 1, 1/2, and 1/3, for an hexagonal cellular topology and assuming eNB with tri-sector antennas.

When reuse factor 1 is applied, in practice no frequency reuse is being used, because all cells are using the total available frequency or subcarriers. In this case all cells interfere with each other, but maximum spectral efficiency can be achieved (if channel conditions allow). Reuse 1/2 means that the total frequency available per cell is only half of the one in Reuse 1. This halves the total number of interfering cells, and also halves the subcarriers available to transmit at each cell. The same applies to Reuse 1/3, where only 1/3 of the available frequency is allocated to each cell, and at a given cell, only 1/3 of the cells are interfering.

This restriction of resources is achieved by limiting the power output for a given set of frequencies, at a given cell, both in time and/or frequency domain. Considering a Reuse 1/3 scheme, the total subcarriers available $f$ will be divided into 3 different subsets: $f1$, $f2$, and $f3$. Assuming that every cell will be assigned to a specific subset $f_n$, the maximum output power for a given frequency is

$$P_{\max}(f,t) = P, \quad f \in f_n$$
$$P_{\max}(f,t) = 0, \quad f \notin f_n. \tag{5.2}$$

The level of interference reduces as we increase the frequency partitioning (especially at cell borders). This occurs in part because the distance between cells using same subcarriers increases and the power received from those interfering cells is

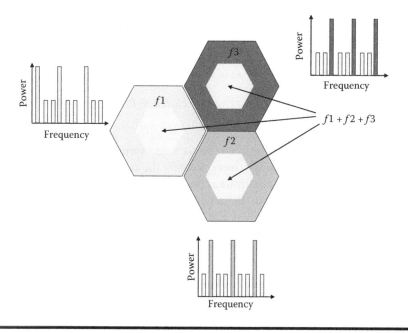

**Figure 5.6    Example of hybrid reuse with Reuse 1 + Reuse 1/3.**

significantly smaller. This is, naturally, an important issue when we aim at lowering inter-cell interference levels at the expense of maximum achievable throughput*. Therefore, only a part of the total throughput can be achieved comparing to using all subcarriers available [Fodor et al. 2009].

### 5.1.3  Soft and Hybrid Reuse

To solve the limitations in terms of spectral efficiency of traditional frequency reuse schemes, hybrid reuse (HR) or soft fractional frequency reuse (SFFR) as it is sometimes called in the literature, was proposed by several authors [Simonsson 2007; Zhou and Zein 2008; Sarperi et al. 2008]. This consists of defining different zones within each cell where different reuse factors are applied. Figure 5.6 is an example of HR 1 + 1/3 (Reuse 1 + Reuse 1/3), where there is an area where all the frequency spectrum is used and another area at the border of the cells where only a fraction of the frequency is used. This way we can achieve maximum spectral efficiency in the center of each cell, and reduced inter-cell interference levels at the border of cells.

---

* Since each cell is using only a part of the total subcarriers available.

This technique is implemented using the same methodology used for normal fractional frequency reuse, but this time different levels of power $P$ are applied to different frequencies

$$P_{\max}(f,t) = P, \quad f \in f_n$$

$$P_{\max}(f,t) = p_{tx} < P, \quad f \notin f_n$$

(5.3)

for each sector of every cell.

The assignment of different power levels to different frequencies is related to the characteristics of radio signals, more precisely the average path losses ($L_{path}$) that radio signals suffer over the air. This has direct impact on the total power received at the UE ($P_{RxUE}$). This way, at the center of the cell, all frequencies transmitted with power $P < p_{tx}$ will be received and, at the edge of the cells, only frequencies transmitted with power $P > p_{tx}$ will be received and will interfere with each other.

To determine the power levels that each frequency should use, we first need to determine the location of the different reuse zones. This can be somewhat tricky, because it is hard for the UE to determine with precision its position, therefore making it hard to know when it should be using one or another reuse. To solve this, we can define certain levels or targets for carrier-to-interference ratio ($C/I$), and based on those target values the UE knows that it should apply for one type of reuse (e.g., if UE $C/I$ is higher than the target $C/I$) or another (e.g., if UE $C/I$ is below target $C/I$). On the other hand, the level of $C/I$ received at UE is associated with a certain distance to the center of the cell (that is, eNB site). This allows us to define distances of reuse (DR) that are directly related to a certain level of $C/I$ and, in turn, these levels of $C/I$ can be determined from the received power of pilot subcarriers at UEs.

Let us analyze the example depicted in Figure 5.7, where HR 1 + 1/3 is considered, meaning that Reuse 1 is used within a certain area (the small inner hexagons)

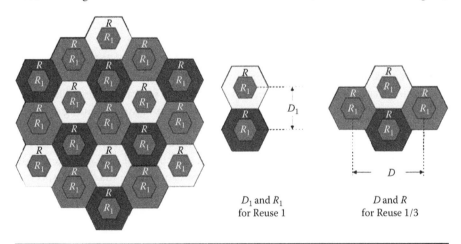

$D_1$ and $R_1$
for Reuse 1

$D$ and $R$
for Reuse 1/3

**Figure 5.7** **Hexagonal cellular topology, using HR 1 + 1/3 scheme.**

and Reuse 1/3 is used outside that area (outer region) for each cell sector. $R$ is the cell radius and $D$ is the distance between the center of cells using the same frequencies, and $R_I$ is the radius of the Reuse 1 zone inside each cell, $D_I$ being the distance between Reuse 1 zones inside each cell.

The carrier-interference ratio ($C/I$) in hexagonal cellular topologies is expressed as

$$\frac{c}{i} = \frac{1}{i_0}\left(\frac{D}{R}\right)^{\alpha}$$

$$\frac{C}{I} = 10\log_{10}\left(\frac{c}{i}\right) \quad [dB]$$

$$(5.4)$$

where $i_0$ is the number of interfering cells interfering at distance $D$, and $\alpha$ is the average exponent of propagation path loss that can take values between 2 and 5 (based on 3GPP specifications, we use $\alpha = 3.488$). According to Figure 5.7, both Reuse 1 and Reuse 1/3 have $i_0 = 6$ interfering cells at distance $D$. If we want the DR to be 1/3 of $R$, for Reuse 1 we have $R_I = DR = R/3$ and $D_I = 2R$. The $C/I$ for users at $DR$ is

$$\frac{c}{i} = \frac{1}{6}\left(\frac{2R}{\frac{R}{3}}\right)^{3.488} = 86.305$$

$$\frac{C}{I} = 10\log_{10}(86.305) = 19.36 \quad [dB]$$

$$(5.5)$$

We know that for a UE located at $DR = R/3$, the $C/I$ target should be 19.36 dB, meaning that UEs with $C/I$ equal or higher than this value will be using Reuse 1, and UEs with $C/I$ lower than that will be using Reuse 1/3.

Next, we must find out what is the power that should be allocated to frequencies meant to be received within Reuse 1, and the power that must be used by frequencies meant to be received on the entire cell (especially in the border of the cell). To do so, we need to first know what is the $C/I$ level received at the edge of the cell; in our example, the edge of the cell at the distance R from the center of cell. We have

$$\frac{c}{i} = \frac{1}{6}\left(\frac{3R}{R}\right)^{3.488} = 7.692$$

$$\frac{C}{I} = 10\log_{10}(7.692) = 8.86 \quad [dB]$$

$$(5.6)$$

The $C/I$ for frequencies supposed to be received by UEs at the edge of the cell will be 8.86 dB. If we consider that for that $C/I$ level the power is $P_{max}$ 46 dBm = 40 W, the frequencies that are not supposed to be received must be transmitted with inferior power $P_{tx}$. This way we have

$$P_{tx} = \frac{40W}{\left(\frac{3R}{R}\right)^{3.488}} = \frac{40W}{46.152} = 0.867[W]. \qquad (5.7)$$

Looking at Figure 5.6, we conclude that, at the center of the cells, frequencies will be transmitted with $P_{max} = 0.867$ [W] and, at the border of the cells, frequencies will be transmitted using $P_{max} = 40$ [W].

The example described can be applied to other configurations such as HR 1 + 1/2 or HR 1/2 + 1/3. Furthermore, more than just two reuse zones can exist. Schemes of HR will be evaluated for various configurations of two and three reuse zones.

## 5.2 Evaluation Methodology

Radio network resources are obviously finite and scarce (that is, limited bandwidth for transmission). As the number of services that use the available bandwidth increases, it is essential to have techniques that optimize the usage and allocation of such limited resources. Scheduling consists of precise allocation of resources, balancing different factors. This allocation is typically performed in the time domain, meaning that scheduling algorithms divide the time domain in small pieces (time slots (TSs)), and then allocate every one of those TSs to different users.

With the introduction of LTE and OFDMA, the paradigm of scheduling is slightly changed since resources can be allocated not only in the time domain but also in the frequency domain. This flexibility of allocation both in time and frequency will, in theory, enable OFDMA-based schedulers to allocate resources more efficiently, for example, by allocating more time and bandwidth to users with more needs.

This section relies on the evaluation of different multi-resolution techniques for E-MBMS services using either MBSFN or SC-PMP, which are PMP broadcasting scenarios. For these two scenarios, the scheduling is straightforward; that is, all users who have subscribed to E-MBMS are sharing the same frequency and physical resource blocks (PRB), so all users receive the same data and are served at the same time. However, with E-MBMS, there is also the possibility of point-to-point (PtP) transmission. In this case, the usage of efficient scheduling algorithms is essential to ensure a correct and efficient usage of resources while guaranteeing that every user is served.

The employed simulators can be divided into two different classes: link-level simulator (LLS) and system-level simulator (SLS). The objective of LSS is to study the characteristics of the radio link that is established over the air between a UE and eNB, more specifically, analyzing the SNR values that a certain link should achieve in order to receive a block of bits, with a certain percentage of bit error rate

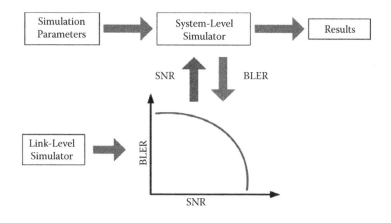

**Figure 5.8 Interaction between LLS and SLS. (Adapted from Correia, A. et al., in *Evolved Cellular Network Planning and Optimization for UMTS and LTE*, Lingyang Song and Jia Shen, Eds.: CRC Press Auerbach Publications, New York, August 2010.)**

(BER) or block error rate (BLER). In the SLS, an operating LTE network is simulated, including several UEs and eNBs, with the objective of understanding how the different UEs/eNBs interfere with each other when transmitting and receiving data. The data collected in LLS (the minimum SNR for each type of link) is used to parameterize the SLS, as indicated by Figure 5.8.

A single simulation that takes into consideration both the link-level and system-level simulation would be preferable, but that implies a significant increase in complexity and huge simulation times. Therefore, two separate but interconnected approaches are the best solution. In the next sections, the results from LLS and SLS will be presented and discussed.

The LLS objective is to model the same conditions experienced in the real world when a radio link between a UE and an eNB is established. The output that is expected consists of one or more figures which show the relation between the minimum required SNR that the link should experience to achieve a given BLER.

The simulator used for this purpose is in essence the same used in [Correia et al. 2010a]. This simulates the radio link between UE and eNB for both uplink and downlink (for this study, only downlink is considered) and considers the specifications of 3GPP Release 7 regarding the signal processing and physical transport channels. The structure of both transmitter and receiver is included in the LLS. Multipath Rayleigh fading channels with additive white Gaussian noise (AWGN) channels are considered, to illustrate a more realistic scenario. Multipath fading plays an important role in the study of the impact as it induces hierarchical modulations. As previously described, the different blocks of bits present in this type of modulation have different sensitivities regarding channel conditions.

**Table 5.1   Link-Level Simulator Parameters**

| Scenario | SC-PMP | MBSFN |
|---|---|---|
| Propagation Channel | Typical Urban | MBSFN |
| Transmission BW | 10 MHz | 10 MHz |
| Sub-carrier spacing | 15 kHz | 15 kHz |
| Subframe duration (TTI) | 0.5 ms | 0.5 ms |
| CP length (samples) | 72 | 256 |
| OFDM symbols per subframe | 7 | 6 |
| FFT size | 1024 | 1024 |
| User mobility | 30 km/h | 30 km/h |
| Number of sub-carriers | 200, 300, 600 | 200, 300, 600 |
| Number of sub-carriers per RB | 12 | 12 |
| Coding rate | 1/2, 3/4 | 1/2, 3/4 |
| Modulations | QPSK, H-16QAM, H-64QAM | QPSK, H-16QAM, H-64QAM |
| Transmitting antennas | 1 | 1 |
| Receiving antennas | 2 | 2 |

As we would expect in a real-life scenario, phenomena like multipath fading are not linearly predictable. Instead, they are nonlinear, assuming random or pseudo-random values and, in order to obtain consistent and reliable data to support our findings, a Monte Carlo methodology is applied to the simulator, where every simulation run is iterated at least 1,000 times. The simulator is written in the Matlab programming language, since it is very efficient for computational processing of mathematical expressions.

For the link-level simulations, the parameters in Table 5.1 were considered. For the typical urban and MBSFN multipath propagation channels, the number of subcarriers simulated was 600, 300, and 200, corresponding to the situations where Reuse 1, Reuse ½, or Reuse 1/3, respectively, are employed. The vehicular A propagation channel was used to simulate the PtP E-MBMS scenario radio link. Since the number of subcarriers allocated to each user will vary along the simulation, an extensive study had to be performed testing all the possible number of subcarriers. The minimum value was 12 subcarriers, and this value was incremented by 12 subcarriers every time (12, 24, 36, 48 subcarriers, and so on) until the maximum subcarrier value of 600 (corresponding to the total usage of the subcarriers available with a 10 MHz transmitting bandwidth) was reached.

The size of the transmitted block depends on the number of subcarriers used and the number of OFDM symbols per subframe, being expressed by

$$L_{TBcod} = N_{symb} \, N_{sc} \tag{5.8}$$

where $L_{TBcod}$ denotes the total size of the block transmitted in a single subframe including coding and cyclic redundancy check (CRC) bits, $N_{symb}$ is the number of OFDM symbols per subframe and per subcarrier, and $N_{sc}$ is the number of occupied subcarriers during one subframe.

The size of the effective data transmitted block (that is, without the coding and correction bits) is inferior to $L_{TBcod}$, being expressed by

$$L_{TB} = L_{TBcod} \, R_c \tag{5.9}$$

where $R_c$ is the respective coding rate.

The hierarchical modulations included in the simulations were the same for all the environments (as indicated by Table 5.1). Nevertheless, for SC-PMP and MBSFN scenarios, the results for QPSK modulation were not taken into consideration in the SLS, since it is not a hierarchical modulation. However, its throughput is the reference for comparison with 16QAM and 64QAM modulations.

For the different combinations of environment (hierarchical modulation, coding rate, etc.), the LLS produced figures with the evolution of the BLER as a function of the link $E_S/N_0$. The $E_S/N_0$ represents the relation between the energy of each symbol ($E_S$), over the spectral noise density ($N_0$). For the PMP modes (SC-PMP and MBSFN scenarios), the target BLER to be achieved is 1% (that is, in every hundred transmitted packets, only one of those is not correctly decoded at the receiver). This occurs because in these scenarios retransmissions are not carried out, thus requiring a very low BLER. The SNR values used in SLS are obtained using

$$SNR = \frac{E_S}{N_0} + 10 \log_{10} \left( \frac{R_b}{\log_2(M) \, B_w} \right) \tag{5.10}$$

where $E_S/N_0$ is obtained from the figures produced by LLS, $R_b$ is the specific bitrate considered in the simulation run, $M$ is the index of the hierarchical modulation used (e.g., for 16QAM, we have the modulation order $M = 16$), and $B_w$ is the total bandwidth available for transmission (that is, 10 MHz for all the cases). The $R_b$ value is obtained by dividing the size of the block of bits being transmitted (before coding) over the duration in seconds that is needed to transmit the entire

**Table 5.2   LLS Results for SC-PMP Scenario, BLER = 1%**

| Hierarchic Modulation | | Coding Rate | [bits] | [bits] | [kbps] | | db/SNR [dB] |
|---|---|---|---|---|---|---|---|
| 16QAM | 200 | 1/2 | 5600 | 2800 | 5,600 | 14.10 | 8.56 |
| | | 3/4 | 5600 | 4200 | 8,400 | 18.76 | 13.23 |
| | 300 | 1/2 | 8400 | 4200 | 8,400 | 13.00 | 9.22 |
| | | 3/4 | 8400 | 6300 | 12,600 | 16.77 | 13.00 |
| | 600 | 1/2 | 16800 | 8400 | 16,800 | 13.98 | 13.21 |
| | | 3/4 | 16800 | 12600 | 25,200 | 16.10 | 15.34 |
| 64QAM | 200 | 1/2 | 8400 | 4200 | 8,400 | 20.52 | 14.98 |
| | | 3/4 | 8400 | 6300 | 12,600 | 24.52 | 18.99 |
| | 300 | 1/2 | 12600 | 6300 | 12,600 | 18.76 | 14.98 |
| | | 3/4 | 12600 | 9450 | 18,900 | 23.22 | 19.45 |
| | 600 | 1/2 | 25200 | 12600 | 25,200 | 18.45 | 17.68 |
| | | 3/4 | 25200 | 18900 | 37,800 | 22.68 | 21.92 |

block of bits (typically this is the subframe duration or time transmission interval [TTI]).

$$R_b = \frac{L_{TB}}{TTI} \tag{5.11}$$

The results from link-level simulations are presented in Tables 5.2 and 5.3 and Figure 5.9 for each of the environments under consideration. The $E_S/N_0$ and SNR values in these tables are those required to receive the totality of the bits for each MCS.[*] The target SNR values are required to receive only a part of those bits (that is, only the strong bits, or strong and average bits) are obviously lower than those presented here.

Figure 5.9 plots the BLER performance of MIMO 2 × 2 with frequency diversity given by complex rotation matrices (CRM) of dimension 2 (that is, CRM2) as a function of the ratio of symbol energy to noise spectral density ($E_S/N_0$) values for different frequency reuse R, antennas A1/A2, and coding rate Cod. Taking as reference BLER = 0.01 (that is, 1%), the corresponding $E_S/N_0$ values would provide the

---

[*] For 16QAM this corresponds to the weak and strong bits, while for 64QAM this corresponds to the weak, average and strong bits.

**Table 5.3  LLS Results for MBSFN Scenario, BLER = 1%**

| Hierarchic Modulation | | Coding Rate | [bits] | [bits] | [kbps] | [dB] | SNR [dB] |
|---|---|---|---|---|---|---|---|
| **16QAM** | 200 | 1/2 | 4800 | 2400 | 4,800 | 13.00 | **6.79** |
| | | 3/4 | 4800 | 3600 | 7,200 | 17.49 | **11.29** |
| | 300 | 1/2 | 7200 | 3600 | 7,200 | 12.48 | **8.03** |
| | | 3/4 | 7200 | 5400 | 10,800 | 16.76 | **12.33** |
| | 600 | 1/2 | 14400 | 7200 | 14,400 | 11.01 | **9.57** |
| | | 3/4 | 14400 | 10800 | 21,600 | 16.06 | **14.63** |
| **64QAM** | 200 | 1/2 | 7200 | 3600 | 7,200 | 19.50 | **13.29** |
| | | 3/4 | 7200 | 5400 | 10,800 | 23.64 | **17.44** |
| | 300 | 1/2 | 10800 | 5400 | 10,800 | 18.76 | **14.31** |
| | | 3/4 | 10800 | 8100 | 16,200 | 22.93 | **18.50** |
| | 600 | 1/2 | 21600 | 10800 | 21,600 | 18.54 | **17.10** |
| | | 3/4 | 21600 | 16200 | 32,400 | 22.93 | **21.50** |

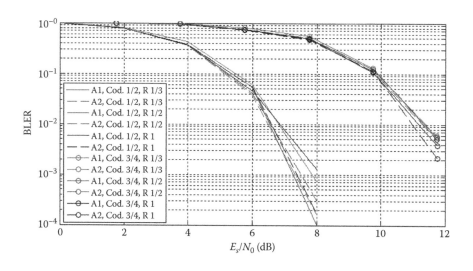

**Figure 5.9    BLER versus Es/No MIMO (2 × 2)+CRM2, QPSK, MBSFN channel.**

SNR values necessary to correctly receive a block of *N* bits being used as an input to run the system-level simulator.

## 5.2.1 System-Level Simulator

The considered system-level simulator was originally developed as part of the work produced in [Soares 2006], for UMTS technology, but has been updated to simulate LTE technology [Rasquete 2009; Correia et al. 2010a]. The simulator was built in the JAVA programming language, due to its characteristics such as portability, multi-platform compatibility, and ease of usage and configuration by users with low experience and familiarization with programming languages.

The core of the SLS is a discrete event generator with some grade of abstraction. The events generated consist of individual tasks such as channel quality indicator (CQI) reporting, packet processing, radio resources management, etc. Propagation, traffic, and mobility models are also part of the system-level simulator and have great impact in the results that will be obtained, especially in terms of coverage and radio link SNR estimation. Moreover, fast-fading and shadowing conditions are emulated, since channel conditions for every eNB/UE combination are time-varying and location dependent.

The geographical environment used in the simulation can be configured manually (that is, setting the geographical position of each eNB). A scenario with 19 sites was configured for the performed simulations and assuming two different configurations: (1) 19 sites corresponding to 19 eNBs with no RN and (2) 19 sites corresponding to 7 eNBs plus 12 RNs. The differences between these two configurations can be seen from Figure 5.10a, where the black triangular shapes and the dark grey

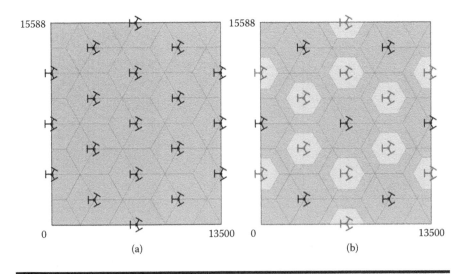

**Figure 5.10** Only eNBs scenario (a); eNBs mixed with h RNs (b).

hexagons represent the location and coverage of eNBs, respectively. In b, the gray triangular shapes and the light gray hexagons represent the location and coverage of RNs. The cellular disposition follows a hexagonal shape pattern (that is, every cell has six adjacent neighbor cells). The inter-site distance (ISD), which represents the distance between the center of two adjacent cells, is represented by the expression

$$ISD = \sqrt{3}R \tag{5.12}$$

where $R$ is the radius of the cells.

Every eNB/RN is made up of three tri-sector antennas, creating three sectors per site (as depicted in Figure 5.10) where the black and dark gray triangular shapes show the orientation of the tri-sector antennas. The general parameterization used for all system-level simulations is shown in Table 5.4 following the 3GPP recommendations [3GPP 2010e; 3GPP 2010f].

The more specific parameters used for point-to-multipoint (SC-PMP and MBSFN scenarios) simulations can be obtained from Table 5.5.

**Table 5.4  General Parameterization for System-Level Simulations**

| Parameter | Values |
|---|---|
| **Simulation Time** | 500 [seg] |
| **Subframe duration (TTI)** | 0.5 [ms] |
| **Carrier Frequency** | 2 [GHz] |
| **Propagation model** | 3GPP extended Okumura-Hata |
| **Distance attenuation (d = distance in kilometers)** | L = 122.23 + 34.88 log(d) [dB] |
| **Cellular layout** | Hexagonal |
| **Number of base station sites** | 19 |
| **User Mobility** | Random walk inside sector |
| **UE antenna height** | 1.5 [m] |
| **eNB Antenna height** | 32 [m] |
| **% of transmitted power by interfering cells** | 90 |
| **Antenna gain of the base station (including feeder loss)** | 15 [dBi] |
| **Width of antenna beam at 3 dB** | 65° |
| **Front/back ratio of the antenna** | 20 [dB] |

**Table 5.5  Parameterization for SC-PMP and MBSFN Simulations**

| Parameter | Values |
|---|---|
| Cell radius | 2250, 3000 [m] |
| Modulations | H-16QAM, H-64QAM |
| Coding rate | 1/2, 3/4 |
| Frequency reuse | • 1/3, 1/2, 1 <br> • HR: (1+1/3), (1+1/2), (1/2+1/3) <br> • HR: (1+1/2+1/3) |
| Number of sectors per base station site | 3 sectors/site |
| Site layout | • 19 eNB <br> • 7 eNB + 12 RN |
| eNB base station power/sector | • 46 [dBm] or 40 [W] |
| RN base station power/sector | • 40 [dBm] or 10 [W] <br> • 34 [dBm] or 2.5 [W] |
| Number of UEs per sector | 20 |

In the following sections, we will analyze the experimental results for every one of the two E-MBMS scenarios exposed, with emphasis on the coverage and throughput results as a function of the percentage of transmitted power per carrier from the base station ($E_C/I_{0r}$). The $E_C/I_{0r}$ is a performance indicator that has a correspondence with the number of OFDMA subcarriers. Having $E_C/I_{0r} = 60\%$ means that, in a certain carrier, we are using 60% of the OFDMA subcarriers (which corresponds to 60% of the carrier power), leaving the remaining 40% to other services. An operator has typically two carriers* per cell sector. Three sectors per cell means that there should be six carriers, each one occupying the maximum 10 MHz carrier bandwidth.

The desired cell coverage for the analyzed scenarios is 95%. It is necessary to perform coverage simulations as a function of $E_C/I_{0r}$ to obtain the percentage of transmitted power values that are necessary to achieve the reference 95% cell coverage. Next, we perform throughput simulations for increasing $E_C/I_{0r}$ to obtain increasing throughputs per carrier. The simulation results presented in this chapter consider not only different scenarios (SC-PMP, MBSFN, PtP) but also different schemes (hierarchical modulations, different MIMO configurations, CRM, relays, etc.). At the end we can find the best compromises between coverage and throughput results for the

---

* With a maximum 10 MHz carrier bandwidth.

different schemes analyzed. The basic idea is to minimize the percentage of transmitted power per carrier from the base station $(E_C/I_{0r})$, and also achieve the reference coverage and find the scheme that provides maximum throughput.

The results presented will only cover the simulations using a cell radius of 2,250 m, since the results obtained for cell radius of 3,000 m are nearly the same, and analysis performed for 2,250 m can be applied to 3,000 m. In the legends of the figures, DR stands as distance of reuse in meters (only applies to hybrid reuse schemes); TD is the output power in watts used by the RNs (only applies to the simulations where RNs where used); in the coverage versus $E_C/I_{0r}$ figures, the following notation is employed:

- For H-16QAM, H1 and H2 represent the Strong and Weak blocks, respectively;
- For H-64-QAM, H1, H2 and H3 represent the Strong, Average, and Weak blocks, respectively;
- For MIMO schemes, A1 and A2 represent Antenna 1 and Antenna 2.

## 5.3 Simulation Results

### 5.3.1 Results for SC-PMP scenario

For the SC-PMP scenario (as well as for the MBSFN scenario), the performed simulations can be categorized in three major groups:

- Case 1—Frequency reuse (reference results): This group includes the conventional reuse schemes using 1, 1/2, and 1/3 partitioning factor. These are the reference results that will serve as a baseline for comparison between the alternatives presented.
- Case 2—Hybrid reuse: This group includes hybrid reuse schemes using two reuse zones (HR 1+1/2, HR 1+1/3, or HR 1/2+1/3) and also three reuse zones (HR 1+1/2+1/3). In the case of two reuse zones, there is an approximate distance from the center of the cell at which the UE switches between one or another reuse scheme. That distance is called distance of reuse or simply DR. When three reuse zones are considered, two of these distances exist, DR1 and DR2, and they tell approximately at which distance the user switches from Reuse 1 to Reuse 1/2, and from Reuse 1/2 to Reuse 1/3, respectively.
- Case 3—Relaying: This group includes a variation of the simulations performed in Case 1 where some of the eNBs are replaced by RNs (as plotted in Figure 5.10b), varying the transmission powers used in the RNs.

When analyzing the obtained results, the values in Table 5.6 should be kept in mind, representing the maximum $R_b$ and target SNR values for each of the reuse, modulation, and coding rates used.

**Table 5.6  Maximum Bit-Rate and Target SNR Values for SC-PMP**

| Hierarchic Modulation | | Coding rate | [kbps] | SNR [dB] |
|---|---|---|---|---|
| 16QAM | Reuse 1/3 | 1/2 | 5,600 | 8.56 |
| | | 3/4 | 8,400 | 13.23 |
| | Reuse 1/2 | 1/2 | 8,400 | 9.22 |
| | | 3/4 | 12,600 | 13.00 |
| | Reuse 1 | 1/2 | 16,800 | 13.21 |
| | | 3/4 | 25,200 | 15.34 |
| 64QAM | Reuse 1/3 | 1/2 | 8,400 | 14.98 |
| | | 3/4 | 12,600 | 18.99 |
| | Reuse 1/2 | 1/2 | 12,600 | 14.98 |
| | | 3/4 | 18,900 | 19.45 |
| | Reuse 1 | 1/2 | 25,200 | 17.68 |
| | | 3/4 | 37,800 | 21.92 |

## 5.3.1.1 Results for SC-PMP with Hierarchical QAM

Figures 5.11 to 5.16 present the system simulation results in terms of coverage versus the percentage of transmitted power from the base station $(E_C/I_{0r})$ for coding rate $(R_c)$ 1/2 and 3/4.

The first conclusion that can be made is about the different $R_c$, where it is clear that $R_c = 1/2$ has better cell coverage (that is, more packets received correctly after decoding with forward error correction) than $R_c = 3/4$, for all block types. This is due to higher forward error correction (FEC) bits when using $R_c = 1/2$, which increases the error detection and correction capabilities of the system. When using $R_c = 3/4$, the best coverage for Strong blocks (H1) is achieved for Reuse 1/3 with relays, but still only 60% of cell coverage is achieved.[*] When using coding rate 1/2 with Reuse 1/3 and RNs transmitting at 2.5 W and with 100% of $E_C/I_{0r}$, 95% of cell coverage is achieved for Strong blocks and, remarkably, 75% of the cell is covered with both Strong and Weak blocks, that is, base and enhancement layer received in 75% of the cell.

When analyzing the throughput results (Figures 5.17 to 5.22), $R_c = 3/4$ has once again worst results than $R_c = 1/2$. This is explained by the lower levels of

---

[*] When at least 95% of the cell should be covered with basic quality, as recommended for E-MBMS.

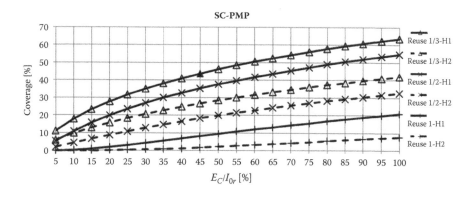

**Figure 5.11** Coverage versus. $E_C/I_{0r}$, H-16QAM, $R_c = 1/2$, for SC-PMP Case 1.

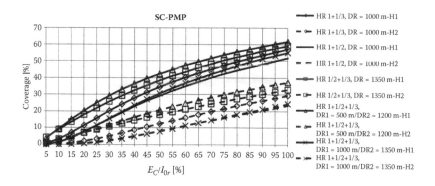

**Figure 5.12** Coverage versus $E_C/I_{0r}$, H-16QAM, $R_c = 1/2$, for SC-PMP Case 2.

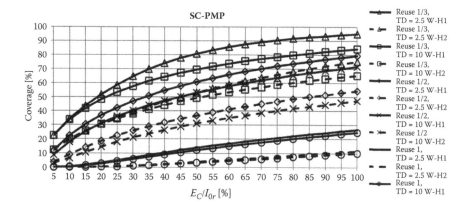

**Figure 5.13** Coverage versus $E_C/I_{0r}$, H-16QAM, $R_c = 1/2$, for SC-PMP Case 3.

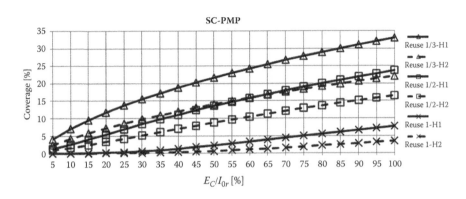

**Figure 5.14** Coverage versus $E_C/I_{0r}$, H-16QAM, $R_c = 3/4$, for SC-PMP Case 1.

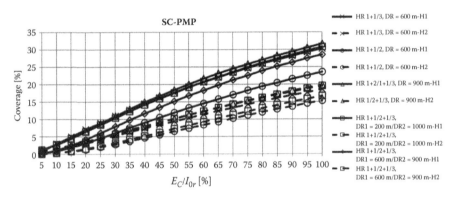

**Figure 5.15** Coverage versus $E_C/I_{0r}$, H-16QAM, $R_c = 3/4$, for SC-PMP Case 2.

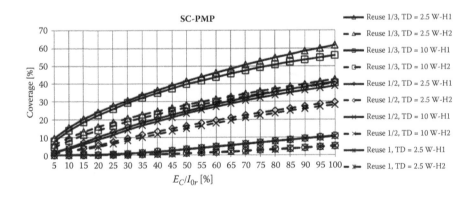

**Figure 5.16** Coverage versus $E_C/I_{0r}$, H-16QAM, $R_c = 3/4$, for SC-PMP Case 3.

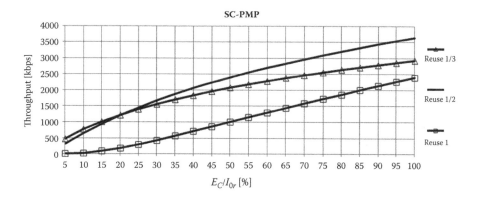

**Figure 5.17** Throughput versus $E_C/I_{0r}$, H-16QAM, $R_c = 1/2$, for SC-PMP Case 1.

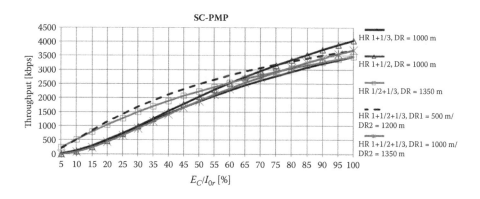

**Figure 5.18** Throughput versus $E_C/I_{0r}$, H-16QAM, $R_c = 1/2$, for SC-PMP Case 2.

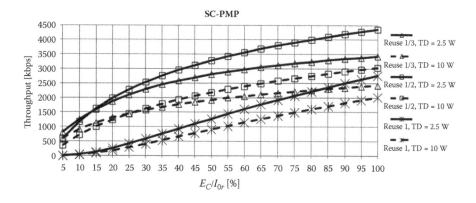

**Figure 5.19** Throughput versus $E_C/I_{0r}$, H-16QAM, $R_c = 1/2$, for SC-PMP Case 3.

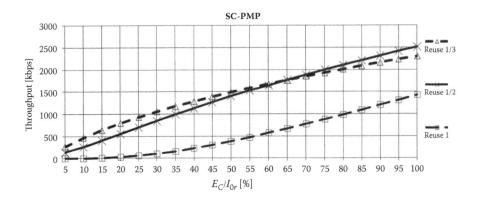

**Figure 5.20  Throughput versus $E_C/I_{0r}$, H-16QAM, $R_c = 3/4$, for SC-PMP Case 1.**

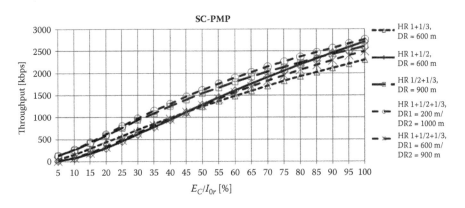

**Figure 5.21  Throughput versus $E_C/I_{0r}$, H-16QAM, $R_c = 3/4$, for SC-PMP Case 2.**

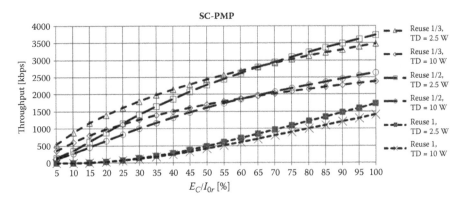

**Figure 5.22  Throughput versus $E_C/I_{0r}$, H-16QAM, $R_c = 3/4$, for SC-PMP Case 3.**

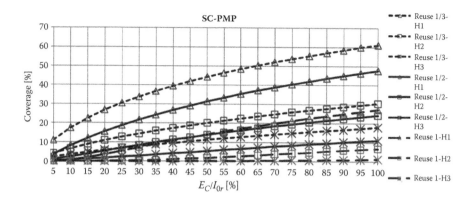

**Figure 5.23   Coverage versus $E_C/I_{0r}$, H-64QAM, $R_c$ = 1/2, for SC-PMP Case 1.**

coverage previously observed for $R_c$ = 3/4, since UEs that cannot receive packets correctly report their throughput as being zero kbps (thus negatively influencing the achieved results). The higher throughput for $R_c$ = 3/4 is achieved for Reuse 1/2 with RNs and TD = 2.5 W (approximately 3700 kbps). The best overall results are achieved when $R_c$ = 1/2, and the highest spectral efficiency is achieved with Reuse 1/2 with RNs and TD = 2.5 W (approximately 4300 kbps). However the results for hybrid reuse (HR) lag behind, achieving throughput values between 3500 kbps and 4000 kbps.

Another interesting conclusion from these results is the fact that the Reuse 1/2 with RNs and TD = 2.5 W has almost 15% less coverage for Strong blocks (H1) than Reuse 1/3 with RNs and TD = 2.5 W (see Figure 5.13), but has almost 1300 kbps more than Reuse 1/3 with RNs and TD = 2.5 W. However, good coverage and good spectral efficiency are not a given. In fact, this suggests the existence of a trade-off between coverage and throughput as we pass from Reuse 1/3 to Reuse 1/2.

When analyzing the coverage results for H-64QAM (Figures 5.23 to 5.32), similar conclusions made for H-16QAM can be drawn. $R_c$ = 1/2 achieves better results than $R_c$ = 3/4. We also note that independently of the $R_c$ used, Reuse 1 (with or without RNs) has the worst results. This is due to the fact that, in the SC-PMP scenario, there is no cell coordination. Therefore, UEs cannot take advantage of spatial diversity or signal diversity due to other adjacent eNBs/RNs transmitting the same content. This is especially true for users at the cell edge suffering large amount of interference from neighboring cells, in contrast with what happens when Reuse 1/3 or even Reuse 1/2 is employed. In this latter case, the interference levels at the edge of cells are greatly reduced, which improves the coverage in those areas. Moreover, H2 of H-64QAM have significantly lower coverage values when compared to H2 of H-16QAM. However, in H-64QAM there are additional H3 blocks that allow UEs to achieve higher throughput values.

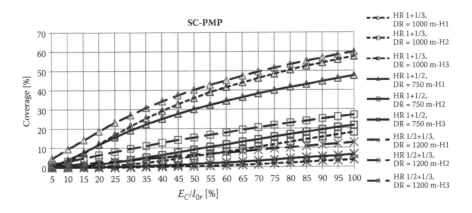

**Figure 5.24  Coverage versus $E_C/I_{0r}$, H-64QAM, $R_c = 1/2$, for SC-PMP Case 2 (1).**

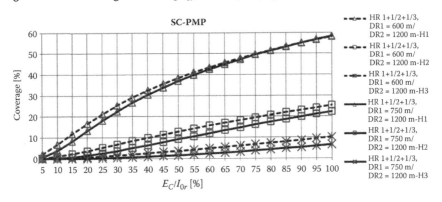

**Figure 5.25  Coverage versus $E_C/I_{0r}$, H-64QAM, $R_c = 1/2$, for SC-PMP Case 2 (2).**

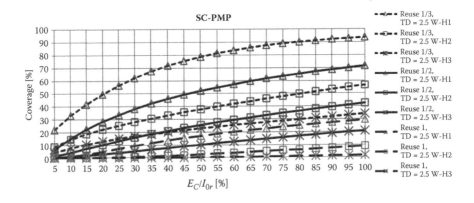

**Figure 5.26  Coverage versus $E_C/I_{0r}$, H-64QAM, $R_c = 1/2$, for SC-PMP Case 3 (1).**

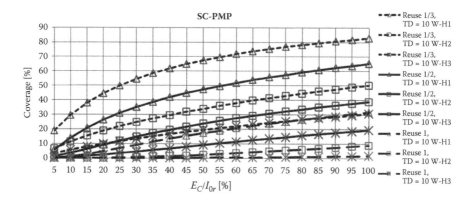

**Figure 5.27  Coverage versus $E_C/I_{0r}$, H-64QAM, $R_c = 1/2$, for SC-PMP Case 3 (2).**

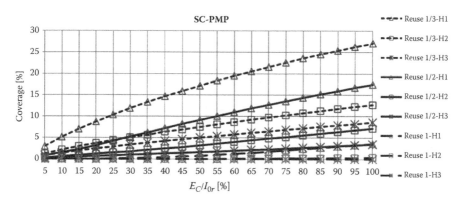

**Figure 5.28  Coverage versus $E_C/I_{0r}$, H-64QAM, $R_c = 3/4$, for SC-PMP Case 1.**

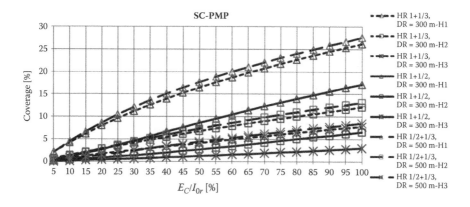

**Figure 5.29  Coverage versus $E_C/I_{0r}$, H-64QAM, $R_c = 3/4$, for SC-PMP Case 2 (1).**

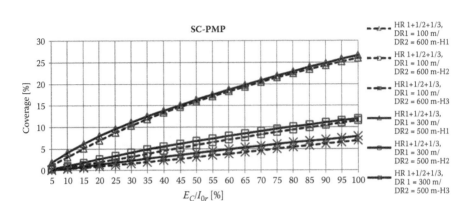

**Figure 5.30**  Coverage versus $E_C/I_{0r}$, H-64QAM, $R_c = 3/4$, for SC-PMP Case 2 (2).

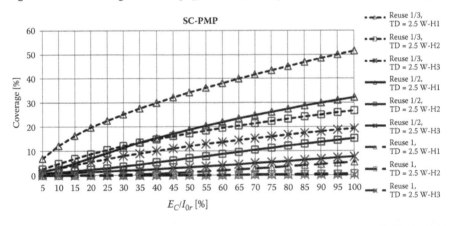

**Figure 5.31**  Coverage versus $E_C/I_{0r}$, H-64QAM, $R_c = 3/4$, for SC-PMP Case 3 (1).

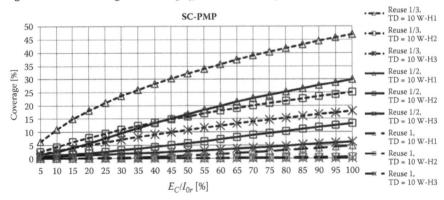

**Figure 5.32**  Coverage versus $E_C/I_{0r}$, H-64QAM, $R_c = 3/4$, for SC-PMP Case 3 (2).

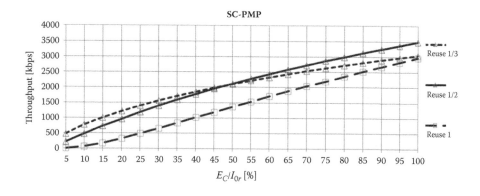

**Figure 5.33** **Throughput versus $E_C/I_{0r}$, H-64QAM, $R_c = 1/2$, for SC-PMP Case 1.**

The throughput results for H-64QAM (Figures 5.33 to 5.38) exhibit approximately the same values and characteristics as the H-16QAM. To explain this, we must keep in mind that the throughput achieved when receiving H1 and H2 blocks in H-16QAM is the same as when receiving Strong (H1) and Average (H2) blocks in H-64QAM. Moreover, since H2 blocks have a lower coverage in H-64QAM than in H-16QAM, the spectral efficiency would be lower. However, in H-64QAM there is a small percentage of area covered where H1+H2+H3 blocks are received. When this occurs, the achieved throughput in those areas is 1.5 times the throughput of H1+H2 in H-16QAM. This can be seen as a trade-off between coverage and throughput (less coverage for H2 and H3 blocks, but increased throughput when receiving H1+H2+H3 blocks).

### 5.3.1.2 Results for SC-PMP with MIMO and CRM

The coverage results for MIMO + CRM with QPSK illustrated in Figures 5.39 and 5.40 are for MIMO $4 \times 4$, QPSK and CRM4 using SC-PMP. As can be seen, better

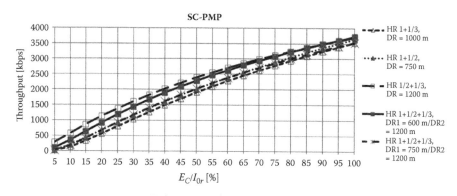

**Figure 5.34** **Throughput versus $E_C/I_{0r}$, H-64QAM, $R_c = 1/2$, for SC-PMP Case 2.**

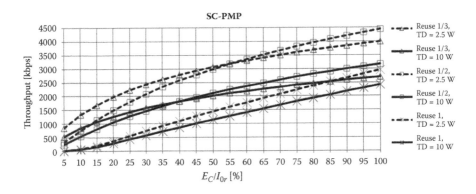

**Figure 5.35** Throughput versus. $E_C/I_{0r}$, H-64QAM, $R_c = 1/2$, for SC-PMP Case 3.

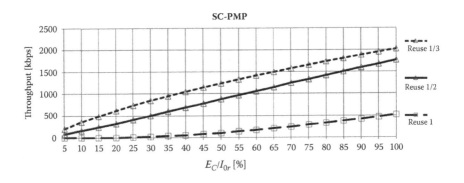

**Figure 5.36** Throughput versus $E_C/I_{0r}$, H-64QAM, $R_c = 3/4$, for SC-PMP Case 1.

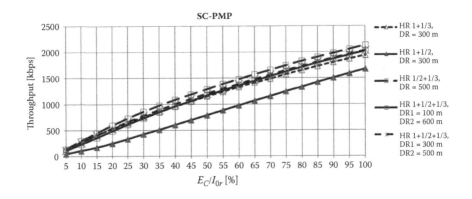

**Figure 5.37** Throughput versus $E_C/I_{0r}$, H-64QAM, $R_c = 3/4$, for SC-PMP Case 2.

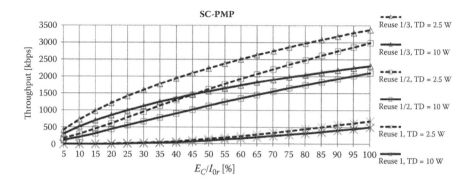

**Figure 5.38    Throughput versus $E_C/I_{0r}$, H-64QAM, $R_c$ = 3/4, for SC-PMP Case 3.**

results are achieved compared to the corresponding H-16QAM and H-64QAM results. The comparison between the two figures indicates a small decrease of coverage when we increase the coding rate. In both figures, the cell radius $R$ is 2,250 m. For a coding rate of 1/2, the only reuse that is close to the reference coverage of 95% is 1/3 (due to small levels of inter-cell interference).

One solution to solve the decrease of coverage might be the reduction of the cell radius to 1,500 m, or the use of MIMO 2 × 2. Recall that the transmitted power per antenna decreases when the number of antennas increases, keeping the total transmitted power constant.

The throughput results for MIMO+CRM with QPSK (Figures 5.41 and 5.42) show better performance compared to the corresponding H-16QAM and H-64QAM results. The comparison between the two figures indicates an increase of throughput

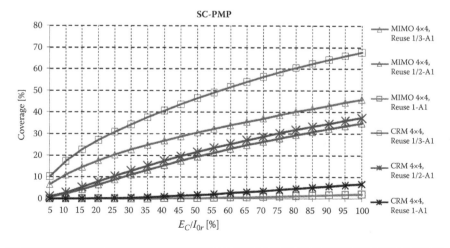

**Figure 5.39    Coverage versus $E_C/I_{0r}$, MIMO (4 × 4) + CRM4, QPSK, SC-PMP, Cod. 3/4.**

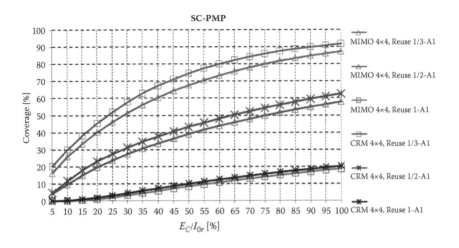

**Figure 5.40** Coverage versus $E_C/I_{0r}$, MIMO (4 × 4)+CRM4, QPSK, SC-PMP, Cod. 1/2.

when we increase the number of antennas. This is due to the higher spectral efficiency of MIMO 4 × 4 compared to MIMO 2 × 2, in spite of its lower coverage.

## 5.3.2 Results for the MBSFN scenario

For the MBSFN scenario, the three cases specified for SC-PMP are also valid. When analyzing the results presented in the following, the values in Table 5.7

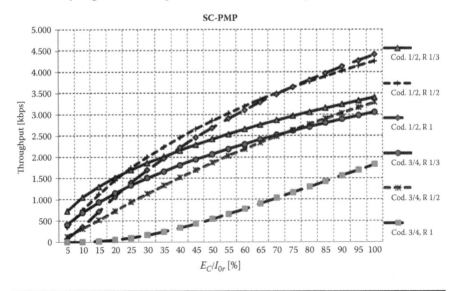

**Figure 5.41** Throughput versus $E_C/I_{0r}$, MIMO (2×2)+CRM2, QPSK, SC-PMP.

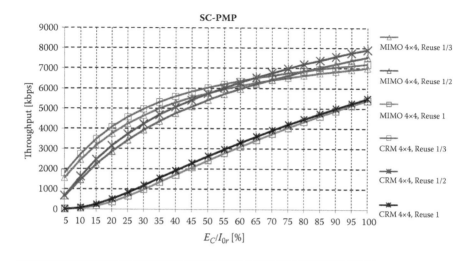

**Figure 5.42** Throughput versus $E_C/I_{0r}$, MIMO (4 × 4)+CRM4, QPSK, SC-PMP, Cod. 1/2.

**Table 5.7 Maximum Bit-Rate and Target SNR Values for MBSFN**

| Hierarchic modulation | | Coding rate | Bit rate [kbps] | SNR [dB] |
|---|---|---|---|---|
| **16QAM** | Reuse 1/3 | 1/2 | 4,800 | **6.79** |
| | | 3/4 | 7,200 | **11.29** |
| | Reuse 1/2 | 1/2 | 7,200 | **8.03** |
| | | 3/4 | 10,800 | **12.33** |
| | Reuse 1 | 1/2 | 14,400 | **9.57** |
| | | 3/4 | 21,600 | **14.63** |
| **64QAM** | Reuse 1/3 | 1/2 | 7,200 | **13.29** |
| | | 3/4 | 10,800 | **17.44** |
| | Reuse 1/2 | 1/2 | 10,800 | **14.31** |
| | | 3/4 | 16,200 | **18.50** |
| | Reuse 1 | 1/2 | 21,600 | **17.10** |
| | | 3/4 | 32,400 | **21.50** |

should be kept in mind, representing the maximum $R_b$ and target SNR values for each of the reuse, modulation, and coding rates used.

Since in MBSFN eNBs transmit coordinately, UEs can take advantage of signal diversity, combining the signals received from various base stations. In this study, each UE combines the three best received signals. The MBSFN channel is configured to use only 6 OFDM symbols per subcarrier (instead, SC-PMP uses 7 OFDM symbols), and that slightly reduces the maximum throughput ($R_b$) that can be achieved in comparison with SC-PMP. Nevertheless, since UEs in MBSFN are expected to achieve better SNR levels (due to additional diversity), coverage results will be significantly better, and there is an increased chance of UEs receiving base and enhancement layers, thus achieving higher spectral efficiency than in the SC-PMP scenario.

### 5.3.2.1 Results for MBSFN with Hierarchical QAM

This section describes the results obtained for the MBSFN environment and hierarchical QAM modulation. As expected, all the coverage results for H-16QAM (Figures 5.43 to 5.50) are better than the results obtained for H-16QAM SC-PMP. Once again, this is explained by the increased SNR ratio that UEs experience in this scenario, due to the exploitation of multiple signal diversity.

When we take a closer look at the results obtained, it can be concluded that when using $R_c = 1/2$, all techniques except Reuse 1 (with or without RNs) achieve or exceed the target value of 95% coverage for the base layer (that is, the Strong blocks - H1). Even more remarkable is the fact that Reuse 1/3 and Reuse 1/2 (without RNs, and with RNs and TD = 2.5 W), using 100% of the total power transmitted, can achieve 95% coverage for Weak Blocks (H2). This means that in those cases at least 95% of the UEs can receive base and enhancement layer (e.g., receiving E-MBMS content with full quality).

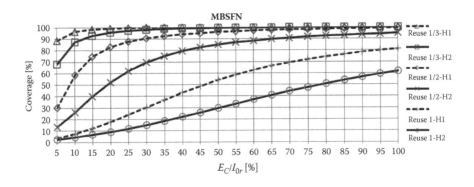

**Figure 5.43** **Coverage versus $E_C/I_{0r}$, H-16QAM, Rc = 1/2, for MBSFN Case 1.**

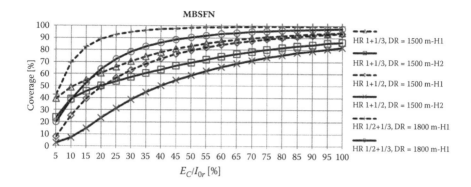

**Figure 5.44** Coverage versus $E_C/I_{0r}$, H-16QAM, $R_c$ = 1/2, for MBSFN Case 2.

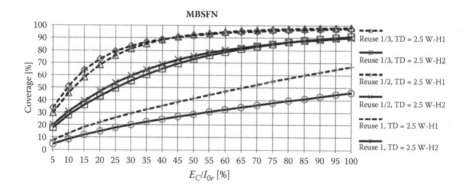

**Figure 5.45** Coverage versus $E_C/I_{0r}$, H-16QAM, $R_c$ = 1/2, for MBSFN Case 3 (1).

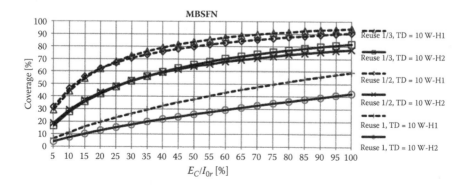

**Figure 5.46** Coverage versus $E_C/I_{0r}$, H-16QAM, $R_c$ = 1/2, for MBSFN Case 3 (2).

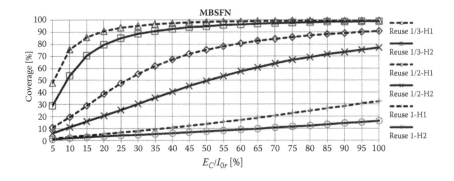

**Figure 5.47  Coverage versus $E_C/I_{0r}$, H-16QAM, $R_c = 3/4$, for MBSFN Case 1.**

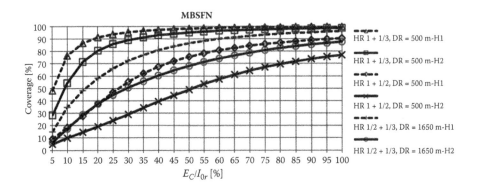

**Figure 5.48  Coverage versus $E_C/I_{0r}$, H-16QAM, $R_c = 3/4$, for MBSFN Case 2.**

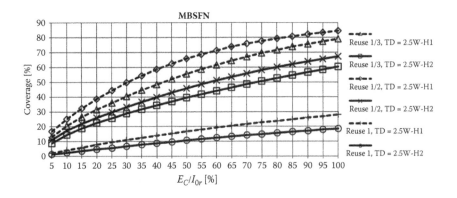

**Figure 5.49  Coverage versus $E_C/I_{0r}$, H-16QAM, $R_c = 3/4$, for MBSFN Case 3 (1).**

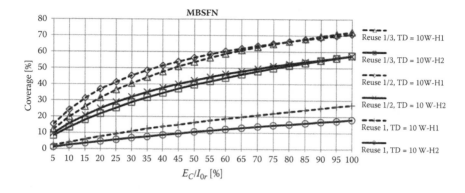

**Figure 5.50** Coverage versus $E_C/I_{0r}$, H-16QAM, $R_c = 3/4$, for MBSFN Case 3 (2).

When Reuse 1 or Reuse 1/2 is used, there is also a saving in power consumption. Figure 5.43 shows that 95% coverage for blocks H1 is achieved using only 5% and 40% of the total transmission power, respectively. For the hybrid reuse schemes, such power saving is more evident for HR 1/2+1/3. The usage of Reuse 1/3 and Reuse 1/2 with RNs and TD = 2.5 W also allows the system to achieve some power saving if we consider that H1 blocks achieve 95% coverage at around 60% of $E_C/I_{0r}$, and that RNs have lower maximum transmission power than eNBs (2.5 W in RN, against 40 W in eNB).

Reuse 1 (with or without RNs) is once again the scheme with worst coverage, due to large inter-cell interference suffered by UEs at the cell edge. However, the results for that technique in MBSFN are significantly better than in SC-PMP, since in MBSFN, UEs at the cell edge explore signal diversity, transforming "interfering signals" into constructive signals, improving the quality of the received signal.

For Rc = 3/4, only Reuse 1/3, HR 1+1/3 and HR 1/2 + 1/3 achieve 95% coverage for H1 blocks.

Figures 5.51 to 5.54 present the throughput results for 16-HQAM and MBSFN network. When comparing the results of MBSFN and SC-PMP, we see that Reuse 1 is now the reuse scheme that achieves higher spectral efficiency with around 10 Mbps of throughput using all the transmission power available. This occurs because the macro-diversity combining existing in MBSFN network allows Reuse 1 to take full advantage of using the total transmission bandwidth available. In MBSFN, the throughput results when RNs are used are (in general) lower than those obtained when no RNs are used. This is explained by the fact that users receiving E-MBMS through RNs experience half of the throughput received by UEs being served directly by eNBs. As expected $R_c = 1/2$ provides higher throughput results than $R_c = 3/4$ due to its higher coverage.

Figures 5.55 to 5.58 present the throughput results for 64-HQAM and MBSFN networks. As can be seen, the maximum throughput values achieved for both $R_c = 1/2$ and $R_c = 3/4$ are around the 10,000 kbps (similar to the best throughput results achieved for H-16QAM MBSFN). There are, however, some substantial

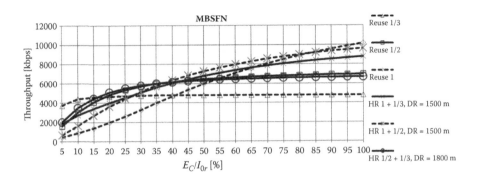

**Figure 5.51** Throughput versus $E_C/I_{0r}$, H-16QAM, $R_c = 1/2$, Case 1 and Case 2.

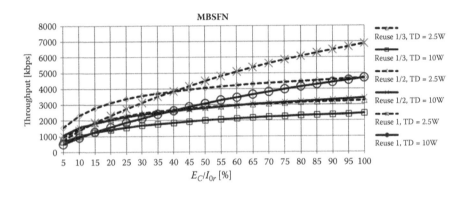

**Figure 5.52** Throughput versus $E_C/I_{0r}$, H-16QAM, $R_c = 1/2$, Case 3.

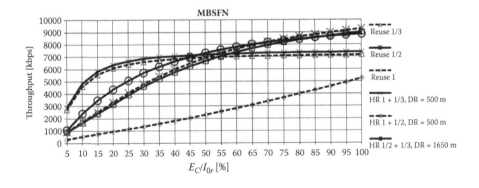

**Figure 5.53** Throughput versus $E_C/I_{0r}$, H-16QAM, $R_c = 3/4$, Case 1 and Case 2.

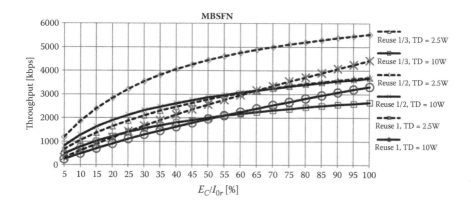

**Figure 5.54    Throughput versus $E_C/I_{0r}$, H-16QAM, $R_c$ = 3/4, Case 3.**

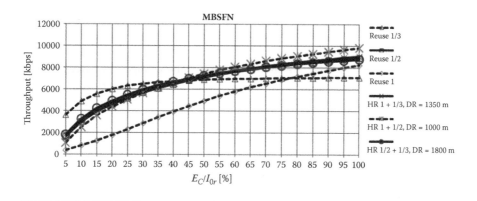

**Figure 5.55    Throughput versus $E_C/I_{0r}$, H-64QAM, $R_c$ = 1/2, Case 1 and Case 2.**

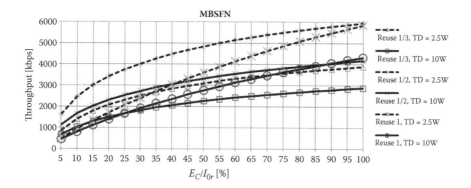

**Figure 5.56    Throughput versus $E_C/I_{0r}$, H-64QAM, $R_c$ = 1/2, Case 3.**

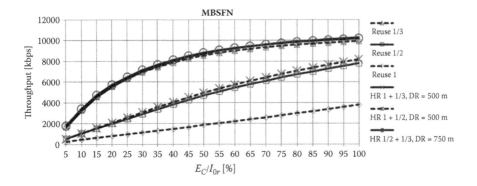

**Figure 5.57    Throughput versus $E_C/I_{0r}$, H-64QAM, $R_c$ = 3/4, Case 1 and Case 2.**

differences. Starting with $R_c = 1/2$, we can see that Reuse 1 in H-64QAM has slightly lower throughput. On the other hand, Reuse 1/3 has 1.5 times more throughput in H-64QAM than with H-16QAM. Also, Reuse 1/2 and HR 1/2+1/3 in H-64QAM achieve approximately 1,800 kbps more of maximum throughput than that achieved in H-16QAM.

In the cases where RNs are used, there are some changes too. Reuse 1 has less 1,000 kbps of maximum throughput, while Reuse 1/3 (TD = 2.5 W and TD = 10 W) and Reuse 1/2 (TD = 10 W) see their maximum values increase up to 800 kbps.

The most spectral efficient technique is HR 1+1/2 (around 9,800 kbps of maximum throughput) and is also the one representing the best compromise between throughput and coverage (coverage – 99.01%; for 100% $E_C/I_{0r}$). Reuse 1/2, HR 1 + 1/3, and HR 1/2 + 1/3 have slightly lower throughput values than HR 1+1/2, but the values they achieve, in conjunction with their coverage results (between 96% and 99%), also make them the right choices for H-64QAM MBSFN, when

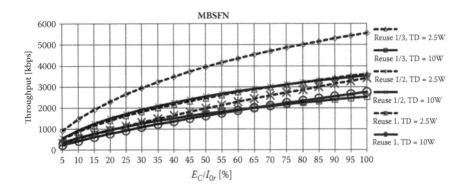

**Figure 5.58    Throughput versus $E_C/I_{0r}$, H-64QAM, $R_c$ = 3/4, Case 3.**

using $R_c = 1$In $R_c = 3/4$, the best results are achieved for HR 1 + 1/3, HR 1/2 + 1/3, and Reuse 1, with values of around 10,000 kbps. A particularity that these two techniques have in common is the fact that all of them have zones where Reuse 1/3 is used (for the HR schemes). This suggests that to achieve good results, some kind of inter-cell interference coordination (such as reuse partitioning or hybrid reuse) should be implemented to minimize the negative effects that high-order modulations and coding schemes have on signals received by UEs at the cell edge. Reuse 1/2 and HR 1+1/2 achieve around 8,000 kbps of throughput, and Reuse 1 has the worst performance, getting only around 4,000 kbps of throughput. In the cases where RNs are employed, the results are not better than those achieved for H-16QAM MBSFN. Reuse 1/3, HR 1+1/3, and HR 1/2+1/3 are all the most spectral-efficient schemes for H-64QAM with $R_c = 3/4$, and are also the best compromise between coverage and throughput (coverage values for those techniques is around 98%).

### 5.3.2.2 Results for MBSFN with MIMO and CRM

In order to increase the spectral efficiency at the cell borders, the MIMO 2 × 2 associated with QPSK modulation and signal space diversity provided by CRM is employed in the MBSFN scenario.

Figures 5.59 to 5.62 present the results for MIMO+CRM and MBSFN networks. Figure 5.59 presents the coverage versus the fraction of the total transmitted power, for MIMO 2X2, cell radius of 1500 m and 2250 m, coding rate 1/2, and the MBSFN scenario. This is the case where there is almost no inter-cell interference due to multipoint MIMO coordination. Again, all potential interfering sites transmit with the maximum power of 90%. Due to coordinated multi-point MIMO transmissions, the coverage has increased substantially. The reference coverage of 95% is reached for $E_C/I_{0r} = 10\%$, independently of the cell radius, for Reuse 1/3. This would allow the

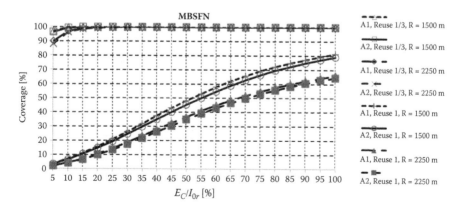

**Figure 5.59  Coverage versus $E_C/I_{0r}$, MIMO (2 × 2)+CRM2, QPSK, MBSFN, Cod. 1/2.**

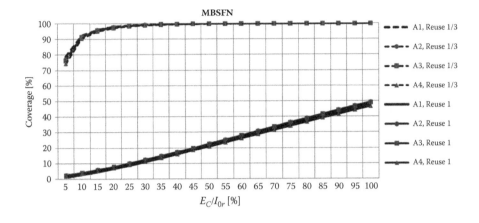

**Figure 5.60  Coverage versus $E_C/I_{0r}$, MIMO (4 × 4)+CRM4, QPSK, MBSFN, Cod. 1/2.**

sharing of the same carrier by different multimedia services. Even Reuse 1 offers 80% coverage if $R = 1{,}500$ m and it decreases to 65% with $R = 2{,}250$ m.

In Figure 5.60, we observe that the reference coverage of 95% is reached for $E_C/I_{0r} = 15\%$ for the cell radius of $R = 2{,}250$ m, for Reuse 1/3 and MIMO 4×4. Reuse 1 assures only 50% for the same radius. The lower coverage values of MIMO $4 \times 4$ compared to MIMO $2 \times 2$ was already expected due to the lower transmitted power per increasing number of antennas.

In Figure 5.61, we observe that the maximum throughput is achieved for Reuse 1, which was expected due to its inherent maximum capacity associated with less inter-cell interference provided by the MIMO coordination of the MBSFN network. In spite of lower coverage, Reuse 1 is capable of providing higher throughput in the

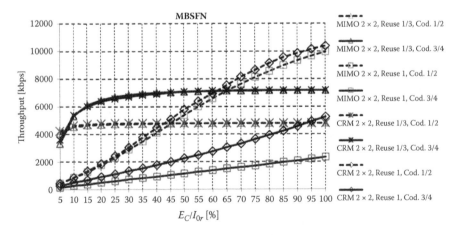

**Figure 5.61  Throughput versus $E_C/I_{0r}$, MIMO (2 × 2)+CRM2, QPSK, MBSFN.**

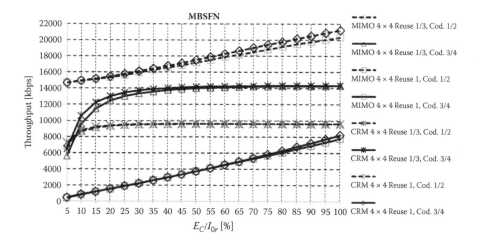

**Figure 5.62**   **Throughput versus** $E_C/I_{0r}$, **MIMO (4 × 4)+CRM4, QPSK, MBSFN.**

cell area. However, the maximum value of throughput is achieved with coding rate 1/2 due to lower sensitivity of the receiver resulting in higher coverage. This means that when the coding rate is 3/4, users located at the cell edge experience much lower throughput than users located close to the base stations.

To further increase the spectral efficiency at cell borders, the number of antennas was increased to MIMO 4 × 4 and the coordinated MIMO transmissions provided by the MBSFN network was used. Figure 5.62 corresponds to Figure 5.61, the difference being the introduction of more antennas. As expected, Reuse 1 offers the highest throughput value. The introduction of CRM4 allows a small increase of throughput compared to the case of MIMO 4 × 4 alone. The comparison between Figures 5.61 and 5.62 indicates that the increase of throughput due to doubling the number of antennas is directly proportional. Using 4 × 4 antennas instead of 2 × 2 allows increasing the maximum throughput from 10,500 kbps to 21,000 kbps. A coding rate 1/2 continues to offer additional average throughput compared to rate 3/4 due to smaller receiver sensitivity (that is, higher coverage), as previously described.

## 5.3.3  Results for Point-to-Point scenario

In the PtP scenario, every UE is served individually, and the link is established by any given UE. Moreover, an eNB is individually allocated with resources, and is also dynamically configured based on the interactivity that exists between UE and eNB (that is, the CQIs sent by the UE to the eNB). This interaction between the UE and eNB (the uplink) is assumed to be always performed under ideal conditions (that is, the CQIs reported by every UE are always correctly received by eNBs).

If a user does not receive a packet properly, there is the option to retransmit the lost packet. Therefore, coverage in this type of system is assured. In this

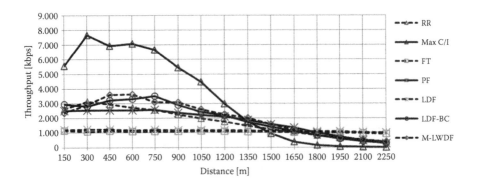

**Figure 5.63** **Distance versus throughput, CBR, 30 UEs per eNB.**

scenario, service delay or outage can be experienced (e.g., due to large waiting times when scheduling); it is one of the aspects that is analyzed here. Another important aspect is the overall system capacity (that is, how many users per cell can the system serve).

Since every UE is individually allocated with resources, and once these are finite, some sort of scheduling mechanism is necessary. Different scheduling mechanisms are tested, using different numbers of UEs in the system to better understand how every scheduling algorithm performs [Gomes 2010]. Moreover, a traffic generation model is employed. The exposed results only cover the traffic model; CBR* @ 37,800 kbps. The Constant Bit-Rate Traffic model, as the name says, generates always the same amount of data, with exactly the same time intervals between consecutive data. This is the traffic model that comes closer to the type of traffic generated in PMP transmissions.

The traffic model used generates a packet of 37,800 bits every 1 ms (millisecond). This represents a traffic generator offering a load of 37,800 kbps per UE.

## 5.3.3.1 Throughput Results

The throughput results plotted in Figures 5.63 and 5.64 for CBR show the limitations that PtP transmission have when compared to PMP. Right from the beginning and having only 30 UEs per eNB, fair throughput (FT) and largest delay first (LDF) schedulers can only achieve 1,000 kbps from the 37,800 kbps offered to each user. For the maximum carrier-to-interference (Max $C/I$) scheduler, the maximum throughput achieved is around 8,000 kbps, but that is only for UEs with good channel conditions, since for users far from the BS the throughput achieved is much less than 1 Mbps. Among the remaining schedulers, round robin (RR), proportional fair (PF), largest delay first with best channel (LDF-BC), only modified largest waiting delay first (M-LWDF) seems to perform slightly better than others,

---

* CBR stands for constant bit rate.

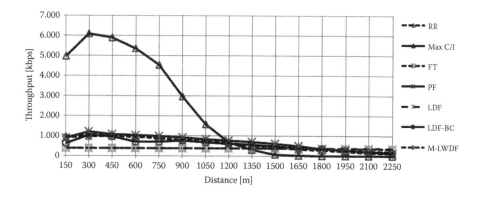

**Figure 5.64    Distance versus throughput, CBR, 90 UEs per eNB.**

mostly because it is a rather complex scheduling algorithm that takes into account several QoS constraints, which can have a particular important effect when the resources are limited and every UE has large amounts of data to transmit.

The situation becomes even more severe as the number of UEs increases to 90 per eNB. There is a clear degradation of the overall system performance in terms of throughput for all schedulers without exception, the Max *C/I* being the scheduler that keeps achieving the highest throughput values, at the expense of close-to-zero throughputs to UEs at cell border.

## 5.3.3.2 Delay Results

Figure 5.65 plots the average delay that packets suffer in each scheduling algorithm for 30 UEs. It is evident that the results of delay for Max *C/I* reach up to 55,000 ms

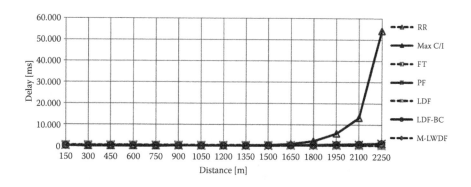

**Figure 5.65    Distance versus average packet delay, CBR, 30 UEs per eNB.**

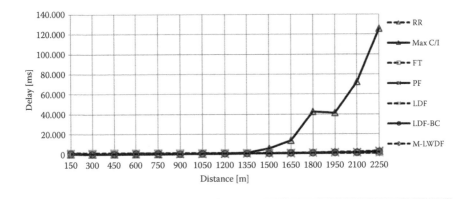

**Figure 5.66** **Distance versus average packet delay, CBR, 90 UEs per eNB.**

(or 55 s) for 30 UEs, and go all the way to 126 s of delay when we have 90 UEs per cell (see Figure 5.66). FT, LDF, and M-LWDF maintain a constant delay at all distances (approximately 1,800 ms). RR, PF, and LDF-BC start with small delays for UEs at close distances (between 600 and 900 ms), reaching up to 5,882 (RR), 3,158 (PF), and 4,769 ms (LDF-BC) of delay to UEs at the cell edge.

Table 5.8 lists several types of average delay that packets suffer for each scheduling algorithm, independently of the distance, for 90 UEs per eNB. It is evident that the coverage is very high for all schedulers, and denied packets (outage measure) is very small, independently of the scheduling algorithm.

## 5.4 Conclusions

The main goal of this chapter was to describe a set of techniques that may be implemented in future LTE-Advanced networks for E-MBMS, using PMP and PtP scenarios. These techniques introduce the concept of multi-resolution, since they allow users to have different types of service quality (or quality of service resolutions).

The PMP scenarios for E-MBMS are the ones creating greater expectations in the LTE community. PMP scenarios, the SC-PMP and MBSFN, have an inherent advantage over point to point, which offers nearly unlimited system capacity. In essence, since PMP transmits data in broadcast/multicast mode, there are no individual retransmissions or significant UE-eNB interaction. Moreover, the channel where data is being transmitted is shared by all, pretty much like television broadcast. The limitations of the PMP scenarios come from other factors such as coverage and packet loss, since in PMP it is not possible to dynamically configure the link between eNB and each UE with the best modulation and coding scheme that achieves best coverage and packet loss rate. This is what multi-resolution techniques are used for.

**Table 5.8 Overview of CBR, for 90 UEs per eNB**

| Scheduler | RR | Max C/I | FT | PF | LDF | LDF-BC | M-LWDF, $\lambda=15070$ ms |
|---|---|---|---|---|---|---|---|
| Avg. Inter serving delay [ms] | 9.39 | 11.16 | 6.27 | 9.63 | 6.95 | 10.63 | 8.71 |
| Avg. packet delay [ms] | 883.53 | 407.13 | 1085.32 | 636.23 | 1093.72 | 843.39 | 744.89 |
| Avg. RLC to MAC delay [ms] | 701.97 | 322.48 | 861.48 | 504.31 | 874.31 | 672.79 | 594.61 |
| Avg. MAC to PHY delay [ms] | 181.56 | 84.65 | 223.84 | 131.92 | 219.41 | 170.60 | 150.28 |
| Coverage (%) | 99.97 | 99.94 | 99.98 | 99.82 | 99.97 | 99.89 | 99.92 |
| HARQ transmissions | | | | | | | |
| 1 | 85.25% | 87.22% | 85.68% | 81.68% | 83.35% | 83.99% | 90.43% |
| 2 | 13.55% | 10.89% | 13.30% | 14.50% | 15.19% | 13.35% | 7.94% |
| 3 | 1.06% | 1.60% | 0.92% | 3.10% | 1.30% | 2.19% | 1.31% |
| 4 | 0.11% | 0.24% | 0.09% | 0.54% | 0.13% | 0.37% | 0.24% |
| Denied packets | 0.03% | 0.06% | 0.02% | 0.18% | 0.03% | 0.11% | 0.08% |

The use of fractional reuse schemes helps reduce inter-cell interference suffered by users, especially those at the cell border. The disadvantage of this technique is that by partitioning the existing frequencies available, every cell will only have access to a part of the total available bandwidth, and this is unfair to UEs closer to the center of the cell (with good channel conditions) that could achieve higher throughputs if using the total bandwidth.

Hybrid reuse consists of applying the concept of fractional reuse to create two or more zones inside a cell, where different reuse schemes are utilized. The advantage of this method is that two zones can be configured: a zone closer to the center of the cell for no fractional reuse, allowing users to achieve maximum throughput, and a second zone close to the cell edge (where users suffer large amounts of inter-cell interference) where some type of fractional reuse is employed, thus reducing the maximum throughput achievable and interference, improving coverage in those zones.

The employment of relay nodes is a new feature introduced in the LTE-Advanced standard. RNs are fixed structures with no backhaul connection, which receive the information transmitted by eNBs and retransmit it to UEs within its area of coverage. RNs can be regarded as a way to extend the coverage of a certain eNB. RNs use lower transmission power than eNBs, which translates into power saving benefits when using these structures, instead of eNBs.

The results obtained for SC-PMP suggest that the best results for both throughput and coverage are achieved when using RNs. Reuse 1/2 with RNs and TD = 2.5 W achieves the best throughput (for both H-16QAM and H-64QAM) and Reuse 1/3 with RNs and TD = 2.5 W for best coverage. The latter configuration should be preferred over the former, since it achieves the requirement of 95% coverage for E-MBMS, but Reuse 1/2 (TD = 2.5 W) can also be an option if a trade-off between coverage and throughput is preferred. The differences between H-16QAM and H-64QAM are close to none in terms of throughput, but in terms of coverage for H1 and H2 blocks, H-64QAM performs worst. The reason why H-64QAM performs similar to H-16QAM results from the fact that H-64QAM has H3 blocks and, in the zones where these are received by UEs, the throughput is 1.5 times the throughput of H-16QAM, receiving H1 and H2 blocks. Increasing the coding rate is not advised since $R_c = 3/4$ performs worst.

MBSFN allows UEs to explore signal diversity, by combining signals transmitted from several eNBs. In this scenario, most of the techniques achieve good results in terms of coverage except Reuse 1. Also, there is no advantage in terms of performance from using RNs, except the power saving achieved. That is due to the fact that when RNs are used instead of eNBs, UEs cannot combine the signals received from RNs and eNBs simultaneously, thus losing the advantage of signal diversity. Fractional reuse and hybrid reuse achieve most of the times the 95% coverage requirement. Moreover, since they achieve the target coverage using only $E_C/I_{0r} = 60\%$ or less, they allow some power saving. The best compromise between coverage and throughput is achieved for H-64QAM using $R_c = 3/4$, for Reuse 1/3, HR 1+1/3, and HR 1/2+1/3. If power saving is required instead, the 95% coverage target can be achieved with good spectral efficiency by using H-64QAM, $R_c = 1/2$, since a low coding rate requires less $E_C/I_{0r}$ to achieve the same coverage results than $R_c = 3/4$.

The introduction of signal space diversity converted to frequency diversity in multipath Rayleigh channels with OFDMA transmission and spatial multiplexing using 4X4 or 2X2 MIMO enables enhancing the spectral efficiency at the cell borders of MBSFN. The use of coding rate 1/2 and Reuse 1 provides the highest spectral efficiency. It is not recommended to increase the coding rate within the MBSFN network to not decrease the throughput at the cell borders.

Finally, it can be concluded that E-MBMS services in LTE-Advanced using PMP scenarios bring great advantages in terms of throughput and system capacity over PtP transmissions. The introduction of MBSFN in E-MBMS significantly improves the system performance, and the different evaluated techniques have enabled multi-resolution for the E-MBMS.

# References

[3GPP 2003] 3GPP, 25.212-v5.2.0, Multiplexing and Channel Coding (FDD), 2003.

[3GPP 2004] 3GPP, TS 25.213-v6.1.0, Spreading and Modulation (FDD), December, 2004.

[3GPP 2005] 3GPP TR 25.913, V7.0.0, Requirements for Evolved UTRA (E-UTRA) and Evolved UTRAN (E-UTRAN), June 2005.

[3GPP 2006] 3GPP TR 25.814, 3rd Generation Partnership Project: Technical Specification Group Radio Access Network; Physical Layers Aspects for Evolved UTRA, 2006.

[3GPP 2007a] 3GPP, Feasibility study for evolved Universal Terrestrial Radio Access (UTRA) and Universal Terrestrial Radio Access Network (UTRAN), Tech rep. 25.912 v7.1.0, http://www.3gpp.org, 2007.

[3GPP 2007b] 3GPP, Improvements of the Multimedia Broadcast Multicast Service (MBMS) in UTRAN, TR 25.905 v7.2.0, December 2007.

[3GPP 2008a] 3GPP TR 36.913, V8.0.0, Requirements for Further Advancement for E-UTRA (LTE-Advanced), June 2008.

[3GPP 2008b] 3GPP, Feasibility Study on Improvement of the Multimedia Broadcast Multicast Service (MBMS), TR 25.905 version 7.2.0 Release 7, 2008.

[3GPP 2009a] 3GPP, Requirements for Evolved UTRA (E-UTRA) and Evolved UTRAN (E-UTRAN), TR 25.913 v9.0.0, December 2009.

[3GPP 2009b] 3GPP, Feasibility Study for Evolved Universal Terrestrial Radio Access (UTRA) and Universal Terrestrial Radio Access Network (UTRAN), TR 25.912 V9.0.0, 2009.

[3GPP 2009c] 3GPP, Technical Specification Group Radio Access Network; MBMS synchronization protocol (SYNC), TS 25.446 v9.0.0, December 2009.

[3GPP 2010a] 3GPP, Evolved Universal Terrestrial Radio Access (E-UTRA) and Evolved Universal Terrestrial Radio Access Network (E-UTRAN); Overall description, TS 36.300 V9.3.0, 2010.

[3GPP 2010b] 3GPP, Feasibility study for Further Advancements for E-UTRA (LTE-Advanced), TR 36.912 v9.2.0, 2010.

[3GPP 2010c] 3GPP, Evolved Universal Terrestrial Radio Access (E-UTRA); Relay architectures for E-UTRA (LTE-Advanced), TR 36.806 v9.0.0, May 2010.

[3GPP 2010d] 3GPP, General aspects and principles for interfaces supporting Multimedia Broadcast Multicast Service (MBMS) within E-UTRAN, TS 36.440 v9.1.0, March 2010.

[3GPP 2010e] 3GPP, Evolved Universal Terrestrial Radio Access (E-UTRA); Radio Frequency (RF) system scenarios, Technical Report TR 36.942 v9.0.1, April 2010.

[3GPP 2010f] 3GPP, Evolved Universal Terrestrial Radio Access (E-UTRA); User Equipment (UE) radio transmission and reception, TS 36.101 v9.3.0, 2010.

[3GPP 2011a] 3GPP, Evolved Universal Terrestrial Radio Access (E-UTRA); Multiplexing and channel coding TS 36.212 v10.4.0, December 2011.

[3GPP 2011b] 3GPP, Evolved Universal Terrestrial Radio Access (E-UTRA); Physical Channels and Modulation, TS 36.211 v10.4.0, December 2011.

[3GPP 2011c] 3GPP, Deployment Aspects, TR 25.943 v10.0.0, April 2011.

[A. Orozco-Lugo et al. 2004] Orozco-Lugo, A., Lara, M., and McLernon, D. Channel estimation using implicit training, *IEEE Transactions on Signal Processing* , vol. 52, no. 1, Jan. 2004.

[Abramowitz and Stegun 1972] Abramowitz, M., and Stegun, I., *Handbook of Mathematical Functions*, New York, Dover Publications, 1972.

[Alamouti 1998] Alamouti, S. M., A Simple transmitter diversity scheme for wireless communications, *IEEE JSAC*, pp. 1451–1458, Oct. 1998.

[Andrews et al. 2007] Andrews, J. G., Gosh, A., and Muhamed, R., *Fundamentals of WiMAX: Understanding Broadband Wireless Networking*, Prentice-Hall, New Jersey, 2007.

[Araújo and Dinis 2006] Araújo, T., and Dinis, R., Efficient detection of zero-padded OFDM signals with large blocks, *IASTED SIP'06*, Honolulu, Hawaii, 2006.

[Araújo and Dinis 2007] Araújo, T., and Dinis, R., Performance evaluation of quantization effects on multicarrier modulated signals, *IEEE Transactions on Vehicular Technology*, vol. 56, no. 5, parte 2, pp. 2922–2930, Sep. 2007.

[Araújo and Dinis 2010] Araújo, T., and Dinis, R., On the accuracy of the Gaussian approximation for the evaluation of nonlinear effects in OFDM signals, *IEEE VTC'2010 (Fall)*, Ottawa, Canada, Sep. 2010.

[Araújo and Dinis 2011] Araújo, T. and Dinis, R., Analytical evaluation of nonlinear effects on OFDMA signals, *IEEE Transactions on Wireless Communications*, vol.9, no.11, pp. 3472–3479, Nov. 2010.

[Araújo and Dinis 2012a] Araújo, T., and Dinis, R., Loading techniques for OFDM systems with nonlinear distortion effects, *European Transactions on Telecommunications*, 2012.

[Araújo and Dinis 2012b] Araújo, T., and Dinis, R., On the accuracy of the Gaussian approximation for the evaluation of nonlinear effects in OFDM signals, *IEEE Transactions on Communications*, 2012.

[Armstrong 2001] Armstrong, J., New OFDM peak-to-average power reduction scheme, *IEEE VTC'2001*(Spring), Rhodes, Greece, May 2001.

[Armstrong 2002] Armstrong, J., Peak-to-average power reduction for OFDM by repeated clipping and frequency-domain filtering, *IEE Electronic Letters*, vol. 38, no. 5, Feb. 2002.

[Astely et al. 2009] Astely, D., Dahlman, E., Furuskar, A., Jading, Y., Lindstrom, M., and Parkvall, S., LTE: The evolution of mobile broadband, *IEEE Communications Magazine*, pp. 44–51, April 2009.

[Bahl et al. 1974] Bahl, L. R., Cocke, J., Jeinek, F., and Raviv, J., Optimal decoding of linear codes for minimizing symbol error rate. *IEEE Transactions on Information Theory*, vol. IT-20, pp. 248–287, March, 1974.

[Benedetto and Montorsi 1997] Benedetto, S., and Montorsi, G., A soft-input soft-output APP module for iterative decoding of concatenated codes, *IEEE Communications Letters*, vol. 1, no. 1, pp. 22–24, January, 1997.

[Benvenuto and Tomasin 2002a] Benvenuto, N., and Tomasin, S., On the comparison between OFDM and single carrier with DFE using a frequency domain feedforward filter, *IEEE Transactions on Communications*, vol. 50, no. 6, pp. 947–955, June 2002.

[Benvenuto and Tomasin 2002b] Benvenuto, N., and Tomasin, S., Block iterative DFE for single carrier modulation, *IEE Electronics Letters*, vol. 39, no. 19, pp. 1144–1145, Sep. 2002.

[Benvenuto and Tomasin 2005] Benvenuto, N., and Tomasin, S., Iterative design and detection of a DFE in the frequency domain, *IEEE Transactions on Communications*, vol. 53, no. 11, pp. 1867–1875, Nov. 2005.

[Benvenuto et al. 2010] Benvenuto, N., Dinis, R., Falconer, D., and Tomasin S., Single carrier modulation with non linear frequency domain equalization: An idea whose time has come—again, *Proceedings of IEEE*, vol. 98, no. 1, pp. 69–96, Jan. 2010.

[Berrou et al. 1993] Berrou, C., Glavieux, A., and Thitimajshima, P., Near Shannon limit error correcting coding and decoding: Turbo-codes, *Proceedings of IEEE International Conference on Communications*, pp. 1064–1070, Geneva, Switzerland, May 23–26, 1993.

[Bhat et al. 2012] Bhat, P., Nagata, S., Campoy, L., Berberana, I., Derham, T., Liu, G., Shen, X., Zong, P., and Yang, J., LTE-advanced: An operator perspective, *IEEE Communications Magazine*, vol. 50, n°. 2, February 2012, pp. 104–114.

[Bolcskei et al. 2002] Bölcskei, H., Gesbert, D., and Paulraj, A. J., On the capacity of OFDM-based spatial multiplexing systems, *IEEE Transactions on Communications*, vol. 50, Feb. 2002, pp. 225–234.

[Boutros and Viterbo 1998] Boutros, J., and Viterbo, E., Signal space diversity: A power- and bandwidth-efficient diversity technique for the Rayleigh fading channel, *IEEE Transactions on Information Theory*, vol. 44, no. 4, July 1998.

[Cavers 1991] Cavers, J. K., An analysis of pilot symbol assisted modulation for Rayleigh fading channels, *IEEE Transactions on Vehicular Technology*, vol. 40, no. 4, pp. 686–693, November 1991.

[Chen and Tsai 2009] Chen, Y., and Tsai, Y., Adaptive resource allocation for multi-resolution multicast services with diversity in OFDM systems, in *IEEE 69th Vehicular Technology Conference*, Barcelona, April 2009, pp. 1–5.

[Chiurtu et al. 2001] Chiurtu, N., Rimoldi, B., and Telatar, E., Dense multiple antenna systems, in *Proceedings of the IEEE Information Theory Workshop*, 2001, pp. 108–109.

[Cimini 1985] Cimini, L., Jr., Analysis and simulation of a digital mobile channel using orthogonal frequency division multiplexing, *IEEE Transactions on Communications*, vol. 33, no. 7, pp. 665–675, 1985.

[Cimini and Sollenberger 1999] Cimini, L. Jr., and Sollenberger, N., Peak-to-average power reduction of an OFDM Signal using partial transmit sequences, *IEEE Communications Letters*, Nov. 1999.

[Cooley and Tukey 1965] Cooley, J. W., and Tukey, J. W., An algorithm for the machine calculation of complex Fourier series, *Mathematics of Computation*, vol. 19, no. 90, 297–301, 1965.

[Correia 2002] Correia, A., Optimised complex constellations for transmitter diversity, *Wireless Personal Communications Journal*, vol. 20, No.3, pp. 267–284, March 2002.

[Correia et al. 2010a] Correia, A., Dinis, R., Souto, N., and Silva, J., LTE E-MBMS capacity and inter-site gains, in *Evolved Cellular Network Planning and Optimization for UMTS and LTE*, Lingyang Song and Jia Shen, Eds.: CRC Press Auerbach Publications, 1st edition, New York, New YorkAugust 2010.

[Correia et al. 2010b] Correia, L., Zeller, D., Blume, O., Ferling, D., Jading, Y., István, G., Auer, G., and Perre, L., Challenges and enabling techniques for energy aware mobile radio networks, *IEEE Communications Magazine*, vol. 48, pp. 66–72, November 2010.

[Cover 1972] Cover T., Broadcast channels, *IEEE Transactions on Information Theory*, vol. IT-18, pp. 2–14, January 1972.

[Cox 1974] Cox, D., Linear amplification with nonlinear components, *IEEE Transactions on Communications*, vol. 22, no. 12, pp. 1942–1945, Dec. 1974.

[Di Zenobio et al. 1995] Di Zenobio D., Santella G., and Mazzenga F., Adaptive linearization of power amplifier in orthogonal multicarrier schemes, *IEEE Wireless Communication System Symposium*, pp 225–230, 1995.

[Dinis and Gusmão 1996a] Dinis, R., and Gusmão, A., Performance evaluation of OFDM transmission with conventional and two-branch combining power amplification schemes, *IEEE GLOBECOM'96*, London, vol. 1, pp. 734–739, Nov. 1996.

[Dinis and Gusmão 1996b] Dinis, R., and Gusmão, A., CEPB-OFDM: A new technique for multicarrier transmission with saturated power amplifiers, *IEEE ICCS'96*, Singapore, Nov. 1996.

[Dinis and Gusmão 1997] Dinis, R., and Gusmão, A., Carrier synchronization with CEPB-OFDM, *IEEE VTC'97*, Phoenix, ArizonaMay 1997.

[Dinis and Gusmão 1998] Dinis, R., and Gusmão, A., Performance evaluation of a multicarrier modulation technique allowing strongly nonlinear amplification, *IEEE ICC'98*, Atlanta, Georgia July 1998.

[Dinis and Gusmão 1999] Dinis, R., and Gusmão, A., On the performance evaluation of OFDM transmission using clipping techniques, *IEEE VTC'99* (Fall), Amsterdam, September 1999.

[Dinis and Gusmão 2000] Dinis, R., and Gusmão, A., A class of signal processing algorithms for good power/bandwidth tradeoffs with OFDM transmission, *IEEE ISIT'2000*, Sorrento, Italy, June 2000.

[Dinis and Gusmão 2001a] Dinis, R., and Gusmão, A., A new class of signal processing schemes for bandwidth-efficient OFDM transmission with low envelope fluctuation, *IEEE VTC'2001* (Spring), Rhodes, May 2001.

[Dinis and Gusmão 2001b] Dinis, R., and Gusmão, A., Performance evaluation of peak cancellation schemes for bandwidth-efficient OFDM transmission with low envelope fluctuation, *IEEE ISCTA'01*, Ambleside, UK, July 2001.

[Dinis and Gusmão 2001c] Dinis, R., and Gusmão, A., Signal processing schemes for power/bandwidth efficient OFDM transmission with conventional or LINC transmitter structures, *IEEE ICC'2001*, Helsinki, Finland vol. 4, pp. 1021–1027, June 2001.

[Dinis and Gusmão 2002] Dinis, R., and Gusmão, A., Comparison of techniques for low-PMEPR OFDM transmission, *IEEE VTC'02* (Spring), Birmingham, Alabama, vol. 4, pp. 1751–1755, May 2002.

[Dinis and Gusmão 2003a] Dinis, R., and Gusmão, A., Performance evaluation of an iterative PMEPR-reducing technique for OFDM transmission, *IEEE GLOBECOM'03*, Vol. 1, pp. 20–24, Dec. 2003.

[Dinis and Gusmão 2003b] Dinis, R., and Gusmão, A., An iterative technique for CEPB-OFDM transmission with low out-of-band radiation, *IEEE VTC'03* (Fall), Orlando, Florida, Oct. 2003.

[Dinis and Gusmão 2004] Dinis, R., and Gusmão, A., A class of nonlinear signal processing schemes for bandwidth-efficient OFDM transmission with low envelope fluctuation, *IEEE Transactions on Communications*, Nov. 2004.

[Dinis and Gusmão 2008] Dinis, R., and Gusmão, A., Nonlinear signal processing schemes for OFDM modulations within conventional or LINC transmitter structures, *European Transactions on Telecommunications*, 2008.

[Dinis and Silva 2006] Dinis, R., and Silva, P., Analytical evaluation of nonlinear effects in MC-CDMA signals, *IEEE Transactions on Wireless Communications*, Ago. 2006.

[Dinis et al. 2003] Dinis, R., Gusmão, A., and Esteves N., On broadband block transmission over strongly frequency-selective fading channels, *Proceedings of Wireless 2003*, Calgary, Canada, July 2003.

[Dinis et al. 2004] Dinis R., Kalbasi R., Falconer D., and Banihashemi A., Iterative layered space-time receivers for single-carrier transmission over severe time-dispersive channels, *IEEE Communications Letters*, vol. 8, no. 9, pp. 579–581, Sep. 2004.

[Dinis et al. 2010] Dinis, R., Araujo, T., Pedrosa, P., and Nunes, F., Joint turbo equalisation and carrier synchronisation for SC-FDE schemes, *European Transactions on Telecommunications*, vol. 21, no. 2, pp. 131–141, March 2010.

[Dinur and Wulich 2001] Dinur, N., and Wulich, D., Peak-to-average power ratio in high-order OFDM, *IEEE Transactions on Communications*, vol. 49, no. 6, June 2001.

[Divsalar and Pollara 1995] Divsalar, D., and Pollara, F., Multiple turbo codes, in *Proceedings of the IEEE Military Communications Conference*, pp. 279–285, San Diego,California, November, 1995.

[Driessen and Foschini 1999] Driessen, P. F., and Foschini, G. J., On the capacity formula for multiple-input multiple-output wireless channels: A geometric interpretation, *IEEE Transactions on Communications*, vol. 47, Feb 1999, pp. 173–176.

[Durgin and Rappaporto 1999] Durgin, G. D., and Rappaport, T. S., Effects of multipath angular spread on the spatial cross-correlation of received voltage envelopes, *49th IEEE Vehicular Technology Conference (VTC) 1999*, vol. 2, pp. 996–1000.

[Eklund et al. 2002] Eklund, C. et al., IEEE Standard 802.16: A technical overview of the WirelessMAN Air interface for broadband wireless access, *IEEE Communications Magazine*, vol. 40, no.6, June 2002, pp. 98–107.

[Elias 1954] Elias, P., Error-free coding, *IRE Transactions*, vol.vIT-4, pp.29–37, 1954.

[Engels and roohling 1998] Engels V., and Rohling, H., Multi-Resolution 64-DAPSK Modulation in a Hierarchical COFDM Transmission System, *IEEE Transactions on Broadcasting*, vol. 44, no.1, pp. 139–149, March, 1998.

[Ericsson 2007] Ericsson, Sustainable Energy use in Mobile Communications, White Paper, August 2007 http://www.ericsson.com, accessed in 23 December 2011.

[ETSI 1998a] ETSI, TR 101 112 v3.2.0, Selection procedures for the choice of radio transmission technologies of UMTS, Sophia Antipolis, France, 1998.

[ETSI 1998b] ETSI, Channel models for HIPERLAN/2 in Different Indoor Scenarios, ETSI EP BRAN 3ERI085B, pp. 1–8, March 1998.

[ETSI 1998c] ETSI, TR 101 112 v3.2.0, Selection procedures for the choice of radio transmission technologies of UMTS, Sophia Antipolis, France, 1998.

[ETSI 2004a] ETSI EN 300 744 V1.5.1; European Standard (Telecommunications series); Digital Video Broadcasting (DVB); Framing structure, channel coding and modulation for digital terrestrial television, June 2004.

[ETSI 2004b] ETSI, ETS 300 744: Digital video broadcasting (DVB) framing structure, channel coding and modulation for digital terrestrial television (DVB-T) V1.5.1, European Telecommunication Standard, November, 2004.

[ETSI 2006] ETSI EN 300 401 V1.4.1; European Standard (Telecommunications series); Radio Broadcasting Systems; Digital Audio Broadcasting (DAB) to mobile, portable and fixed receivers, January 2006.

[ETSI 2009a] ETSI EN 302 307 V1.2.1: Digital Video Broadcasting (DVB); Second generation framing structure, channel coding and modulation systems for Broadcasting, Interactive Services, News Gathering and other broadband satellite applications (DVB-S2), 2009.

[ETSI 2009b] ETSI EN 302 755 V1.1.1 Frame structure channel coding and modulation for a second generation digital terrestrial television broadcasting system (DVB-T2), 2009.

[Falconer et al. 2002] Falconer D., Ariyavisitakul S., Benyamin-Seeyar A., and Eidson B., Frequency domain equalization for single-carrier broadband wireless systems, *IEEE Communications Magazine*, vol. 4, no. 4, pp. 58–66, Apr. 2002.

[Ferling et al. 2010] Ferling, D., Bohn, T., Zeller, D., Frenger, P., Gódor, I., Jading, Y., and Tomaselli, W., Energy efficiency approaches for radio nodes, *Proceedings of Future Networks Mobile Summit 2010*, Florence, Italy, June 2010.

[Fodor et al. 2009] Fodor, G. et al., Intercell interference coordination in OFDMA networks and in the 3GPP long term evolution system, *Journal of Communications*, vol. 4, pp. 445–453, August 2009.

[Forney 1966] Forney, J.D. Jr., *Concatenated Codes*, MIT Press Research Monograph 37, 1966.

[Foschini 1996] Foschini, G. J., Layered space-time architecture for wireless communication in a fading environment when using multiple antennas, *Bell Laboratories Technical Journal*, vol. 1, no. 2, Autumn, 1996, pp. 41–59.

[Foschini and Gans 1998] Foschini, G. J., and Gans, M. J., On limits of wireless communications in fading environments when using multiple antennas, *Wireless Personal Communications*, vol. 6, pp. 315–335, March 1998.

[Fossorier et al. 1998] Fossorier, M. P. C., Burkert, F., Lin, S., and Hagenauer, J., On the equivalence between SOVA and Max-Log-MAP decodings, *IEEE Communications Letters*, vol. 2, no. 5, May, 1998.

[Gallager 1962] Gallager, R. G., Low-density parity-check codes, *IRE Transactions on Information Theory*, vol. IT-8, pp. 21-28, January 1962.

[Gesbert et al. 2002] Gesbert, D., Christophersen, N., and Ekman, T., Capacity limits of dense palm-sized MIMO arrays, in *Proceedings of the Global Communications Conference*, 2002.

[Ghogho et al. 2005] Ghogho, M., McLernon, D., A-Hernandez, E., and Swami, A., Channel estimation and symbol detection for block transmission using data-dependent superimposed training, *IEEE Signal Processing Letters*, vol. 12, no. 3, pp. 226–229, March 2005.

[Goldsmith and Chua 1997] Goldsmith, A., and Chua, S. G., Variable-rate variable power M-QAM for fading channels, *IEEE Transactions on Communications*, vol. 45, no. 10, pp. 1218–1230, October 1997.

[Gomes 2010] Gomes, P. S., Scheduling techniques to transmit multi-resolution in E-MBMS services of LTE-advanced, Ph.D. thesis, ISCTE-IUL, September 2010.

[Gusmao et al. 2000] Gusmão A., Dinis R., Conceição J., and N. Esteves, Comparison of two modulation choices for broadband wireless communications, *IEEE VTC'00* (Spring), Tokyo, Japan, May 2000.

[Gusmão et al. 2003] Gusmão, A., Dinis, R., and Esteves, N., On frequency-domain equalization and diversity combining for broadband wireless communications, *IEEE Transactions on Communications*, vol. 51, no. 7, pp. 1029–1033, July 2003.

[Gusmão et al. 2006] Gusmão, A., Torres, P., Dinis, R., and Esteves, N., A class of iterative FDE techniques for reduced-CP SC-based block transmission, *International Symposium on Turbo Codes*, April 2006.

[Gusmão et al. 2007] Gusmão, A., Torres, P., Dinis, R., and Esteves, N., A turbo FDE technique for reduced-CP SC-based block transmission systems, *IEEE Transactions on Communications*, vol. 55, no. 1, pp. 16–20, Jan. 2007.

[Hagenauer and Hoher 1989] Hagenauer, J., and Hoher, P., A Viterbi algorithm with soft-decision outputs and its applications, *Proceedings of IEEE GLOBECOM'89*, Dallas, U.S.A, 1989.

[Heegard and Wicker 1999] Heegard, C., and Wicker, S. B., *Turbo Coding*, Kluwer Academic Publishers, 1999.

[Ho et al. 2001] Ho, C., Farhang-Boroujeny, B., and Chin, F., Added pilot semi-blind channel estimation scheme for ofdm in fading channels', *IEEE GLOBECOM'01*, Nov. 2001.

[Hochwald et al. 2001] Hochwald, B., Marzetta, T., and Papadias, C., A transmitter diversity scheme for wideband CDMA systems based on space–time spreading, *IEEE Journal on Selected Area in Communications*. 19(1), pp. 48–60, Jan. 2001.

[Hoher et al. 1997] Hoher, P., Kaiser, S., and Robertson, P., Pilot-symbol-aided channel estimation in time and frequency, IEEE Communication Theory Mini-Conference (CTMC), *IEEE GLOBECOM'97*, pp. 90–96, 1997.

[Hottinen et al. 2003] Hottinen, A., Tirkkonen, O., and Wichman, R., Multi-antenna transceiver techniques for 3G and beyond, Chichester, UK, 2003.

[IEEE 2006] IEEE, IEEE Standard for Local and metropolitan area networks—Part 16: Air interface for broadband wireless access systems, IEEE Std 802.16e-2005, February 2006.

[IEEE 2011a] IEEE, IEEE Standard for Local and metropolitan area networks—Part 16: Air interface for broadband wireless access systems, IEEE Std 802.16m-2011, May 2011.

[IEEE 2011b] IEEE, IEEE Standard for Information technology—Telecommunications and information exchange between systems—Local and metropolitan area networks—specific requirements—Part 11: Wireless LAN Medium Access Control (MAC) and Physical Layer (PHY) Specifications. Amendment 5: Enhancements for Higher Throughput, IEEE Std 802.11n-2009, October 2011.

[IEEE 802.16] IEEE 802.16 Relay Task Group, 2008; http://www.ieee802.org/16/relay, accessed 3 January 2012.

[IEEE 802.16-2004] IEEE Standard 802.16-2004, Air interface for fixed broadband wireless access systems, 2004.

[IEEE 802.16e] IEEE Standard 802.16e-2005, Amendment to air interface for fixed and mobile broadband wireless access systems—Physical and medium access control layers for combined fixed and mobile operations in licensed bands, 2005.

[IEEE 802.16m] IEEE Standard 802.16m-07/002r8, IEEE 802.16m System Requirements, Jan. 2009. http://ieee802.org/16/tgm/index.html, accessed 5 January 2012.

[ITU 2010] ITU Paves Way for Next-Generation 4G Mobile Technologies, ITU press release, 21 October 2010.

[ITU-R 2008] ITU-R Recommendation M.2133—Requirements, evaluation criteria and submission templates for the development of IMT-Advanced, 2008.

[Jiang and Wilford 2005] Jiang, H., and Wilford, P., A Hierarchical modulation for upgrading digital, *IEEE Transactions on Broadcasting*, vol. 51, no. 2, pp. 223–229, June 2005.

[Jones and Wilkinson 1996] Jones, A., and Wilkinson, T., Combined coding for error control and increased robustness to system nonlinearities in OFDM, *IEEE VTC'96*, Atlanta, May 1996.

[Josiam and Rajan 2007] Josiam, K., and Rajan, D., Bandwidth efficient channel estimation using super-imposed pilots in OFDM systems, *IEEE Transactions Wireless Communications*, vol. 6, no. 6, pp. 2234–2245, June, 2007.

[Kim et al. 1997] Kim, Y.-S., Kim, C.-J., Jeong, G.-Y., Bang, Y.-J., Park, H.-K., and Choi, S. S., New Rayleigh fading channel estimator based on PSAM channel sounding technique, in *Proceedings of IEEE International Conference on Communications*, pp. 1518–1520, Montreal, Canada, June 1997.

[Kudoh and Adachi 2003] Kudoh, E., and Adachi, F., Transmit power efficiency of a multi-hop virtual cellular system, *Proceedings of IEEE Vehicular Technology Conference, 2003* (VTC2003-Fall), vol. 5, pp. 2910–2914, October 6–9, 2003.

[Lam et al. 2006] Lam, C., Falconer, D., Danilo-Lemoine, F., and Dinis, R., Channel estimation for SC-FDE systems using frequency domain multiplexed pilots, *IEEE VTC'06* (Fall) , Sep. 2006.

[Lam et al. 2008] Lam, C., Falconer, D., Danilo-Lemoine, F., and Dinis, R., Iterative frequency domain channel estimation for DFT-Precoded OFDM systems using in-band pilots, *IEEE Journal on Selected Areas in Communication*, vol. 26, No. 2, pp. 348–358, Feb. 2008.

[Li and Cimini 1998] Li, X., and Cimini, L. J. Jr., Effects of clipping and filtering on the performance of OFDM, IEEE Comm. Letters, Vol. 2, No. 5, pp. 131--133, May 1998.

[Liu 2012] Liu, L., Chen, R., Geirhofer, S., Sayana, K., Shi, Z., and Zhou, Y., Downlink MIMO in LTE-advanced: SU-MIMO vs. MU-MIMO, *IEEE Communications Magazine*, vol. 50, no. 2, pp. 140–147, February 2012.

[Liu and Li 2005] Liu, H., and Li, G. *OFDM-Based Broadband Wireless Networks*, New Jersey: John Wiley & Sons, 2005.

[Lozano et al. 2001] Lozano, A., Farrokhi, F. R., and R. A. Valenzuela, Lifting the limits on high-speed wireless data access using antenna arrays, *IEEE Communications Magazine*, vol. 39, Sept. 2001, pp. 156–162.

[Lugo et al. 2004] Orozco-Lugo, A., Lara, M., and McLernon, D., Channel estimation using implicit training, *IEEE Transactions on Signal Processing*, vol. 52, no. 1, Jan. 2004.

[Mackay and Neal 1995] David J. C. MacKay and Radford M. Neal, Good codes based on very sparse matrices, in BOYD, c. (editor), *Cryptography and Coding. 5th IMA Conference, 1995* (Springer), pp. 100–111 (number 1025 in Lecture Notes in Computer Science).

[Mackay and Neal 1996] David J. C. MacKay and Radford M. Neal, Near Shannon limit performance of low density parity check codes, *Electronics Letters*, July 1996.

[Marques da Silva 2012] Marques da Silva, M., *Multimedia Communications and Networking*, CRC Press, 1st edition, ISBN: 9781439874844, New York, New YorkMarch 2012.

[Marques da Silva and Correia 2001] Marques da Silva, M., and Correia, A., Space Time Diversity for the Downlink of WCDMA, *IEEE—Wireless Personal and Mobile Communications—WPMC'01* (Aalborg–Denmark), September 9–12, 2001.

[Marques da Silva and Correia 2002a] Marques da Silva, M., and Correia, A., Space Time Block Coding for 4 antennas with coding rate 1, *IEEE—International Symposium on Spread Spectrum Techniques and Application-ISSSTA (Prague–Check Republic)*, September 2–5, 2002.

[Marques da Silva and Correia 2002b] Marques da Silva, M., and Correia, A., Space time coding schemes for 4 or more antennas, *IEEE—International Symposium on Personal Indoor and Mobile Radio Communications—PIMRC'02 (Lisbon–Portugal)*, September 16–18, 2002.

[Marques da Silva and Correia 2003] Marques da Silva, M., and Correia, A., Combined transmit diversity and beamforming for WCDMA, *IEEE EPMCC'03*, April 2003, Glasgow, Scotland.

[Marques da Silva et al. 2004] Marques da Silva, M., Correia, A., Silva, J. C., and Souto, N., Joint MIMO and parallel interference cancellation for the HSDPA, *Proceedings of IEEE International Symposium on Spread Spectrum Techniques and Applications 2004 (ISSSTA'04)*, Sydney, Australia, September 2004.

[Marques da Silva et al. 2005] Marques da Silva, M., Dinis, R., and Correia, A., A V-BLAST detector approach for W-CDMA signals with frequency-selective fading, *Proceeding of the 16th IEEE Personal Indoor and Mobile Radio Communications 2005 (PIMRC'05)*, Berlin, Germany, 11-14 September 2005.

[Marques da Silva et al. 2008] Marques da Silva, M., Correia, A., and Dinis, R., On pre-processing for mimo w-CDMA, *11th International Symposium on Wireless Personal Multimedia Communications (WPMC 2008)*, Lapland, Finland, September 2008.

[Marques da Silva et al. 2009a] Marques da Silva, M., Correia, A., and Dinis, R., On transmission techniques for multi-antenna W-CDMA systems, *European Transactions on Telecommunications*, vol. 20, Issue 1, pp. 107–121, January 2009.

[Marques da Silva et al. 2009b] Marques da Silva, M., Correia, A., and Dinis, R., On transmission techniques for multi-antenna W-CDMA systems, *European Transactions on Telecommunications*, vol. 20, Issue 1, pp. 107–121, January 2009.

[Marques da Silva et al. 2009c] Marques da Silva, M., Dinis, R., and Correia, A., Iterative frequency-domain receivers for STBC schemes, *2009 IEEE 70th Vehicular Technology Conference 2009* (VTC2009—Fall), Anchorage, Alaska, September 2009.

[Marques da Silva et al. 2010] Marques da Silva, M., Correia, A., Dinis, R., Souto, N., and Silva, J.C., *Transmission Techniques for Emergent Multicast and Broadcast Systems*, CRC Press Auerbach Publications, 1st edition, ISBN: 9781439815939, New York, New YorkMay 2010.

[May and Rohling 1998] May, T., and Rohling, H., Reducing the peak-to-average power ratio in OFDM radio transmission systems, *IEEE VTC'98*, Ottawa, May 1998.

[Meng et al. 2007] Meng, X., Tugnait, J., and He, S., Iterative joint channel estimation and data detection using superimposed training: Algorithms and performance analysis, *IEEE Transactions on Vehicular Technology*, vol. 56, no. 4, pp. 1873–1880, July 2007.

[Montezuma and Gusmão 2001] Montezuma, P., and Gusmão, A., On analytically described Trellis-coded modulation schemes, *ISCTA'01*, Ambleside, UK, ,July 2001.

[Müller and Huber 1997] Müller, S. and Huber, J., A comparison of peak reduction schemes for OFDM, *IEEE GLOBECOM'97*, Phoenix, Arizona, May 1997.

[Müller et al. 1997] Müller, S., Bräuml, R., Fischer, R., and Huber, J., OFDM with reduced peak-to-average power ratio by multiple signal representation, *Annales of Telecommunications*, vol. 52, Feb. 1997.

[Muquet et al. 2000a] Muquet, B., Courville, M., Dunamel, P., and Giannakis, G., OFDM with trailing zeros versus OFDM with cyclic prefix: Links, comparisons and application to the HiperLAN/2 system, *IEEE ICC'00*, pp. 1049–1053, Santa-Barbara, California, June 2000.

[Muquet et al. 2000b] Muquet, B., Courville, M., Giannakis, G., Wang, Z., and Dunamel, P., Reduced complexity equalizers for zero-padded OFDM transmission, *IEEE ICASSP'00*, pp. 2973–2976, Istanbul, Turkey, June 2000.

[Nee and Prasad 2000] Van Nee, R. and Prasad, R., *OFDM for Wireless Multimedia Communications*, Artech House Publ., 2000.

[Ochiai and Imai 2000] Ochiai, H., and Imai, H., Performance of deliberate clipping with adaptive symbol selection for strictly band-limited OFDM systems, *IEEE Journal on Selected Areas in Communications*, vol. 18, no. 11, Nov. 2000.

[Ochiai and Imai 2002] Ochiai, H. and Imai, H., Performance analysis of deliberately clipped OFDM signals, *IEEE Transactions on Communications*, vol. 50, no. 1, Jan. 2002, pp. 89–101.

[Ohno and Giannakis 2002] Ohno, S., and Giannakis, G., Optimal training and redundant precoding for block transmissions with application to wireless OFDM', *IEEE Transactions on Communications*, vol. 50, no. 12, pp. 2113–2123, Dec., 2002.

[Ohrtman 2008] Ohrtman, F., *WiMAX Handbook*, McGraw-Hill Communications, 2008.

[O'Neill and Lopes 1995] O'Neill, R., and Lopes, L., Envelope variations and spectral splatter in clipped multicarrier signals, *IEEE PIMRC'95*, Sep. 1995.

[Oyman et al. 2007] Oyman, O., Laneman, J. N., and Sandhu, S., Multihop relaying for broadband wireless mesh networks: From theory to practice, *IEEE Communications Magazine*, vol. 45, no. 11, pp. 116–122, November 2007.

[Peters and Heath 2009] Peters, S. W., and Heath, R. W., The future of WiMAX: Multihop relaying with IEEE 802.16j, *IEEE Communications Magazine*, pp. 104–111, January 2009.

[Proakis 1995] Proakis, J., *Digital Communications*, McGraw-Hill, New York, New York 1995.

[Proakis 2001] Proakis, J., *Digital Communications*, 4th edition, McGraw-Hill, New York, New York 2001.

[Pursley and Shea 1999] Pursley, M. B., and Shea, J. M., Non-uniform phase-shift-key modulation for multimedia multicast transmission in mobile wireless networks, *IEEE Journal on Selected Areas in Communications*, vol. 17, no. 5, pp. 774–783, May 1999.

[Rainish 1996] Rainish, D., Diversity transform for fading channels, *IEEE Transactions on Communications*, vol. 44, no. 12, pp. 1653–1661, December 1996.

[Raleigh and Cioffi 1998] Raleigh, G., and Cioffi, J. Spatio-temporal coding for wireless communication', *IEEE Transactions on Communications*, vol. 46, no. 3, pp. 357–366, March 1998.

[Ramchandran et al. 1993] Ramchandran, K., Ortega, A., Uz, K. M., and Vetterli, M., Multiresolution broadcast for digital HDTV using joint source/channel coding, *IEEE Journal on Selected Areas in Communications*, vol. 11, no. 1, pp. 6–23, January 1993.

[Rapajic and Popescu 2000] Rapajic, P. B., and Popescu, D., Information capacity of a random signature multiple-input multiple-output channel, *IEEE Transactions on Communications*, vol. 48, pp. 1245–1248, Aug 2000.

[Rasquete 2009] Rasquete, G., Transmissão de Serviços Multimédia Através do HSDPA nas Futuras Redes LTE, ISCTE/IUL- DCTI, Dissertation presented in partial fulfillment of the Requirements for the Degree of Master of Computer Science and Telecommunications Engineering June 2009.

[Richardson and Urbanke 2001a] Richardson, T., and Urbanke, R., The capacity of low-density parity-check codes under message-passing decoding, *IEEE Transactions on Information Theory*, vol. 47, no. 2, pp. 599–618, February 2001.

[Richardson and Urbanke 2001b] Richardson, T., and Urbanke, R., Efficient encoding of low-density parity check codes, *IEEE Transactions on Information Theory*, vol. 47, no. 2, pp. 638–656, Feb. 2001.

[Robertson and Worz 1995] Robertson, P., Villebrun, E., and Hoeher, P., A comparison of optimal and sub-optimal MAP decoding algorithms operating in the log domain, *Proceedings of the IEEE International Conference on Communications (ICC '95)*, vol. 2, pp. 1009–1013, Seattle, 1995.

[Robertson and Worz 1998] Robertson, P., and Worz, T., Bandwidth-efficient Turbo Trellis-coded modulation using punctured component codes, *IEEE Journal on Selected Areas in Communications*, vol. 16, no.2, pp. 206–218, February,1998.

[Rooyen et al. 2000] Rooyen, P. V., Lötter, M., and Wyk, D., *Space-Time Processing for CDMA Mobile Communications*, Kluwer Academic Publishers, Boston, 2000.

[Rowe 1982] Rowe, H., Memoryless nonlinearities with gaussian input: Elementary results, *Bell System Technical Journal*, vol. 61, Sep. 1982.

[Saleh 1981] Saleh, A., Frequency-independent and frequency-dependent nonlinear models of TWT amplifiers, *IEEE Transactions on Communications*, vol. 29, no. 11, Nov. 1981.

[Sari et al. 1994] Sari H., Karam G., and Jeanclaude I., An analysis of orthogonal frequency-division multiplexing for mobile radio applications, in *Proceedings of the IEEE Vehicular Technology Conference*, VTC'94, pp. 1635–1639, Stockholm, June 1994.

[Sarperi et al. 2008] Sarperi, L., Hunukumbure, M., and Vadgama, S., Simulation study of fractional frequency reuse in WiMAX networks, *FUJITSU Scientific and Technical Journal (FSTJ)*, vol. 44, pp. 318–324, July 2008.

[Sengupta and Mitra 2000] Sengupta, M., and Mitra, P. P., Capacity of multivariate channels with multiplicative noise: I. Random matrix techniques and large-n expansions for full transfer matrices, *Phy. Arch.*, no. 0 010 081, 2000.

[Shannon 1948] Shannon, C. E., A mathematical theory of communication, *Bell System Technical Journal*, October 1948.

[Shiu 1999] Shiu, D., *Wireless Communication Using Dual Antenna Arrays*, ser. International Series in Engineering and Computer Science. Norwell, MA: Kluwer, 1999.

[Shiu et al. 2000] Shiu, C., Foschini, G. J., Gans, M. J., and Kahn, J. M., Fading correlation and its effect on the capacity of multi-element antenna systems, *IEEE Transactions on Communications*, vol. 48, no. 3, pp. 502–513, March 2000.

[Silva et al. 2005a] Silva, J. C., Souto, N. S., Correia, A., Rodrigues, A. J., and Cercas, F. C., Enhanced MMSE detection for MIMO systems, *Proceedings of ConfTele'2005*, Tomar, Portugal, April 2005.

[Silva et al. 2005b] Silva, J. C., Souto, N. S., Cercas, F. C., Correia, A., and Rodrigues, A. J., Equalization based receivers for Wideband MIMO/BAST Systems, *Proceedings of WPMC'2005*, Aalgorg, Denmark, September 2005.

[Silverstein 1995] Silverstein, J. W., Strong convergence of the empirical distribution of eigenvalues of large dimensional random matrices, *Journal of Multivariate Analysis*, vol. 55, no. 2, pp. 331–339, 1995.

[Simonsson 2007] Simonsson, A., Frequency reuse and intercell interference co-ordination in E-UTRA, in *IEEE 65th Vehicular Technology Conference*, Dublin, pp. 3091–3095, April 2007.

[Soares 2006] Soares, A. B., Técnicas para a Optimização da Transmissão de Serviços Multimédia em Modos Multicast/Broadcast nas Redes UMTS, ISCTE/IUL—DCTI, Dissertation presented in partial fulfillment of the Requirements for the Degree of Master in Computer Science and Telecommunications Engineering, July 2006.

[Souto et al. 2005a] Souto, N. S., Silva, J. C., and Cercas, F. C., Iterative turbo multipath interference cancellation for WCDMA systems with non-uniform modulations, *Proceedings of the IEEE Vehicular Technology Conference*, VTC2005-Spring, Stockholm, Sweden, May–June, 2005.

[Souto et al. 2005b] Souto, N. S., Silva, J. C., Cercas, F. C., Rodrigues, A. J., and Correia, A., Non-uniform constellations for broadcasting and multicasting services in WCDMA systems, *Proceedins of the IEEE IST Mobile and Wireless Communications Summit*, Dresden, Germany, June 19–23, 2005.

[Souto et al. 2007a] Souto, N., Cercas, F., Dinis, R., and Silva, J. C., On the BER performance of hierarchical M-QAM constellations with diversity and imperfect channel estimation, *IEEE Transactions on Communications*, vol. 55, no. 10, pp. 1852–1856, October 2007.

[Souto et al. 2007b] Souto, N., Dinis, N., and Silva, J. C., Iterative decoding and channel estimation of MIMO-OFDM transmissions with hierarchical constellations and implicit pilots, *Proceedings of the IEEE International Conference on Signal Processing and Communications*, Nov. 2007, Dubai, EAU.

[Souto et al. 2010] Souto N., Dinis R., and Silva J., Analytical matched filter bound for M-QAM hierarchical constellations with diversity reception in multipath Rayleigh fading channels, *IEEE Transactions on Communications*, vol. 58, no. 3, pp. 737–741, Mar 2010.

[Stette 1974] Stette, G., Calculation of intermodulation from a single carrier amplitude characteristic, *IEEE Transactions on Communications*, vol. 22, no 3, March 1974.

[Stuber 2001] Stuber, G., *Principles of Mobile Communication, Second Edition.* Kluwer Academic Publishers, Boston, Mass. 2001.

[Sydir and Taori 2009] Sydir, J. and Taori, R., An evolved cellular system architecture incorporating relay stations, *IEEE Communications Magazine*, vol. 47, no. 6, pp. 150–155, June 2009.

[Tanner 1981] Tanner, R. M., A recursive approach to low complexity codes, *IEEE Transactions on Information Theory*, vol. 27, 1981.

[Tarokh et al. 1999] Tarokh, V., Jafarkhani, H., and Calderbank, A. R., Space-time block codes from orthogonal designs, *IEEE Transactions on Information Theory*, pp. 1456–1467, July 1999.

[Telatar 1995] Telatar, I. E., Capacity of multiantenna Gaussian channels, AT&T Bell Laboratories, Tech. Memo., June 1995.

[Tüchler and Hagenauer 2001] Tüchler, M. and Hagenauer, J., Linear time and frequency domain turbo equalization, *IEEE VTC'01* (Fall), pp. 2773–2777, Atlantic City, New Jersey, 2001.

[Tugnait and He 2006] Tugnait, J., and He, S., Doubly-selective channel estimation using data-dependent superimposed training and exponential bases models, *Proceedings of the 40th Annual Conf. on Information Sciences and Systems*, 2006.

[Tugnait and Meng 2006] Tugnait, J., and Meng, X., Superimposed training for channel estimation: Performance, analysis, training power allocation and frame synchronization,

[Van Eetvelt et al. 1996] Van Eetvelt P., Wade G., and Tomlinson M., Peak to average power reduction for OFDM schemes by selective scrambling, *IEE Electronics Letters*, vol. 32 , no. 21, pp. 1963–1964, 1996.

[Viterbi 1967] Viterbi, A. J., Error bounds for convolutional codes and an asymptotically optimum decoding algorithm, *IEEE Transactions on Informations Theory*, vol. IT-13, pp. 260–269, April 1967.

[Vitthaladevuni and Alouini 2003] Vitthaladevuni, P. K., and Alouini, M.-S., A recursive algorithm for the exact BER computation of generalized hierarchical QAM constellations, *IEEE Transactions on Information Theory*, vol. 49, no. 1, pp. 297–307, January, 2003.

[Vitthaladevuni and Alouini 2004] Vitthaladevuni, P. K., and Alouini, M.-S., A closed-form expression for the exact BER of generalized PAM and QAM constellations, *IEEE Transactions on Communications*, vol. 52, no. 5, pp. 698–700, May, 2004.

[Vitthaladevuni and Alouini 2001] Vitthaladevuni, P. K., and Alouini, M.-S., BER computation of 4/M-QAM hierarchical constellations, *IEEE Transactions on Broadcasting*, vol. 47, no.3, September 2001.

[Vucetic and Yuan 2002] Vucetic, B., and Yuan, J., *Turbo Codes: Principles and Applications*, Kluwer Academic Publ., 2002.

[Webb and Hanzo 1994] Webb, W. T., and Hanzo, L., Modern quadrature amplitude modulation: Principles and applications for fixed and wireless channels, New York: IEEE Press, 1994.

[Webb and Steele 1995] Webb, W. T., and Steele, R., Variable rate QAM for mobile radio, *IEEE Transactions on Communications*, vol. 43, no. 7, pp. 2223–2230, July, 1995.

[Wei 1993] Wei, L.-F., Coded modulation with unequal error protection, *IEEE Transactions on Communications*, vol. 41, no. 10, pp. 1439–1449, October 1993.

[Wei et al. 2002] Wei, S., Goeckel, D., and Janaswami, R., On the capacity of fixed length linear arrays under bandlimited correlated fading, in *Proceedings of CISS*, Princeton, NJ, Apr. 2002.

[Wu et al. 2008] Wu, Z., Qiu, R., and Zhu, S., MIMO-OFDM PAPR reduction by space-frequency permutation and inversion, *Wireless Communications, Networking and Mobile Computing*, pp. 1–4, 2008.

[Yarali and Rahman 2008] Yarali, A., and Rahman, S., WiMAX broadband wireless access technology: services, architecture and deployment models, *IEEE CCECE2008*, Niagara Falls, Ontario, pp. 77–82, May 4–7, 2008.

[Yu et al. 2001] Yu, K., Bengtsson, M., Ottersten, B., McNamara, D., Karlsson, P., and Beach, M., Second order statistics of NLOS indoor MIMO channels based on 5.2 GHz measurements, *IEEE GLOBECOM 2001*, vol. 1, pp. 156–160.

[Zhou and Zein 2008] Zhou, Y., and Zein, N., Simulation study of fractional frequency reuse for mobile WiMAX, in *IEEE Vehicular Technology Conference*, Singapore, May 2008, pp. 2592–2595.

[Zhu et al. 2003] Zhu, H., Farhang-Boroujeny, B., and Schlegel, C., Pilot embedding for joint channel estimation and data detection in mimo communication systems, *IEEE Communications Letters*, Jan. 2003.

# Index